Self-Assembly of Atomically Precise Nanoclusters: From Irregular Assembly to Crystalline Assembly

Self-Assembly of Atomically Precise Nanoclusters: From Irregular Assembly to Crystalline Assembly

Guest Editor

Rodolphe Antoine

MDPI

Basel • Beijing • Wuhan • Barcelona • Belgrade • Novi Sad • Cluj • Manchester

Guest Editor
Rodolphe Antoine
Institut Lumière Matière
Université Claude Bernard
Lyon 1 and CNRS
Lyon
France

Editorial Office
MDPI AG
Grosspeteranlage 5
4052 Basel, Switzerland

This is a reprint of the Special Issue, published open access by the journal *Nanomaterials* (ISSN 2079-4991), freely accessible at: https://www.mdpi.com/journal/nanomaterials/special_issues/assembly_nanoclusters.

For citation purposes, cite each article independently as indicated on the article page online and as indicated below:

Lastname, A.A.; Lastname, B.B. Article Title. *Journal Name* **Year**, *Volume Number*, Page Range.

ISBN 978-3-7258-4225-4 (Hbk)
ISBN 978-3-7258-4226-1 (PDF)
https://doi.org/10.3390/books978-3-7258-4226-1

Contents

About the Editor

Rodolphe Antoine

Rodolphe Antoine received his Ph.D. in Molecular Physics from The University of Lyon (Michel Broyer). He was previously a postdoctoral researcher at the Swiss Federal Institute of Technology Lausanne, EPFL, in nonlinear optics at interfaces (Hubert H. Girault). His background spans atomic and molecular physics, laser spectroscopy, and physical chemistry. He has broad, multi -disciplinary interests in both experimental and computational avenues of research related to nanoclusters. He is currently a research group leader, focusing on the structure and dynamics of proteins, nanoclusters, and nanoparticles at the Institut Lumière Matière at the University of Lyon and CNRS.

nanomaterials

Editorial

Self-Assembly of Atomically Precise Nanoclusters: From Irregular Assembly to Crystalline Assembly

Rodolphe Antoine

Institut Lumière Matière UMR 5306, Université Claude Bernard Lyon 1, CNRS, Université Lyon, F-69100 Villeurbanne, France; rodolphe.antoine@univ-lyon1.fr; Tel.: +33-(0)-787-098-059

The persistent efforts toward achieving superior properties for assembled nanoscale particles have been held back due to the resulting polydispersity associated with colloidal routes of synthesis [1]. The emergence of ligand-protected atomic clusters seems to be a solution to this limitation [2]. In this case, the ligands stabilizing the clusters are highly reactive in nature and thus provide a facile avenue for the "ligand-mediated spatial organization of nanoclusters" [3]. Further, the most important characteristic of nanoclusters, which distinguishes them from other classes of nanomaterials, is that atomic clusters are imbued with structural integrity. Even polydispersed nanoclusters may be purified following regular purification techniques, and the precise chemical formula of the nanoclusters may be deciphered thanks to mass spectrometry [4]. Thus, unlike other forms of nanoscale particles, a dispersion of atomic clusters typically constitutes of structurally and chemically related species. Controlling the size of nano-constructs is of the utmost importance to obtain a nano-theranostic device, as outlined by Mellor et al. [5]. Hence, the study of chemical reactions toward achieving complex nanostructures in a controllable manner could yield self-assembled nanoclusters with multiple functions and collective properties [6], widening their application potential [7].

The aim of this Special Issue on "Self-Assembly of Atomically Precise Nanoclusters" was to provide a unique international forum aimed at covering a broad description of the various approaches developed for assembling nanostructures (in particular nanoclusters) into higher ordered structures in various dimensions.

Synthetic routes are at the heart of supramolecular chemistry. Innovative strategies, inspired by colloidal routes, lead to 3D assemblies of nanoscale materials. Kim et al. [8] used the evaporation of a fine fountain pen to form 3D colloidal assemblies composed of nanoparticles. This route could be widely applied as a simple fabrication tool in order to explore complex metamaterials constructed of nanoparticles, as this method is highly flexible in varying the shape as well as the composition ratio of self-assembled structures. Also, noble metal nanoparticles provide a reaction platform that plays dual roles in the formation of ligand-protected gold nanoclusters. Cheng et al. used the surface of nanoparticles, with four different shapes, to reduce gold ions and to attract capping ligands [9]. Alternative synthetic routes can use a pulsed plasma approach for the production and surface deposition of silicon nanoclusters (SiNCs) for a size of about one to two nanometers. The as-produced one-to-two-nanometer SiNCs can assemble to form much larger "superclusters" with a size of tens of nanometers. These superclusters possess extremely high permanent electric dipole moments that can be exploited to orient and guide these clusters with external electric fields, opening the path to the controlled architecture of silicon nanostructures [10]. Amorphous or irregular assemblies are also of great interest since they are found in natural processes. The growth of so-called amorphous calcium carbonate nanoparticles into micro-to-millimeter scale crystals is the key step in the formation of the shells and exoskeletons of most marine animals. Understanding the mechanism by which these nanoparticles self-assemble is essential to better understand the effects of climate change on marine life. In this spirit, Clark et al. explored the kinetics of aragonite formation from solution via amorphous calcium carbonate [11].

Ligands, which protect the nanoclusters, play a vital role for stability and for enhancing their properties. In particular, ligands may affect the intramolecular configuration, intermolecular packing, and optical properties of metal nanoclusters. Wu et al. [12] used a Ag_{22} nanocluster template to address the effects of surface modification on intracluster constructions and intercluster packing modes, as well as the properties of nanoclusters or cluster-based crystallographic assemblies. At the supramolecular level, the regulation of intramolecular and intermolecular interactions in nanocluster crystallographic assemblies rendered them CIEE (crystallization-induced emission enhancement) active or inactive nanomaterials. Silver nanoclusters (Ag NCs), as a material with good aggregation-induced emission and biocompatibility, have been widely used in the field of luminescence. However, there are few studies on inducing Ag NCs to obtain chirality through supramolecular self-assembly. Therefore, realizing the chiral self-assembly of Ag NCs is still an urgent problem to be solved. Wang et al. [13] used silver NCs (Ag_9-NCs, $[Ag_9(mba)_9]$, where H2mba = 2-mercaptobenzoic acid) and peptide DD-5 (polymerization of five aspartic acids) to obtain highly ordered fluorescent nanotubes through supramolecular self-assembly. With the introduction of DD-5, the Ag9-NCs AIE effect was triggered and the chirality of the peptide was successfully transferred to the supramolecular assembly, resulting in an assembly that possessed chirality and good circularly polarized luminescence characteristics. Proteins, and in particular bovine serum, are good templates to stabilize and provide highly fluorescent bioconjugates [14]. It is well-known that the deuterium–hydrogen isotope effect causes significant changes in the folding–unfolding processes of proteins. Fehér et al. [15] showed that heavy water, compared to normal water, induces stronger effects in both global and fine structures of bioconjugates and that these changes bring a significantly increased red fluorescence than that observed in normal water.

To conclude this overview on the papers published in the Special Issue "Self-Assembly of Atomically Precise Nanoclusters", I am confident that the readers will enjoy these contributions and may be able to find inspiration for their own research within this Special Issue. This series of manuscripts on this topic will give maximum impact and allow workers in other research areas to apply the same methodologies in understanding the mechanisms of self-assembly in their systems.

Acknowledgments: I am grateful to all the authors for submitting their studies to the present Special Issue and for its successful completion. I deeply acknowledge the Nanomaterials reviewers for enhancing the quality and impact of all submitted papers. Finally, I sincerely and warmly thank the editorial staff of Nanomaterials for their stunning support during the development and publication of this Special Issue. Moreover, the Shanghai Science and Technology Innovation Program (22520712500) is gratefully acknowledged. CNRS is acknowledged for funding through International Emerging Actions between Institut Lumière Matière, CNRS, France, and Mahatma Gandhi University, India.

Conflicts of Interest: The author declares no conflict of interest.

References

1. Kotnala, A.; Zheng, Y. Digital Assembly of Colloidal Particles for Nanoscale Manufacturing. *Part. Part. Syst. Charact.* **2019**, *36*, 1900152. [CrossRef] [PubMed]
2. Antoine, R. Supramolecular Gold Chemistry: From Atomically Precise Thiolate-Protected Gold Nanoclusters to Gold-Thiolate Nanostructures. *Nanomaterials* **2020**, *10*, 377. [CrossRef] [PubMed]
3. Basu, S.; Paul, A.; Antoine, R. Controlling the Chemistry of Nanoclusters: From Atomic Precision to Controlled Assembly. *Nanomaterials* **2022**, *12*, 62. [CrossRef] [PubMed]
4. Comby-Zerbino, C.; Dagany, X.; Chirot, F.; Dugourd, P.; Antoine, R. The emergence of mass spectrometry for characterizing nanomaterials. Atomically precise nanoclusters and beyond. *Mater. Adv.* **2021**, *2*, 4896–4913. [CrossRef]
5. Mellor, R.D.; Uchegbu, I.F. Ultrasmall-in-Nano: Why Size Matters. *Nanomaterials* **2022**, *12*, 2476. [CrossRef] [PubMed]
6. Kolay, S.; Bain, D.; Maity, S.; Devi, A.; Patra, A.; Antoine, R. Self-Assembled Metal Nanoclusters: Driving Forces and Structural Correlation with Optical Properties. *Nanomaterials* **2022**, *12*, 544. [CrossRef] [PubMed]
7. Li, T.; Zhu, H.; Wu, Z. Viewing Aggregation-Induced Emission of Metal Nanoclusters from Design Strategies to Applications. *Nanomaterials* **2023**, *13*, 470. [CrossRef] [PubMed]

8. Kim, S.-J.; Lee, I.-H.; Kim, W.-G.; Hwang, Y.-H.; Oh, J.-W. Fountain Pen-Inspired 3D Colloidal Assembly, Consisting of Metallic Nanoparticles on a Femtoliter Scale. *Nanomaterials* **2023**, *13*, 2403. [CrossRef] [PubMed]
9. Chang, C.-Y.; Wu, Y.-R.; Tseng, T.-H.; Su, J.-H.; Wang, Y.-S.; Jen, F.-Y.; Chen, B.-R.; Huang, C.-L.; Chen, J.-C. Shape Dependence of Silver-Nanoparticle-Mediated Synthesis of Gold Nanoclusters with Small Molecules as Capping Ligands. *Nanomaterials* **2023**, *13*, 2338. [CrossRef] [PubMed]
10. Jardali, F.; Tran, J.; Liège, F.; Florea, I.; Leulmi, M.E.; Vach, H. Electric Field-Induced Nano-Assembly Formation: First Evidence of Silicon Superclusters with a Giant Permanent Dipole Moment. *Nanomaterials* **2023**, *13*, 2169. [CrossRef] [PubMed]
11. Clark, S.M.; Grigorova, V.; Colas, B.; Darwish, T.A.; Wood, K.; Neuefeind, J.; Jacob, D.E. The Kinetics of Aragonite Formation from Solution via Amorphous Calcium Carbonate. *Nanomaterials* **2022**, *12*, 4151. [CrossRef] [PubMed]
12. Wu, S.; Wei, X.; Li, H.; Shen, H.; Han, J.; Kang, X.; Zhu, M. Ligand Effects on Intramolecular Configuration, Intermolecular Packing, and Optical Properties of Metal Nanoclusters. *Nanomaterials* **2021**, *11*, 2655. [CrossRef] [PubMed]
13. Wang, W.; Wang, Z.; Sun, D.; Li, S.; Deng, Q.; Xin, X. Supramolecular Self-Assembly of Atomically Precise Silver Nanoclusters with Chiral Peptide for Temperature Sensing and Detection of Arginine. *Nanomaterials* **2022**, *12*, 424. [CrossRef] [PubMed]
14. Zare, I.; Chevrier, D.M.; Cifuentes-Rius, A.; Moradi, N.; Xianyu, Y.; Ghosh, S.; Trapiella-Alfonso, L.; Tian, Y.; Shourangiz-Haghighi, A.; Mukherjee, S.; et al. Protein-protected metal nanoclusters as diagnostic and therapeutic platforms for biomedical applications. *Mater. Today* **2023**, *66*, 159–193. [CrossRef]
15. Fehér, B.; Mihály, J.; Demeter, A.; Almásy, L.; Wacha, A.; Varga, Z.; Varga, I.; Pedersen, J.S.; Bóta, A. Advancement of Fluorescent and Structural Properties of Bovine Serum Albumin-Gold Bioconjugates in Normal and Heavy Water with pH Conditioning and Ageing. *Nanomaterials* **2022**, *12*, 390. [CrossRef] [PubMed]

nanomaterials

MDPI

Article

Fountain Pen-Inspired 3D Colloidal Assembly, Consisting of Metallic Nanoparticles on a Femtoliter Scale

Sung-Jo Kim [1], Il-Hyun Lee [2,3,4], Won-Geun Kim [1,5], Yoon-Hwae Hwang [6,*] and Jin-Woo Oh [1,6,*]

1 BIT Fusion Technology Center, Pusan National University, Busan 46241, Republic of Korea; sungjokim84@pusan.ac.kr (S.-J.K.); kim1guen@postech.ac.kr (W.-G.K.)
2 Department of Nano Fusion Technology, Pusan National University, Busan 46241, Republic of Korea; 2023710262@skku.edu
3 Department of Nano Engineering, SKKU Advanced Institute of Nanotechnology (SAINT), Sungkyunkwan University (SKKU), Suwon 16419, Republic of Korea
4 Department of Nano Science and Technology, SKKU Advanced Institute of Nanotechnology (SAINT), Sungkyunkwan University (SKKU), Suwon 16419, Republic of Korea
5 Department of Mechanical Engineering, Pohang University of Science and Technology (POSTECH), Pohang 37673, Republic of Korea
6 Department of Nano Energy Engineering, Pusan National University, Busan 46241, Republic of Korea
* Correspondence: yhwang@pusan.ac.kr (Y.-H.H.); ojw@pusan.ac.kr (J.-W.O.)

Abstract: The 3D colloidal assemblies composed of nanoparticles (NPs) are closely associated with optical properties such as photonic crystals, localized surface plasmon resonance, and surface-enhanced Raman scattering. However, research on their fabrication remains insufficient. Here, the femtoliter volume of a 3D colloidal assembly is shown, using the evaporation of a fine fountain pen. A nano-fountain pen (NPF) with a micrometer-level tip inner diameter was adopted for the fine evaporation control of the ink solvent. The picoliters of the evaporation occurring at the NFP tip and femtoliter volume of the 3D colloidal assembly were analyzed using a diffusion equation. The shape of the 3D colloidal assembly was dependent on the evaporation regarding the accumulation time and tip size, and they exhibited random close packing. Using gold-, silver-, and platinum-NPs and mixing ratios of them, diverse 3D colloidal assemblies were formed. The spectra regarding a localized surface plasmon resonance of them were changed according to composition and mixing ratio. We expect that this could be widely applied as a simple fabrication tool in order to explore complex metamaterials constructed of nanoparticles, as this method is highly flexible in varying the shape as well as composition ratio of self-assembled structures.

Keywords: colloidal assembly; nanoparticles; 3D nano-cluster; localized surface plasmon resonance; small structure

1. Introduction

Small objects such as nanoparticles (NPs) have gained interest due to their utility in localized surface plasmon resonance (LSPR) [1–5]. The incident light to subwavelength scales of metallic NPs exhibits LSPR which is a useful optical phenomenon in diverse practical applications such as biosensors [4–6], energy devices [7–9], chemical synthesis [10–12] and surface-enhanced Raman spectroscopy [13–15]. To achieve the fabrication of high-performance functional optical devices, it is necessary to be able to design and manipulate LSPRs. Through the assembly of metallic NPs, plasmonic modes were able to be induced in individual NPs [16,17]. In the superradiant mode in which these dipole plasmons vibrate in phase, the radiation damping increases and the spectrum broadens, and when they vibrate out-of-phase, the radiation damping decreases, causing a spectral dip. These ensembles of plasmon modes depend on the NPs cluster and determine the LSPR characteristics. The plasmon modes related to NPs cluster can be tuned by the shape, size, dielectric surroundings, and interparticle distances of NPs [16,17], and this tunability opens the possibility for sensor applications. Recently, it has also been demonstrated that LSPR can be enhanced or manipulated by changing the geometrical packing structure or composition of NPs clusters [18–21].

The lithography with top-down fabrication has long been adopted to develop plasmon structures [16,17,22,23]. Electron- or ion-beam lithography is capable of fabricating nanometer-sized metallic objects by adjusting the shapes and gap distances; however, it suffers from cost inefficiency and low throughput. In addition, the fabricated structures are restricted to two-dimensions, and expanding to three-dimensions through vertical stacking is challenging. Using colloidal assembly has emerged as an alternative method to make 3D plasmon clusters instead of top-down fabrication [24–26]. This method utilizing solution-processed self-assembly provides advantages in terms of simplicity, scalable, and high-throughput route for fabricating 2D as well as 3D plasmonic clusters. Marangoni flow caused by nonuniform evaporation-induced surface tension gradient contributes to the assembly of NPs resulting in densely packed clusters at a two- or three-phase interface, and this solution-mediated mass transfer process such as dip coating [24], the Langmuir–Blodgett process [25], and template-assisted self-assembly [26] have been devised for fabricating NP clusters. Although many colloidal assembly techniques have been developed, constraints on the free 3D structure creation and the change in component composition ratio of plasmon clusters consisting of NPs remain challenging issues.

In this work, we propose an evaporation-induced fabrication and working mechanism for the 3D colloidal assembly consisting of metallic NPs. In order to precisely control the evaporation of the ink solvent, a micrometer-scale fountain pen tip is introduced into the 3D colloidal assembly process. Nanoparticles dispersed in ink using picoliter volume level evaporation are used to fabricate a 3D NP cluster with a femtoliter volume at the micrometer-scale level, and the packing structure is analyzed. Finally, we investigate the LSPR properties of a variety of NP clusters created using NPs composed of gold, silver, and platinum.

2. Materials and Methods

2.1. Fabrication of Nano-Fountain Pen

To make a nano-fountain pen (NFP), which has an inner tip diameter of several micrometer levels, we used a micropipette puller (SU-P97 FLAMING/BROWN PIPETTE PULLER, Sarasota, FL, USA), where a standard glass capillary tube consisting of borosilicate glass (ITEM No. 1B100F-6, outer diameter = 1.0 mm) was used to fabricate NFP. These were purchased from World Precision Instruments. We adjusted the pulling conditions of the puller to 520–530 °C and 10–11 levels of pulling velocity, where the pulling velocity is a set unit used in the pulling machine representing the rate of separation of the puller bars when the glass capillary tube starts to melt. In this condition, the part of the melting capillary tube was pulled and changed to the hourglass neck. In the end, the neck is broken to create two NFPs with a narrow inner diameter tip ranging between 3–5 μm.

2.2. Preparation of Nanoparticle-Dispersed Ink

The ink inserted into the NFP used for 3D colloidal assembly was a colloid consisting of liquid water (dispersion medium) and metallic NPs (dispersed phase). Metallic nanoparticles composed of gold (Au), silver (Ag), and platinum (Pt), respectively, were used as building blocks for the 3D colloidal assembly, and these nanoparticles were purchased from Nanocomposix (San Diego, CA, USA) and coated with polyvinylpyrrolidone (PVP) for dispersion. The diameters (D) of metallic NPs used are as follows; AuNPs (77 ± 9 nm and 20 ± 2 nm), AgNPs (71 ± 7 nm and 26 ± 4 nm), and PtNP (72 ± 4 nm). Using the mean diameters of NPs, we assigned the symbols to G $\equiv$ $AuNP_{77}$, g $\equiv$ $AuNP_{20}$, S $\equiv$ $AgNP_{71}$, s $\equiv$ $AgNP_{26}$, and P $\equiv$ $PtNP_{72}$ where uppercase and lowercase symbols represent relatively large and small nanoparticles. The concentrations (c in the unit of mg/mL) of the nanoparticles dispersed in the ink were defined as c_G, c_g, c_S, c_s, and c_P according to the sizes and compositions of nanoparticles. In the case of the number density ($n = \frac{6c}{\pi D^3 \tilde{\rho}}$, $\tilde{\rho}$ is a mass density of nanoparticle depending on the composition), we defined to n_G, n_g, n_S, n_s, and n_P.

2.3. Motorized Printing Process

Adjusting the 3D colloidal assembly was controlled by distance changes between the tip and the substrate using a 3-axis motorized stage operated by a controller (XPS-D4, Newport, Irvine, CA, USA), and the position of the initial NFP is adjusted using a 3-axis manual stage (PT3A/M, Thorlabs, Newton, NJ, USA). The height change on the z-axis of the NFP and substrate (Si wafer mounted on the 3-axis motorized stage) was measured using a side-view optical microscopy system consisting of a halogen fiber illuminator (OSL2, Thorlabs). Three steps were set for the stamping process, Accumulation–Contact–Detachment, while observing using an objective lens (MY20X-804, Mitutoyo, Kawasaki, Japan), and a charge-coupled device (CCD) camera. After a stamping process, moving the substrate along the x- or y-axis enables the formation of a regular pattern of 3D colloidal assembly structures on the substrate.

2.4. Structure and Optical Spectra of 3D Colloidal Assembly

The structure of the 3D colloidal assembly was measured using a scanning electron microscope (JSM-7900F, JEOL Ltd., Tokyo, Japan). We manually marked the center of mass of the NPs composing the structure using ImageJ software and found their center coordinates using the function of Analyze Particles. To investigate the optical properties of 3D colloidal assemblies, we used an optical microscope (BX53M, Olympus, Tokyo, Japan) in dark-field mode using an objective lens (LMPlanFL N, 100×, NA = 0.8, Olympus). The spectral characteristics of the scattered light collected by the microscope were measured with a spectrometer mounted on the microscope (US/USB4000, Ocean Optics, Duiven, The Netherlands), and the color from the measured spectra was evaluated using CIE 1931 color space plugged in Origin 2021.

3. Results and Discussion

3.1. 3D Colloidal Assembly Using a Nano-Fountain Pen

Water evaporation is ubiquitous, but occurs too slowly for us to perceive it visually. In particular, controlling the evaporation of liquid droplets in the aerosol state is a field of great interest as it occupies a very important part of engine technology, where the time required for complete evaporation is proportional to the radius (R) of a spherical droplet: this is known as the Radius-Square-Law [27]. At this time, the evaporating mass of liquid per unit of time from the surface of the droplet is proportional to R. When the liquid is continuously supplied to the droplet and R does not change, the mass loss is proportional to R and evaporation time (t_e), and a similar situation occurs at the tip of a fountain pen. In the ink inside the fountain pen, nanometer-level dye particles are dispersed in a solvent to express the color of the writing, and the nanoparticles gradually accumulate at the tip due to evaporation of the solvent occurring at the tip. If the tip with accumulated nanoparticles is stamped on a hard substrate such as glass where solvent absorption does not occur, the accumulated nanoparticles might move to the substrate leaving traces of a 3D colloidal assembly with the base area and height. The volume of the 3D colloidal assembly created through this stamping process is proportional to the inner diameter of the tip and the evaporation time. Since the scale of the 3D colloidal assembly's base area is proportional to R^2, its height is proportional to R^{-1} and t_e. Through this dimensional analysis, we reached the conclusion that it can be used to create the 3D colloidal assembly through the stamping process by reducing the size of the fountain pen tip or increasing the accumulation time. In particular, in order to prepare for the fast manufacturing speed for large-area processes in the future, it is more reasonable to reduce the inner diameter of the tip of the fountain pen than to increase the accumulation time. Therefore, we prepared a nano-fountain pen (NFP) with micrometer-sized levels of inner tip diameter for the 3D colloidal assembly (refer to Materials and Methods).

In order to precisely control the 3D colloidal assembly using the nano-fountain pen (NFP), it was important to control the evaporation of water in the ink that occurs at the NFP tip. The Si wafer was used as a substrate mounting on the 3-axis motorized stage (A of Figure 1a), and the motorized stage was utilized to adjust the stamping process of the 3D colloidal assembly. The 3-axis manual stage (B of Figure 1a) was used to set the initial position of the NFP hanging on the manual stage (C of Figure 1a). The motorized stage setting and assembly process was observed by the side-view microscope system that consists of the light source, objective lens, convex lens, and CCD camera (D–G of Figure 1a). We note that the stamping process setup was isolated from the external environment with an acrylic box to control humidity and temperature. In order to achieve 3D colloidal assembly using the NFP, the cartridge was filled with ink above a critical height (Figure 1b). The inner diameter of the NFP tip was $2R_1$, and the thick inner diameter of the ink cartridge of the NFP was $2R_2$, and R_2 was about 100 times larger than R_1. Following this, the ink volume filled in the NFP would be approximate to $V_{\text{ink}} \approx \pi R_2^2 H$ where H is the filling height of the ink in the cartridge. The ink filled in the NFP experiences three forces; Two capillary forces known as Young-Laplace Equation ($F_1 = \pi R_1^2 \times \frac{2\gamma\cos\theta}{R_1}$ and $F_2 = \pi R_2^2 \times \frac{2\gamma\cos\theta}{R_2}$) regarding the two water-air interfaces, and gravity $F_3 = \rho g V_{\text{ink}}$. Here, $\gamma = 72$ mN/m is the surface tension for the air-water interface [28], θ is the contact angle for the air-water-glass surface [29,30], ρ is the mass density of water, and g is the gravitational acceleration. So, the net force parallel to the z-axis is expressed as $F = -F_1 + F_2 - F_3$ and is approximate to $F \approx F_2 - F_3$ due to $R_2 \gg R_1$. In order for the protruding hemispherical ink droplets to form on the NFP tip, F_3 must be greater than F_2, and the height of ink that must be filled in the cartridge to meet the condition expressed as $H > H_C$, where a critical height, $H_C = \frac{2\gamma\cos\theta}{\rho g R_2}$ coincides with a well-known capillary rise of liquid in a capillary [28]. Therefore, the ink filled the cartridge over 17 mm for easy contact of the ink droplet to the substrate.

In order to achieve a stable state with high entropy, A- (green spheres) and B-NPs (red spheres) preferentially mix evenly with each other (Figure 1b). Therefore, when the ink is mixed via sufficient vortexing (30 s), the A- and B-NPs remain in an evenly mixed and dispersed state. Moreover, the sedimentation of metallic NPs dispersed in the ink is negligible. The metallic NPs dispersed in the ink sink by receiving the buoyancy, $F_B = v(\tilde{\rho} - \rho)g$, due to the difference in mass density with the dispersion medium (water), and at this experience the Stokes drag force, $F_D = -3\pi D\eta u_D$, due to the sinking speed u_D. Here, $v = \frac{\pi D^3}{6}$ is a volume of NP with a diameter D. Since the system corresponds to the Low Reynolds number ($\text{Re} = \frac{\rho D u_D}{\eta} \ll 1$), their sinking speed is determined by the balance of the buoyancy and the Stokes drag force, resulting in $u_D = \frac{v(\tilde{\rho}-\rho)g}{3\pi D\eta}$. Where η is the viscosity of water. Metallic NPs also have a speed $u_B = \sqrt{\frac{3k_B T}{v\tilde{\rho}}}$ due to Brownian motion associated with thermal fluctuations, and thus the sedimentation is negligible due to $|\frac{u_D}{u_B}| \ll 1$. Here, k_B and T are Boltzmann constant and absolute temperature, respectively. In other words, the dispersion state of the metallic NPs in the ink is maintained uniformly during the fabrication process.

The 3D colloidal assembly was achieved by a stamping process consisting of three steps, Accumulation–Contact–Detachment (Figure 1c). Adjusting the accumulation time (t_A) in the first step is crucial for precise volume control of the 3D colloidal assembly where the NFP tip is waiting at a sufficiently far distance from the substrate. In this step, the water of the ink evaporates continuously at the NFP tip, and NPs are gradually accumulated on the tip. The accumulated NPs are transported on the substrate through the capillary bridge which connects the tip and substrate in the contact step for 1 s. In the last detachment step, the substrate moves rapidly downward and the capillary bridge is disconnected leaving NPs on the substrate. The NPs stacked on the substrate exhibit interesting 3D colloidal assembled structures which are dependent on accumulation time, inner diameter of the NFP tip, and mixing state of A- and B-NPs.

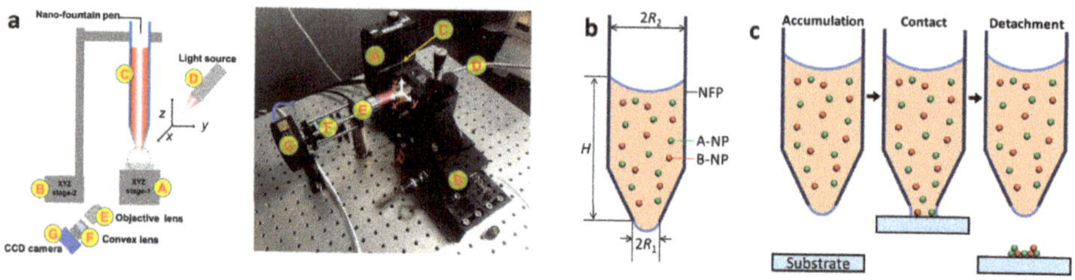

Figure 1. Fabrication of 3D colloidal assembly using NFP. (**a**) Experimental setup for 3D colloidal assembly: schematic (left) and picture (right). (**b**) Schematic of NFP and NPs dispersed in the ink. (**c**) Stamping process to form the 3D colloidal assembly.

3.2. Fabrication Mechanism of a Femtoliter Level of 3D Colloidal Assembly

In order to more precisely control the 3D colloidal assembly at the femtoliter level using the nano-fountain pen (NFP), an analytic solution from the evaporation of the ink solvent was found. We assumed the shape of a protruding water droplet to be a hemisphere with R_1 to simplify the calculation when the ink is filled in the cartridge higher than the critical height (H_C) as explained in Figure 1b. The mass loss of the solvent due to evaporation occurring on the surface of the droplet can be calculated by the diffusion equation for solvent expressed as $\frac{\partial \tilde{c}}{\partial t} = C_D \nabla^2 \tilde{c}$, and the mass loss generated here is filled with the ink in the cartridge as soon as it is supplied so that the size of the droplet always remains the same (blue arrow in Figure 2a). Here, $C_D \sim 2.6 \times 10^{-5}\ \mathrm{m^2/s}$ is the diffusion constant and $\tilde{c}$ is the vapor concentration for water [31]. In addition, due to the very small droplet size and the relatively large reservoir size, the diffusion of the solvent quickly reaches a steady condition and the diffusion equation becomes the Laplace equation expressed as $C_D \nabla^2 \tilde{c} = 0$. Since the concentration function must be continuous, the vapor concentration of the solvent near the droplet surface becomes saturated (defined as $\tilde{c}_s \sim 19.96\ \mathrm{g/m^3}$) for a given temperature, and the vapor concentration corresponds to the relative humidity of the reservoir (defined as $\tilde{c}_0 \sim 7.99\ \mathrm{g/m^3}$) when the distance is far enough away. Under these conditions, the distribution of vapor concentration can be represented by $\tilde{c} = \tilde{c}_0 - (\tilde{c}_s - \tilde{c}_0)R_1/r$ as shown in Figure 2b, and the flux of the solvent by evaporation can be expressed as $\vec{J} = -C_D \nabla \tilde{c}$ from the spatial gradient of the vapor concentration. By integrating the flux passing through the closed surface surrounding the droplet, the mass loss rate of the solvent ($\dot{M} = dM/dt$) due to evaporation can be obtained as $\dot{M} = 1/2 \times \oint \vec{J} \cdot d\vec{a}$ where the one-half multiplied before the integral is due to only the hemisphere surface contributing to evaporation. Consequently, the volume loss rate of the ink solvent can be calculated as $\dot{V} = \dot{M}/\rho$ (Figure 2c).

The picoliter levels of volume loss per unit of time occurring at the NFP tip leads results in a volume change of the 3D colloidal assembly (Figure 2a). The volume loss of the ink solvent is proportional to the evaporation time (t_e) in the accumulation step which is now defined as the term accumulation time (t_A). The NPs are accumulated continuously at the NFP tip by evaporation, and then the accumulated NPs volume should also increase gradually. The concentration of NPs dispersed in the ink was $c = 5\ \mathrm{mg/mL}$, and the mass of accumulated NPs ($M_A = c\dot{V}t_A$) was yielded to the accumulated volume of NPs ($V_A = M_A/\tilde{\rho}$) to approximate 1 fL for $t_A = 5$ s and the mass density of $AuNP_{77}$, $\tilde{\rho} = 19.32\ \mathrm{g/cm^3}$. This estimation coincides with the result as shown in Figure 2d. The volume of the 3D colloidal assembly is monotonically increased according to the accumulation time. Interestingly, as the accumulation time increases, the empty space of the half-torus is gradually filled and the shape of the 3D colloidal assembly is changed to that of a short pillar. This confirms that the shape, as well as the volume of the 3D colloidal assembly, can be changed by adjusting the accumulation time. The inner diameter size of the NFP tip also influences the volume of the 3D colloidal assembly, with the results shown in Figure 2e,f exhibiting that the larger inner diameter forms a larger volume for the same accumulation time. Moreover, the ink mixed with A- and B-NPs also forms the 3D colloidal assembly well (Figure 2g) where the bright and dark NPs represent $AuNP_{77}$ and $AgNP_{71}$, respectively. The packing structures of the 3D colloidal assemblies as shown in the bottom images of Figure 2e–g seem like random closed packing and this is analyzed later in greater detail.

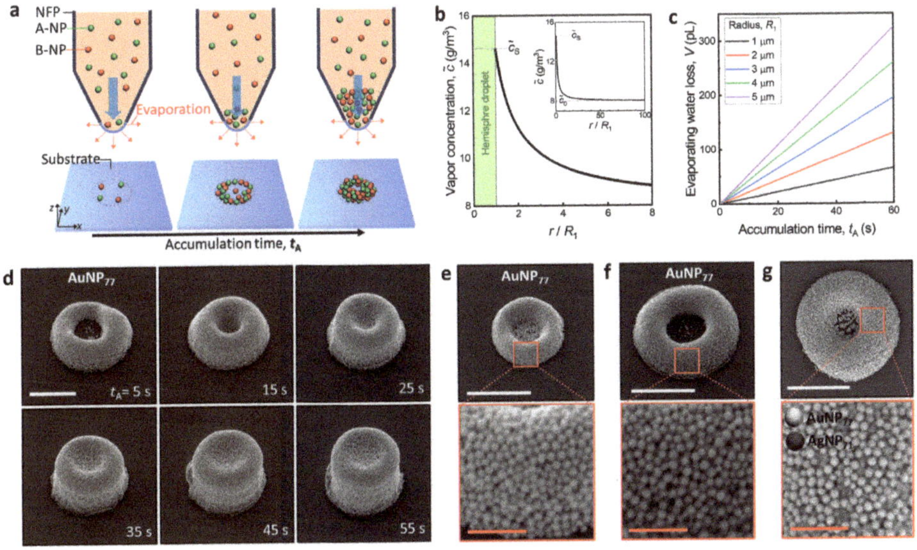

Figure 2. Volume and shape depending on evaporation. (**a**) Volume increment of 3D colloidal assembly regarding the accumulation time, t_A. (**b**) Distribution of vapor concentration induced by evaporation at the NFP tip. (**c**) Evaporation-induced mass loss of water according to the tip inner radius. (**d**) Scanning electron microscope (SEM) images of 3D colloidal assemblies consisting of $AuNP_{77}$. The scale bar represents 2 μm. SEM images of $AuNP_{77}$ assemblies for $t_A = 10$ s: (**e**) $R_1 \sim 3$ μm and (**f**) $R_1 \sim 5$ μm. (**g**) Binary assembly consisting of $AuNP_{77}$ (bright particles) and $AgNP_{71}$ (dark particles). In Figure 2e–g, the scale bars colored white (top) and red (bottom) represent 3 μm and 0.5 μm, respectively.

3.3. Packing Structures of 3D Colloidal Assembly

The characteristics of the packing structure were found from the position of the center of mass of the NPs in the 3D colloidal assembly. Each NP can be identified in the SEM image, and their center of mass ($\vec{r}_i$) was manually located which is marked in red dots (Figure 3a), and the center positions are plotted as shown in Figure 3b, and using these, we found the distances between two NPs ($r_{ij} = |\vec{r}_j - \vec{r}_i|$, for $i \neq j$). We found the number of the first nearest neighbors from a certain NP as marked in red and blue circles.

The 3D colloidal assembly exhibited a random close packing. The distribution of distance (black dots) and their cumulative percent (red solid line) were counted for the distance interval of $\Delta r_{ij} = 10$ nm (Figure 3c). Since the measuring area is finite, the tail of the distribution of distance converges to zero. To find the packing distance, we focused on the distribution for $r_{ij} < 0.2$ μm with smoothing (red dots of Figure 3d). The result showed two peaks at 73 nm and 141 nm, which correspond with the distances for the first (FNNs) and third nearest neighbors (TNNs), while the peak corresponding with the second nearest neighbors (SNNs) is absent. Since the centers do not lie in one plane, the distance of the FNN measures slightly less than 77 nm. In the case of the TNNs, the distance is slightly less than twice that of FNNs, meaning that the particles in the TNNs are not lined up exactly, but rather slightly deviated. The absence of the peak for SNNs also indicates the structure is not hexatic packing and we investigated the number of FNNs and their ratio regarding the number (Figure 3e). The number of FNNs was counted for each NP located within a certain radius that is set to 1.5 times the distance for FNNs. The highest ratio occurred at 6, 5, and 7 follows suggesting that the structure is not hexatic. Therefore, we confirmed that the 3D colloidal assembly is random close packing through comprehensive structural analysis. For random close packing, the volume fraction of the 3D colloidal assembly is $\alpha \approx 0.64$ [32], and thus the 3D colloidal assembly volume (V_{3D}) and accumulated volume can be expressed as $\alpha V_{3D} = V_A$. In the cases of 1, 2, and 3, the number is influenced by the NPs' position where the NPs are located near the border of the measuring area while the 8 and 9 are observed because of the NPs located beneath the outer layer.

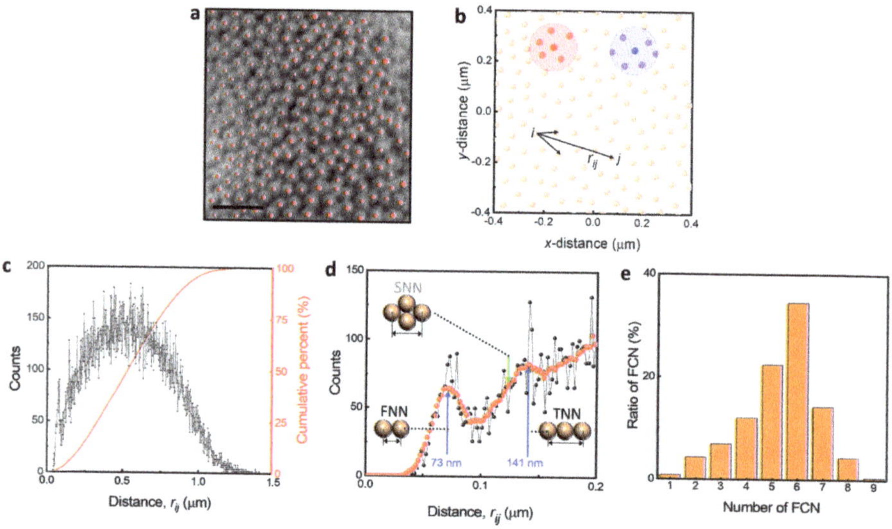

Figure 3. Packing structure of 3D colloidal assembly. (**a**) Center of mass of $AuNP_{77}$. The scale bar represents 0.3 μm. (**b**) Characterization of packing structure using the distance between two NPs (r_{ij}) and the number of first nearest neighbors (FNNs). (**c**) Distribution of r_{ij}. The counts represent the number of r_{ij} existing within distance increment for each $\Delta r_{ij} = 10$ nm. (**d**) Packing distance of 3D colloidal assembly. (**e**) Ratio of FNNs. The number of FNNs was found by counting the NPs existing within 1.5 times the FNN distance.

3.4. Optical Properties Regarding the Structural Changes of 3D Colloidal Assembly

Clusters of metallic NPs created by the 3D colloidal assembly can be used for developing optical metamaterials via wavelength-selective light scattering due to localized surface plasmon resonance (LSPR) [18–21]. In order to tune LSPR, the structure of the cluster needs to be controlled, since the packing structure and mixing of metallic NPs are responsible for the LSPR change. The advantage of using NFP is that the mixing and dispersion state of the NPs dispersed in the ink determines the packing characteristics of the 3D colloidal assembly. It is possible to study the optical properties of optical metamaterials more easily by tuning and fabricating 3D colloidal assemblies of single or binary configurations.

The strategy for tuning LSPR was approached by changing the composition of the NPs and changing the packing structure. For changing the composition of the NPs, the NPs composed of gold (G and g), silver (S and s), and platinum (P) are used, and the large and small NPs are also adopted for changing packing structure (Figure 4a). In order to tune the dispersion state of the 3D colloidal assembly, it is important to adjust the mixing ratio of the ink regarding the NP size. The dispersion state of the 3D colloidal assembly composed of two types of A-NPs and B-NPs is largely determined by the size difference between them. We considered two major states: (I) The size of A-NPs and B-NPs are similar, and (II) A-NP is approximately four-times larger than B-NP. Firstly, we set the conditions so that

B-NPs enclose A-NPs or are enclosed by A-NPs for 3D colloidal assembly, and the number density ratio was estimated from the radii of A-NPs and B-NPs, R_A and R_B. The spherical surface area $S = 4\pi(R_A + R_B)^2$, and the approximate number of B-NPs needed to wrap A-NP in closed packing is $\frac{1}{2}\frac{S}{\pi R_B^2}$. Here, $\frac{1}{2}$ was multiplied as when A-NPs are enclosed, they share B-NPs. Therefore, the number density ratio of 10:1 and 1:10 for case (I) and 1:50 for case (II) is sufficient for the enclosing condition in 3D colloidal assembly. Considering these points, we finally chose the number density ratio of A-NPs and B-NPs, ($n_A : n_B$) as follows: (1:0, 10:1, 1:1, 1:10, 0:1) for case (I) and (1:0, 1:1, 1:50, 0:1) for case (II). The packing structure changes following the yellow arrow for case (I) and the red arrow for case (II) as shown in Figure 4b. We note that approximately 1–50 nanoparticles are dispersed in a 1 μm cube box in the ink in a state that satisfies the experimental conditions described previously.

The spectrum of scattered light is changed by the 3D colloidal assembly of various structures made using NFP. For case (I) of $PtNP_{72}$ and $AgNP_{71}$, the number density ratio ($n_P : n_G$) is (1:0, 10:1, 1:1, 1:10, 0:1), their scattered light spectrum changes (Figure 4c), and the RBG value changes as shown in Figure 4d when these spectra are converted using the CIE 1931 color space. The 3D colloidal assemblies for different four kinds of NPs as shown in Figure 4a exhibited color changes according to the mixing ratio (Figure 4e). The dark star (P) represents the spectrum obtained from the single-component 3D colloidal assembly consisting of PtNPs. The red stars exhibit the spectra measured by the single-component 3D colloidal assembly for $AuNP_{77}$ (G), $AuNP_{20}$ (g), $AgNP_{71}$(S), and $AgNP_{26}$(s). The yellow and red arrows in Figure 4e indicate the structure change as shown in Figure 4b. Their color changes occurred with the signals of two single-component 3D colloidal assemblies. We anticipate that by carefully using metal nanoparticles whose scattering spectra occur in parts distinct from the color map, a wide color distribution between the two parts can be achieved. Moreover, it is expected that the fabrication of a simple 3D colloidal assembly using NFP and the adjustment of the mixing composition ratio will be applicable to the study of micrometer-scale optical metamaterials.

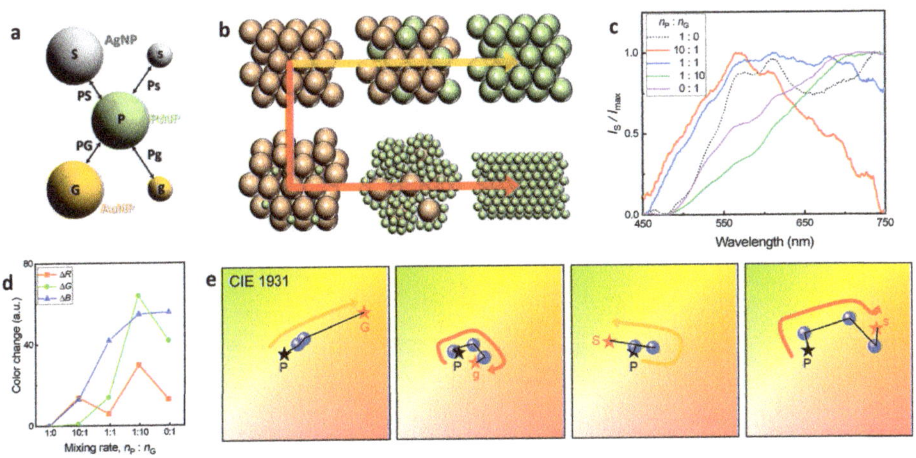

Figure 4. Optical property of the 3D colloidal assembly regarding the diverse structures and mixing compositions. Schematics for (**a**) Mixing combinations of metallic NPs and their (**b**) Packing structures of 3D colloidal assemblies regarding composition and size. The colors represent the composition of NP. The yellow and red arrows show the changes in packing structures when the mixed NP sizes are similar and different, respectively. (**c**) Light scattering spectra of 3D colloidal assemblies consisting of $AuNP_{77}$ and $AuNP_{20}$. (**d**) RGB color changes depending on the mixing ratio. (**e**) Color shift of diverse 3D colloidal assemblies on the CIE 1931 color space. The yellow and red arrows represent packing structure changes as described in Figure 4b.

4. Conclusions

We investigated the fabrication and principles of a 3D colloidal assembly inspired by a nano-fountain pen (NFP). When the inner diameter of the NFP tip was reduced to a micrometer-size scale, 1 pL of water per second was evaporated with the ink solvent generated at the tip. This evaporation caused the metal nanoparticles (NPs) dispersed in the ink to accumulate on the NFP tip. The accumulated volume increased in proportion to the accumulation time with a rate of increase of 1 fL per second. Therefore, through the accumulation phenomenon and the stamping process, a 3D colloidal assembly with a volume of 1 fL could be fabricated on a Si wafer substrate. Depending on the accumulation time, the shape of the manufactured 3D colloidal assembly changed from half-torus to short-pillar. Through examining the location of the center of mass of the NPs constituting the 3D colloidal assembly, the structure was found to have a random close packing. In addition, by changing the mixing of metal nanoparticles dispersed in the ink, 3D colloidal assemblies of various sizes and material compositions were fabricated, and the possibility of localized surface plasmon resonance tuning was confirmed. It is expected that the fabrication mechanism presented here, as well as the simple binary cluster fabrication method, can be used not only for research on various combinations of optical metamaterials but also for the development of optical sensor devices and nanocatalysis using this structure [33–35].

Author Contributions: Conceptualization, S.-J.K., Y.-H.H. and J.-W.O.; methodology, S.-J.K., I.-H.L.and W.-G.K.; formal analysis, S.-J.K.; investigation, I.-H.L. and W.-G.K.; data curation, S.-J.K.; writing—original draft preparation, S.-J.K.; writing—review and editing, S.-J.K., Y.-H.H. and J.-W.O. All authors have read and agreed to the published version of the manuscript.

Funding: This work was supported by a 2-Year Research Grant of Pusan National University.

Data Availability Statement: The data presented in this study are available on request from the first author.

Conflicts of Interest: The authors declare no conflict of interest.

Abbreviations

The following abbreviations are used in this manuscript:

NP	Nanoparticle
NFP	Nano-Fountain Pen
SEM	Scanning Electron Microscope
FNN	First Nearest Neighbor
SNN	Second Nearest Neighbor
TNN	Third Nearest Neighbor
LSPR	Localized Surface Plasmon Resonance

References

1. Hutter, E.; Fendler, J.H. Exploitation of localized surface plasmon resonance. *Adv. Mater.* **2004**, *16*, 1685–1706. [CrossRef]
2. Jensen, T.R.; Malinsky, M.D.; Haynes, C.L.; Van Duyne, R.P. Nanosphere lithography: Tunable localized surface plasmon resonance spectra of silver nanoparticles. *J. Phys. Chem.* **2000**, *104*, 10549–10556. [CrossRef]
3. Ringe, E.; McMahon, J.M.; Sohn, K.; Cobley, C.; Xia, Y.; Huang, J.; Schatz, G.C.; Marks, L.D.; Van Duyne, R.P. Unraveling the effects of size, composition, and substrate on the localized surface plasmon resonance frequencies of gold and silver nanocubes: A systematic single-particle approach. *J. Phys. Chem. C* **2010**, *114*, 12511–12516. [CrossRef]
4. Mayer, K.M.; Hafner, J.H. Localized surface plasmon resonance sensors. *Chem. Rev.* **2011**, *111*, 3828–3857. [CrossRef] [PubMed]
5. Petryayeva, E.; Krull, U.J. Localized surface plasmon resonance: Nanostructures, bioassays and biosensing—A review. *Anal. Chim. Acta* **2011**, *706*, 8–24. [CrossRef] [PubMed]
6. Cao, J.; Sun, T.; Grattan, K.T. Gold nanorod-based localized surface plasmon resonance biosensors: A review. *Sens. Actuators B Chem.* **2014**, *195*, 332–351. [CrossRef]
7. Wu, J.L.; Chen, F.C.; Hsiao, Y.S.; Chien, F.C.; Chen, P.; Kuo, C.H.; Huang, M.H.; Hsu, C.S. Surface plasmonic effects of metallic nanoparticles on the performance of polymer bulk heterojunction solar cells. *ACS Nano* **2011**, *5*, 959–967. [CrossRef]
8. Du, P.; Jing, P.; Li, D.; Cao, Y.; Liu, Z.; Sun, Z. Plasmonic Ag@ oxide nanoprisms for enhanced performance of organic solar cells. *Small* **2015**, *11*, 2454–2462. [CrossRef]

9. Lee, H.B.; Kim, W.G.; Lee, M.; Lee, J.M.; He, S.; Kumar, N.; Devaraj, V.; Choi, E.J.; Jeon, I.; Song, M.; et al. Gap Plasmon of Virus-Templated Biohybrid Nanostructures Uplifting the Performance of Organic Optoelectronic Devices. *Adv. Opt. Mater.* **2020**, *8*, 1902080. [CrossRef]
10. Xiao, Q.; Jaatinen, E.; Zhu, H. Direct Photocatalysis for Organic Synthesis by Using Plasmonic-Metal Nanoparticles Irradiated with Visible Light. *Chem. Asian J.* **2014**, *9*, 3046–3064. [CrossRef]
11. Kolemen, S.; Ozdemir, T.; Lee, D.; Kim, G.M.; Karatas, T.; Yoon, J.; Akkaya, E.U. Remote-controlled release of singlet oxygen by the plasmonic heating of endoperoxide-modified gold nanorods: Towards a paradigm change in photodynamic therapy. *Angew. Chem.* **2016**, *128*, 3670–3674. [CrossRef]
12. Hung, W.H.; Aykol, M.; Valley, D.; Hou, W.; Cronin, S.B. Plasmon resonant enhancement of carbon monoxide catalysis. *Nano Lett. Mater.* **2010**, *10*, 1314–1318. [CrossRef] [PubMed]
13. Mao, P.; Liu, C.; Favraud, G.; Chen, Q.; Han, M.; Fratalocchi, A.; Zhang, S. Broadband single molecule SERS detection designed by warped optical spaces. *Nat. Commun.* **2018**, *9*, 5428. [CrossRef] [PubMed]
14. Kneipp, J.; Kneipp, H.; Kneipp, K. SERS—A single-molecule and nanoscale tool for bioanalytics. *Chem. Soc. Rev.* **2008**, *37*, 1052–1060. [CrossRef] [PubMed]
15. Alvarez-Puebla, R.A.; Liz-Marzán, L.M. SERS detection of small inorganic molecules and ions. *Angew. Chem. Int. Ed.* **2012**, *51*, 11214–11223. [CrossRef] [PubMed]
16. Sonnefraud, Y.; Verellen, N.; Sobhani, H.; Vandenbosch, G.A.; Moshchalkov, V.V.; Van Dorpe, P.; Nordlander, P.; Maier, S.A. Experimental realization of subradiant, superradiant, and Fano resonances in ring/disk plasmonic nanocavities. *ACS Nano* **2010**, *4*, 1664–1670. [CrossRef] [PubMed]
17. Fan, J.A.; Bao, K.; Wu, C.; Bao, J.; Bardhan, R.; Halas, N.J.; Manoharan, V.N.; Shvets, G.; Nordlander, P.; Capasso, F. Fano-like interference in self-assembled plasmonic quadrumer clusters. *Nano Lett.* **2010**, *10*, 4680–4685. [CrossRef] [PubMed]
18. Shao, T.; Wang, X.; Dong, H.; Liu, S.; Duan, D.; Li, Y.; Song, P.; Jiang, H.; Hou, Z.; Gao, C.; et al. A Stacked Plasmonic Metamaterial with Strong Localized Electric Field Enables Highly Efficient Broadband Light-Driven CO_2 Hydrogenation. *Adv. Mater.* **2022**, *34*, 2202367. [CrossRef]
19. Ji, T.; Peng, L.; Zhu, Y.; Yang, F.; Cui, Y.; Wu, X.; Liu, L.; He, S.; Zhu, F.; Hao, Y. Plasmonic broadband absorber by stacking multiple metallic nanoparticle layers. *Appl. Phys. Lett.* **2015**, *106*, 161107. [CrossRef]
20. Saeed, A.; Panaro, S.; Zaccaria, R.P.; Raja, W.; Liberale, C.; Dipalo, M.; Messina, G.C.; Wang, H.; De Angelis, F.; Toma, A. Stacked optical antennas for plasmon propagation in a 5 nm-confined cavity. *Sci. Rep.* **2015**, *5*, 11237. [CrossRef]
21. Kim, W.G.; Lee, J.M.; Yang, Y.; Kim, H.; Devaraj, V.; Kim, M.; Jeong, H.; Choi, E.J.; Yang, J.; Jang, Y.; et al. Three-dimensional plasmonic nanocluster-driven light–matter interaction for photoluminescence enhancement and picomolar-level biosensing. *Nano Lett.* **2022**, *22*, 4702–4711. [CrossRef] [PubMed]
22. Zhang, Y.; Zhen, Y.R.; Neumann, O.; Day, J.K.; Nordlander, P.; Halas, N.J. Coherent anti-Stokes Raman scattering with single-molecule sensitivity using a plasmonic Fano resonance. *Nat. Commun.* **2014**, *5*, 4424. [CrossRef] [PubMed]
23. Chu, M.W.; Myroshnychenko, V.; Chen, C.H.; Deng, J.P.; Mou, C.Y.; de Abajo, F.J.G. Probing bright and dark surface-plasmon modes in individual and coupled noble metal nanoparticles using an electron beam. *Nano Lett.* **2009**, *9*, 399–404. [CrossRef] [PubMed]
24. Lee, J.; You, E.A.; Hwang, D.W.; Kang, S.; Wi, J.S. Active accumulation of spherical analytes on plasmonic hot spots of double-bent Au strip arrays by multiple dip-coating. *Nanomaterials* **2019**, *9*, 660. [CrossRef] [PubMed]
25. Tahghighi, M.; Mannelli, I.; Janner, D.; Ignés-Mullol, J. Tailoring plasmonic response by Langmuir-Blodgett gold nanoparticle templating for the fabrication of SERS substrates. *Appl. Surf. Sci.* **2018**, *447*, 416–422. [CrossRef]
26. Hanske, C.; Tebbe, M.; Kuttner, C.; Bieber, V.; Tsukruk, V.V.; Chanana, M.; König, T.A.; Fery, A. Strongly coupled plasmonic modes on macroscopic areas via template-assisted colloidal self-assembly. *Nano Lett.* **2014**, *14*, 6863–6871. [CrossRef] [PubMed]
27. Rana, A.S.; Lockerby, D.A.; Sprittles, J.E. Lifetime of a nanodroplet: Kinetic effects and regime transitions. *Phys. Rev. Lett.* **2019**, *123*, 154501. [CrossRef] [PubMed]
28. CVR. *Lange's Handbook of Chemistry*, Revised 10th Edition; McGraw-Hill: New York, NY, USA, 1968; p. 449.
29. Cui, J.; Ju, Y.; Liang, K.; Ejima, H.; Lörcher, S.; Gause, K.T.; Richardson, J.J.; Caruso, F. Nanoscale engineering of low-fouling surfaces through polydopamine immobilisation of zwitterionic peptides. *Soft Matter* **2014**, *10*, 2656–2663. [CrossRef]
30. Shuturminska, K.; Tarakina, N.V.; Azevedo, H.S.; Bushby, A.J.; Mata, A.; Anderson, P.; Al-Jawad, M. Elastin-like protein, with statherin derived peptide, controls fluorapatite formation and morphology. *Front. Physiol.* **2017**, *8*, 368. [CrossRef]
31. Cussler, E.L. *Diffusion: Mass Transfer in Fluid Systems*; Cambridge University Press: Cambridge, UK, 2009.
32. Berryman, J.G. Random close packing of hard spheres and disks. *Phys. Rev. A* **1983**, *27*, 1053. [CrossRef]
33. Qin, Z.; Li, Z.; Sharma, S.; Peng, Y.; Jin, R.; Li, G. Self-assembly of silver clusters into one-and two-dimensional structures and highly selective methanol sensing. *Research* **2022**, *2022*, 0018. [CrossRef]

34. Zhang, Y.; Zhang, J.; Li, Z.; Qin, Z.; Sharma, S.; Li, G. Atomically precise copper dopants in metal clusters boost up stability, fluorescence, and photocatalytic activity. *Commun. Chem.* **2023**, *6*, 24. [CrossRef]
35. Shi, Q.; Qin, Z.; Sharma, S.; Li, G. Recent progress in heterogeneous catalysis by atomically and structurally precise metal nanoclusters. *Chem. Rec.* **2021**, *21*, 879–892. [CrossRef]

nanomaterials

Article

Shape Dependence of Silver-Nanoparticle-Mediated Synthesis of Gold Nanoclusters with Small Molecules as Capping Ligands

Cheng-Yeh Chang †, Yi-Ru Wu †, Tzu-Hsien Tseng †, Jun-Hao Su, Yu-Shan Wang, Fang-Yi Jen, Bo-Ru Chen, Cheng-Liang Huang and Jui-Chang Chen *

Department of Applied Chemistry, National Chiayi University, Chiayi City 600355, Taiwan; sss1072727@gmail.com (C.-Y.C.); jerry19970228@gmail.com (Y.-R.W.); tzuhome50@gmail.com (T.-H.T.); andsonhao@gmail.com (J.-H.S.); wang30534@gmail.com (Y.-S.W.); sabrinajengg@gmail.com (F.-Y.J.); s1090251@mail.ncyu.edu.tw (B.-R.C.); clhuang@mail.ncyu.edu.tw (C.-L.H.)
* Correspondence: chenjc@mail.ncyu.edu.tw; Tel.: +886-5-2717968; Fax: +886-5-2717901
† These authors contributed equally to this work.

Abstract: In this study, differently shaped silver nanoparticles used for the synthesis of gold nanoclusters with small capping ligands were demonstrated. Silver nanoparticles provide a reaction platform that plays dual roles in the formation of Au NCs. One is to reduce gold ions and the other is to attract capping ligands to the surface of nanoparticles. The binding of capping ligands to the AgNP surface creates a restricted space on the surface while gold ions are being reduced by the particles. Four different shapes of AgNPs were prepared and used to examine whether or not this approach is dependent on the morphology of AgNPs. Quasi-spherical AgNPs and silver nanoplates showed excellent results when they were used to synthesize Au NCs. Spherical AgNPs and triangular nanoplates exhibited limited synthesis of Au NCs. TEM images demonstrated that Au NCs were transiently assembled on the surface of silver nanoparticles in the method. The formation of Au NCs was observed on the whole surface of the QS-AgNPs if the synthesis of Au NCs was mediated by QS-AgNPs. In contrast, formation of Au NCs was only observed on the edges and corners of AgNPts if the synthesis of Au NCs was mediated by AgNPts. All of the synthesized Au NCs emitted bright red fluorescence under UV-box irradiation. The synthesized Au NCs displayed similar fluorescent properties, including quantum yields and excitation and emission wavelengths.

Keywords: gold nanoclusters; silver nanoparticles; quantum yields; fluorescence; TEM

1. Introduction

Over the past decades, the nanoscience of noble metals has drawn tremendous attention and become a fascinating research subject. Amongst the noble metals, gold is considered as one of the most extensively studied elements, and a myriad of related articles have been produced due to its chemical stability, specific photoproperties, nontoxicity, and biocompatibility. It is involved in various research areas and applications, including chemical sensing [1], medicine [2], and catalysis [3].

Gold exhibits characteristic properties that are influenced by its particle size. When the particles are prepared to be 100 nm or less in diameter, they are termed "nanomaterials". These small-sized gold nanomaterials are further classified into two different groups based on their properties and/or dimensions, which are gold nanoparticles (AuNPs) and gold nanoclusters (Au NCs). Au NCs usually consist of several to tens of gold atoms. Structurally, gold clusters are divided into two parts, the outer shell and the inner core. The outer shell is composed of protecting ligands or capping ligands coordinated with the inner core, the aggregate of gold atoms. The inner core of the cluster has to be segregated by the capping ligands to prevent further aggregation into larger nanoparticles. The sizes of gold nanoclusters are typically less than 5 nm in diameter, exhibiting intrinsically unique and size-specific properties. The ultra-small nanoclusters are in a group of their own, as their unique properties are much different from those of Au nanoparticles. For instance, AuNPs absorb visible wavelengths, but Au NCs do not. Unlike AuNPs, Au NCs neither appear in any specific geometrical shape under the detection of TEM or SEM, nor do they display surface plasmon resonance (SPR) under UV-Vis spectroscopy [4]. The most important characteristics distinguishing Au NCs from AuNPs is that the smaller Au NCs exhibit a molecular-like luminescence property [2,5]. This unique phenomenon is attributed to the size of these nanoclusters, similar to an electron Fermi wavelength of gold metal, leading to its continuous band structure, which becomes quantum-like and breaks into discrete energy states [5]. As a result, these Au NCs behave like molecules and emit fluorescence even though there are no conjugated double bonds in a rigid ring [6].

The preparation method of Au NCs is classified into two approaches, "top-down" and "bottom up". Both synthetic procedures require the protection of capping ligands during synthesis. The top-down approach synthesizes smaller-sized Au NCs from larger-sized nanoparticles assisted with itching reagents and solvents [1,7,8]. However, this approach is time-consuming, and isolation is typically required [9]. It also requires large nanoparticles as the precursors, which limits its further development. The "bottom-up" approach usually prepares Au clusters through the reduction of Au ions into atoms, followed by the spontaneous nucleation of the Au atoms. However, Au ions can be reduced into Au atoms and then form nanoparticles, instead of nanoclusters, because of the potential aggregation of Au atoms in the absence of capping ligands. Too many factors have to be taken into consideration for the successful synthesis of Au NCs. In general, the reaction condition for each synthesis of Au NCs is always dependent on the properties of the capping ligands. Selecting suitable capping ligands has become a challenge for the "bottom-up" synthesis of Au NCs. In the past two decades, the synthesis of Au NCs using macromolecules as the capping ligands has attracted the most attention, because macromolecules are capable of sequestering limited Au ions in a restricted space so that fewer Au nanoparticles are produced. Researchers have used dendrimers and proteins as capping ligands for the preparation of Au NCs [10–13]. However, applications of Au NCs are always dictated by the properties or functions of capping ligands. The synthesis of Au NCs with small molecules is frequently required. For instance, Au NCs synthesized with cysteamine or 4-aminothiophenol can associate with monoamine oxidase B (MAO) and inhibit its activity (MAOI). These MAOI-like Au NCs can mimic drugs to treat patients with depression or anxiety [14]. Siddiqui et al. synthesized GST-Au NCs and utilized fluorescence resonance energy transfer (FRET) to detect melanin. The phenomenon of FRET was achieved by bringing melanin into the proximity of GST-Au NCs through the interaction of GST with melanin [15]. However, Au NCs with small molecules as capping agents are more difficult to synthesize than those with polymer molecules as capping ligands. There is no facile method that can be generally adopted for the synthesis of Au NCs with small molecules, although several small capping ligands have been used to synthesize Au NCs, such as glutathione disulfide (GSSG) [16]. However, this chemical reduction method is time-consuming (about 2 days) and produces unwanted large nanoparticles as byproducts. Several researchers have also reported the synthesis of GST-Au NCs [17–19]. Nevertheless, the synthesis process requires high temperatures or further purification.

Our lab reported a new synthesis approach that could be applied to the synthesis of fluorescent Au NCs using small molecules as capping ligands [20]. The synthesis of Au NCs was completed under mild conditions within one day at 37 °C when quasi-spherical AgNPs were available. No Au nanoparticles were produced as byproducts after Au NC synthesis. The new method was demonstrated to synthesize Au NCs with two different molecules as the capping ligands, GSSG and the reduced form of DTT. As we previously suggested, quasi-spherical AgNPs play two important roles in Au NC synthesis: they act as a reducing agent for gold ions and serve as a platform to accumulate capping ligands on the surface of AgNPs. Au ions on the surface of quasi-spherical AgNPs are trapped by capping ligands and reduced by the AgNPs, while the capping ligands are attracted to the surface of the AgNPs. The assembly of Au ions and capping ligands on the surface of AgNPs turns on the synthesis of Au NCs. One advantage of this method is that the capping ligands can be variable if they can associate with AgNPs. However, the morphology of the AgNPs could influence their ability to attract capping ligands to the surface. Thus, it is not clear whether or not the shape of AgNPs is critical for their synthesis. In this report, four different shapes of AgNPs, quasi-spherical silver nanoparticles (QS-AgNPs), spherical silver nanoparticles (S-AgNPs), triangular silver nanoplates (T-AgNPts) and silver nanoplates (AgNPts), were prepared and used in the experiments to examine the effect of AgNP morphologies on the synthesis of Au NCs. Three small molecules, DTT, oxidized DTT (DTTox), and GSSG were used to synthesize Au NCs as capping ligands to examine whether the synthesis of Au NCs is dependent of the shape of the AgNPs.

2. Materials and Methods

2.1. Materials

Chloroauric acid trihydrate ($HAuCl_4 \bullet 3H_2O$) (99.99%) was obtained from Alpha Aesar (Thermo Fisher Scientific, Ward Hill, MA, USA). Sodium citrate dihydrate ($C_6H_5Na_3O_7 \bullet 2H_2O$) (>99.0%) was purchased from J. T. Baker (J.T. Baker, Philipsburg, NJ, USA). DL, 1, 4-dithiothreitol (DTT) (99%) was purchased from Acros Organics (Geel, Belgium). Silver nitrate (≥99%), glutathione disulfide (GSSG) (≥99%), sodium borohydride (99%), and oxidized DTT (≥98%) were all purchased from Sigma Aldrich (Sigma Aldrich Inc. St. Louis, MO, USA).

2.2. Synthesis of AgNPs

2.2.1. Synthesis of Quasi-Spherical AgNPs

The procedure for synthesizing quasi-spherical AgNPs followed the method described previously, except that the violet LEDs (the average intensity 40 mW/cm^2 and wavelength 405 nm) were used to replace the UVB lamp in the irradiation chamber for the reactions [21]. In general, a reaction bottle containing 40 mL ddH_2O was mixed with the silver nitrate solution (0.5 mL, 0.01 M). Sodium citrate (0.066 g) was added to the solution and mixed thoroughly. The reaction solution was added with ddH_2O to a total of 50 mL to reach the final concentrations of 0.1 mM $AgNO_3$ and 4.5 mM sodium citrate. The reaction solution was then irradiated under violet LEDs in the chamber for 2.5 h (Figure S1).

2.2.2. Synthesis of Spherical AgNP

In brief, silver nitrate solution (0.5 mL, 0.01 M) and sodium citrate (0.1 mL, 0.03 M) were sequentially added into a beaker containing 48.5 mL ddH_2O and mixed thoroughly. Freshly prepared sodium borohydride (0.5mL, 0.01 M) was then placed into the solution. Double distilled H_2O was placed into the solution to make a total of 50 mL to obtain final concentrations of 0.1 mM $AgNO_3$, 0.06 mM sodium citrate, and 0.1 mM $NaBH_4$. The solution was then mixed with stirring for 10 min. The mixture was then irradiated under violet LEDs (405 nm) in the chamber for 80 min.

2.2.3. Synthesis of Silver Nanoplates (AgNPts)

The steps used for the synthesis of silver nanoplates followed the procedure that was previously reported [22]. In brief, the synthesis of AgNPts begins with the same protocol as the preparation of quasi-spherical AgNPs. The quasi-spherical AgNP colloid solution was used to undergo shape transformation to form AgNPts by the following steps. The quasi-spherical AgNPs were irradiated under a green LED set-up in a chamber (λ_{max} 520 nm, average power 48 mW/cm^2) for 2 h. The solution was then changed to another chamber for the irradiation of red LEDs (620 ± 18 nm, average power 116 mW/cm^2) for 3 h (Figure S1). The solution was then removed from the LED chamber to stop the reaction. A UV-Visible spectrophotometer was used to confirm whether the surface plasmon resonance absorption band was formed at approximately 700 nm, indicating the synthesis of silver nanoplates.

2.2.4. Synthesis of Triangular Nanoplates (T-AgNPts)

We prepared a solution (50 mL) containing the final concentrations of 0.1 mM $AgNO_3$, 4.5 mM sodium citrate, and 0.1 mM $NaBH_4$. The solution was mixed by stirring for 5 min. The mixture was then irradiated under sodium LEDs (589 nm) for 1.5 h. A UV-Visible spectrophotometer was employed to verify whether the surface plasmon resonance absorption band was observed at approximately 700 nm, an indication that silver triangular nanoplates had been synthesized.

2.3. Synthesis of Au NCs

Typically, a reaction aqueous solution (1 mL) was formed with final concentrations of $HAuCl_4$ and capping ligands of 0.4 mM and 15 mM, respectively. The total silver concentration of AgNPs was 4.8×10^{-2} mM. The final concentration of the capping ligands may be changed, and it will be specified in the discussion. The solution was then incubated at 37 °C for 7 h or longer. The reaction solution was monitored under a UV-box and a fluorescent spectrophotometer. Dialysis was employed to remove silver ions using a 2.0 kDa MWCO dialysis bag, if necessary.

2.4. Fluorescence Spectroscopy

For a typical fluorescent measurement, a quartz cuvette (4 mm × 4 mm) was used and placed with fluorescent samples (250 µL). A Teflon-coated magnetic stir flea (2 mm × 2 mm) was used to facilitate mixing, if necessary. Each fluorescence sample was measured by a Fluorolog®-3 spectrophotometer (HORIBA Jobin Yvon Inc, Edison, NJ, USA). The temperature of the sample chamber was maintained at room temperature. Each measurement was subtracted by a blank sample (ddH_2O). The excitation shutter was kept closed, except during measurements, to avoid the continuous irradiation of light.

2.5. TEM Imaging of AgNPs and Au NCs

A transmission electron microscope (TEM), Joel JEM-2100 (purchased from Japan Electron Optics Laboratory Co., Ltd., Tokyo, Japan), was used to detect Au NCs and AgNPs. The TEM instrument was operated under air-conditioning at 100 kV and 200 kV. To prepare samples for the TEM detection, the as-prepared Au NCs or AgNP colloidal solutions were placed onto a carbon-coated copper grid. The samples were then air-dried at ambient temperature before examination.

2.6. UV-Visible Spectroscopy

The UV-Vis extinction spectra of all samples were measured at room temperature by the Hitachi U-5100 spectrophotometer (Hitachi Science & Technology, Tokyo, Japan) using a quartz cuvette (1.0 cm × 1.0 cm).

2.7. Quantum Yields

The quantum yield was calculated to estimate the fluorescent efficiency of a newly synthesized Au NC sample. It was defined and calculated as follows:

Quantum yield = number of protons emitted/number of protons absorbed, or

$$Q_S = Q_R(I_S \times OD_R \times n_S^2)/(I_R \times OD_S \times n_R^2) \quad (1)$$

where Q_S is the sample's calculated quantum yield; Q_R is the standard's quantum yield (ethidium bromide, 20%); I_S is the sample's total fluorescent intensity; I_R is the standard's total integrated fluorescent intensity (ethidium bromide, EtBr); OD_S is the optical density for the sample at the absorption (or excitation) wavelength; OD_R is the optical density for EtBr as the reference at the absorption wavelength; n_S is the refraction index of the sample solution; and n_R is the refraction index of the EtBr solution. The values of n_S and n_R were very close and were regarded as the same because both solvents were ddH_2O. The quantum yields of the fluorescent sample (Au NCs) were calculated by Equation (1).

3. Results

3.1. Characterization of Silver Nanoparticles

In a previous report, it was demonstrated that AgNPs play important roles in the synthesis of Au NCs using small molecules as capping ligands [20]. However, differently shaped AgNPs could exhibit capping ligands with different binding abilities. In this study, four different shapes of AgNPs were synthesized: quasi-spherical silver nanoparticles (QS-AgNPs), spherical silver nanoparticles (S-AgNPs), triangular silver nanoplates (T-AgNPts), and nanoplates (AgNPts). After AgNP synthesis, the UV-Visible spectrophotometer was employed to examine whether the SPR absorption band was observed and located at its characteristic wavelengths. Figure 1A–D depict the UV-Vis spectra and TEM images of the as-prepared colloidal silver nanoparticles with different shapes. QS-AgNPs, shown in Figure 1A, exhibit a symmetric LSPR band that has the highest absorption at approximately 408 nm. Silver nanoplates synthesized using a three-stage photochemical reaction (Figure 1B) show bands peaking at 672 nm and 343 nm, correctly corresponding to the in-plane dipolar LSPR and out-of-plane quadrupole modes, respectively. The TEM image showed that the colloids consist of triangular- and round-shaped AgNPts. Triangular silver nanoplates were prepared using the plasmon-mediated process involved in the irradiation of the sodium lamp. As shown in Figure 1C, the UV-Vis spectra of T-AgNPts showed two characteristic bands peaking at 702 nm and 340 nm, corresponding to the in-plane dipolar LSPR mode and out-of-plane quadrupole mode, respectively. Spherical silver nanoparticles (S-AgNPs), shown in Figure 1D, exhibit a symmetric LSPR band that peaks at around 408 nm, similar to the UV-Vis absorption of quasi-spherical silver nanoparticles. The TEM image exhibits the round-shaped morphology of the colloids, which could be either spherical or plated. However, since no peak was detected at 330–340 nm in the UV-Vis spectra, the products are indeed spherical AgNPs.

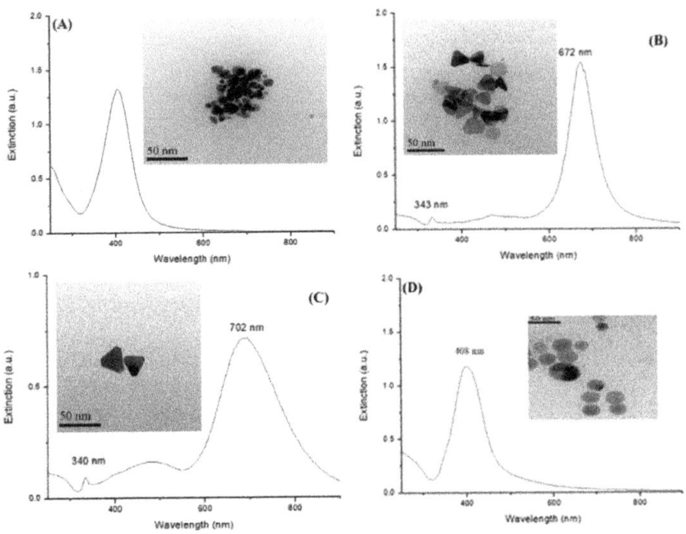

Figure 1. UV-Vis spectra and TEM images of colloidal silver nanoparticles with different shapes. (**A**) Quasi-spherical silver nanoparticles (QS-AgNPs), (**B**) silver nanoplates (AgNPts), (**C**) triangular silver nanoplates (T-AgNPts) and (**D**) spherical silver nanoparticles (S-AgNPs).

3.2. Synthesis of Gold Nanoclusters Mediated by Silver Nanoparticles

A bottom-up (atoms to clusters) strategy was employed to synthesize Au NCs with small molecules as capping ligands. Three capping ligands (DTT, oxidized DTT, and GSSG) and four different shapes of as-prepared AgNPs were used in the synthesis of Au NCs. In general, three required components have to be present in the Au NC synthesis solution: Au^{3+}, Au^{3+}-reducing agents (AgNPs), and capping ligands. In the typical synthesis of Au NCs, AgNP colloids, $HAuCl_4$, and capping ligands are sequentially placed in the reaction solution and mixed. The mixture is then incubated at 37 °C. Figure 2A (odd numbers for QS-AgNPs, even numbers for T-AgNPts) and 2B (odd numbers for AgNPts, even numbers for S-AgNPs) show the results for when GSSG molecules were used as capping ligands and all four different shapes of AgNPs were used as reducing agents. After several hours of incubation, the reaction solution gradually turned bright red under the irradiation of the UV-box equipped with dual wavelengths, 254 and 365 nm. These results suggest that all four different shapes of AgNPs could be used to successfully synthesize Au NCs. To examine whether different concentrations of AgNPs influence the synthesis of Au NCs, different concentrations of QS-AgNPs (0.0, 3.2, 4.8, 6.4, and 8.0 mM) were used in the synthesis. All of the results, except those for 0.0 mM, showed a bright red color under UV-Box when 0.4 mM $HAuCl_4$ and 5 mM GSSG were used. The synthesized GSSG-Au NCs (labelled temporarily) were further investigated to determine their maximal excitation and emission wavelengths using fluorescence spectroscopy. All four synthesized GSSG-Au NCs mediated by four different shapes of AgNPs showed the same maximal excitation and emission wavelengths, at 354 and 623 nm, respectively. In the absence of GSSG, no fluorescent emission was detected (Figure 2A,B, lanes 1 and 2), which revealed that capping ligands are required to produce Au NCs.

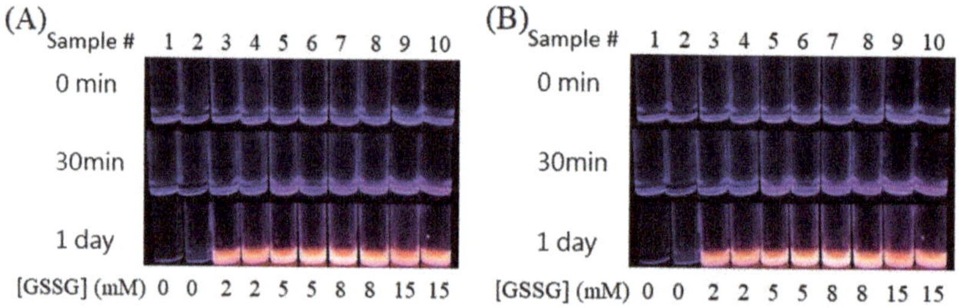

Figure 2. Synthesis of GSSG-Au NCs using 0.4 mM $HAuCl_4$ and different shapes of 4.8×10^{-2} mM AgNPs. Concentrations of GSSG are indicated at the bottom of each figure. The images of the as-prepared GSSG-Au NCs solutions were taken under UV-box detection. Sample # (numbers) are indicated above. (**A**) Odd numbers (3, 5, 7, and 9) indicate that QS-AgNPs were used as the reductants and even numbers (4, 6, 8, and 10) indicate that T-AgNPts were used as the reductants. (**B**) Odd numbers indicate that AgNPts were used as the reductants and even numbers indicate that S-AgNPs were used as the reductants.

To verify whether all four different shapes of Ag NPs are effective for the synthesis of Au NCs with other capping ligands, DTT was used to replace GSSG. The as-prepared reaction solutions of DTT-Au NCs (labeled temporarily), under the detection of the UV-box, are shown in Figure 3A,B. Interestingly, the results clearly show that DTT-Au NCs can be successfully synthesized when QS-AgNPs (#3, 5, 7, and 9 of Figure 3A) and AgNPts (#4, 6, 8, and 10 of Figure 3B) are used in the reaction. In contrast, much less red fluorescence was observed from the solutions when T-AgNPts (#4, 6, 8, and 10 of Figure 3A) and S-AgNPs (#3, 5, 7, and 9 of Figure 3A,B) were used in the synthesis of DTT-Au NCs, indicating that T-AgNPs and S-AgNPs exhibit less efficiency during the synthesis of DTT-Au NCs. To further understand whether more capping ligands or AgNPs are required to optimize the synthesis of Au NCs, higher concentrations of DTT and AgNPs were used to synthesize DTT-Au NCs. However, fluorescent intensities did not increase significantly, suggesting that higher concentrations of capping ligands or AgNPs are not able to increase the percent yield of Au NCs.

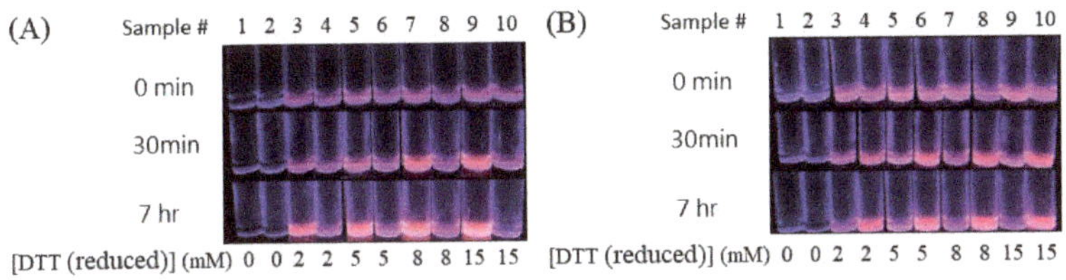

Figure 3. Synthesis of DTT-Au NCs using 0.4 mM $HAuCl_4$ and different shapes of 4.8×10^{-2} mM AgNPs. Concentrations of DTT are indicated on the bottom of each figure. The images of aqueous solutions of the as-prepared DTT-Au NCs are taken under UV-box detection. Sample # (numbers) are indicated above. (**A**) Odd numbers (3, 5, 7, and 9) indicate that QS-AgNPs were used as the reductants and even numbers (4, 6, 8, and 10) indicate that T-AgNPs were used as the reductants. (**B**) Odd numbers indicate S-AgNPs were used as the reductants and even numbers indicate that AgNPts were used as the reductants.

We further used an oxidized form of DTT (DTTox) to synthesize DTTox-Au NCs (labeled temporarily). However, the results are very similar to those using DTT as the capping ligands (Figure 4). Therefore, DTTox-Au NCs can be successfully synthesized when QS-AgNPs and AgNPts are used in the reaction. Less efficient Au NC synthesis was observed when S-AgNPs or T-AgNPs were used to synthesize DTT-Au NCs or DTTox-Au NCs. The difference in efficiency of Au NC synthesis could be attributed to the discrepancy in the association of AgNPs with capping ligands.

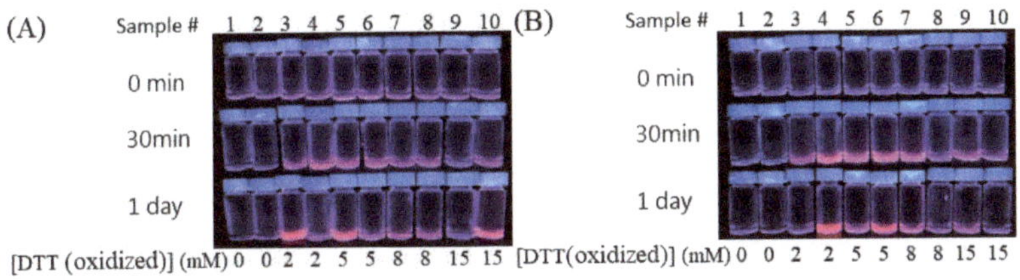

Figure 4. Synthesis of DTTox-Au NCs using 0.4 mM $HAuCl_4$ and different shapes of 4.8×10^{-2} mM AgNPs. Concentrations of DTTox are indicated on the bottom of each figure. The images of aqueous solutions of the as-prepared DTTox-Au NCs are shown under UV-box detection. Sample # (numbers) are indicated above. (**A**) Odd numbers (3, 5, 7, and 9) indicate that QS-AgNPs were used as the reductants and even numbers (4, 6, 8, and 10) indicate that T-AgNPs were used as the reductants. (**B**) Odd numbers indicate S-AgNPs were used as the reductants and even numbers indicate that AgNPts were used as the reductants.

It is conceivable that the association of capping ligands with these AgNPs could play an important role in the synthesis of Au NCs. To evaluate whether GSSG, DTT, and DTTox can associate with AgNPs, differently shaped AgNPs were mixed with these capping ligands. The mixtures were detected by UV-Vis spectrophotometry. Figure 5 demonstrates the time course of the UV-Vis spectra after the DTT molecules were mixed with AgNPs. All of the maximal absorption wavelengths were red-shifted after DTT was associated with various AgNPs, indicating that aggregates of AgNPs were formed. In addition, the intensities decreased significantly. To confirm the aggregation of AgNPs, TEM images (upper panels of Figure 5A,D) of AgNPs interacting with DTT were also taken. The images show that QS-AgNPs (upper panel of Figure 5A) aggregated slightly after interacting with DTT, while T-AgNPs (upper panel of Figure 5D) aggregated aggressively after interacting with DTT. To further understand the kinetics of AgNP aggregation, the figures were replotted to show red-shift wavelength vs. interaction times. The aggregation time can be estimated from the new figures (Figure 5E–H). The results show that the extent of the red-shift wavelengths differs. To investigate the aggregation kinetics of AgNPs in the presence of DTT, the red-shift curves of the samples were analyzed. Exponential functions were employed to fit the curves. The fitting parameters are shown in the panel of each figure. The calculated aggregation times were 213 min for QS-AgNPs (highlighted in red), 18.4 min for S-AgNPs, 166 min for AgNPts, and 22.1 min for T-AgNPts. The aggregation time of QS-AgNPs was longer than that of T-AgNPts, consistent with the observations of the TEM images shown in Figure 5A,D. The aggregation time of S-AgNPs was also much shorter than that of the QS-AgNPs. After comparing the aggregation time with the results of Au NC synthesis, it was found that the longer the aggregations time, the better the Au NC synthesis. These data suggest that the synthesis of Au NCs is dependent on the aggregation rate of AgNPs after being mixed with capping ligands.

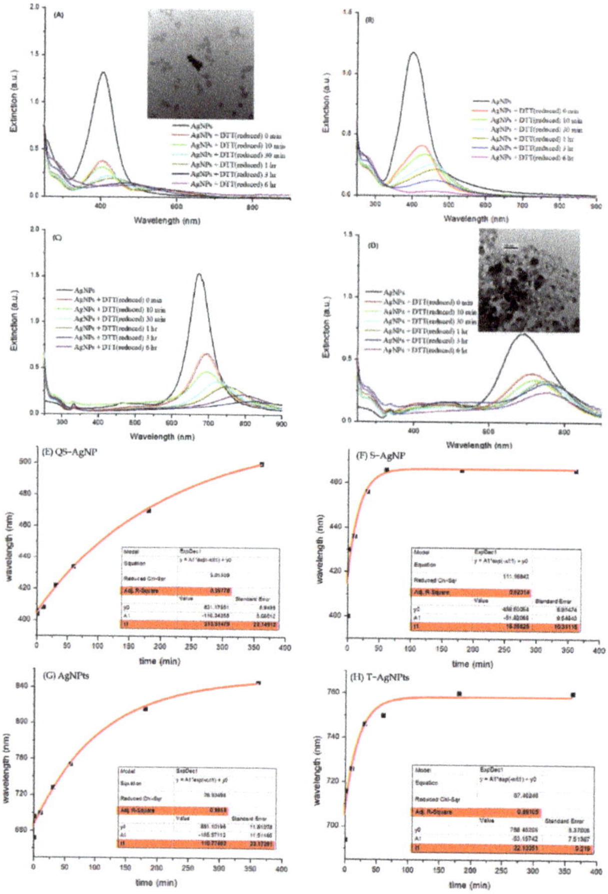

Figure 5. UV-Vis spectra of the association of AgNPs with DTT (**A–D**) and re-plots of red-shift wavelengths versus interaction times (**E–H**). (**A**) QS-AgNP colloids, (**B**) S-AgNP colloids, (**C**) AgNPts colloids, and (**D**) T-AgNPts colloids at various interaction times. Time-dependent AgNP aggregations were shown in (**E**) QS-AgNP colloids, (**F**) S-AgNP colloids, (**G**) AgNPts colloids and (**H**) T-AgNPts colloids.

Based on the current data and results shown for the as-prepared Au NCs, it is concluded that QS-AgNPs and AgNPts are the best silver nanoparticles for mediating the synthesis of Au NCs for all three small molecules as the capping ligands (Table 1). "Good" or "fair" is classified based on whether the as-prepared Au NCs solution can be loaded onto the electrophoretic mobility shift assay (EMSA) and detected by the UV-box without further concentrating the process. In this study, GSSG molecules could be used as capping ligands to synthesize GSSG-Au NCs no matter what shape of AgNPs were used for synthesis.

Table 1. Synthesis efficiency of Au NCs using various AgNPs and capping ligands.

Capping Ligands	QS-AgNPs	T-AgNPts	AgNPts	S-AgNPs
DTTox	good	fair	good	fair
DTT	good	fair	good	fair
GSSG	good	good	good	good

3.3. Characterization of Gold Nanoclusters

To verify that Au NCs had been successfully synthesized in the presence of AgNPs, the as-prepared Au NCs were observed under TEM. The TEM images showed that synthesized particles were approximately 5–6 nm or smaller (Figure 6A–D). The size of the DTT-Au NCs particles was less than 5 nm when QS-AgNPs or AgNPts were used to mediate the synthesis. However, the average size of the GSSG-Au NC particles was approximately 5–6 nm. These results confirm that the presented method can successfully synthesize Au NCs. To confirm that the images of the particles detected by TEM were not smaller-sized AgNPs, an energy dispersive X-ray spectrometer (EDS) was employed to analyze the compositions of the particles. The EDS results for DTT-Au NCs synthesized by QS-AgNPs and GSSG-Au NCs synthesized by S-AgNPs are shown in Figure 7A,B. The EDS data show that the major components of the synthesized particles were Au, C, and S in both cases. These results suggest that Au is from the core of the cluster, and S is from the capping ligands (DTT and GSSG molecules). No silver elements were detected in these EDS analyses. Therefore, these data confirm that the particles are indeed DTT-Au NCs and GSSG-Au NCs, respectively.

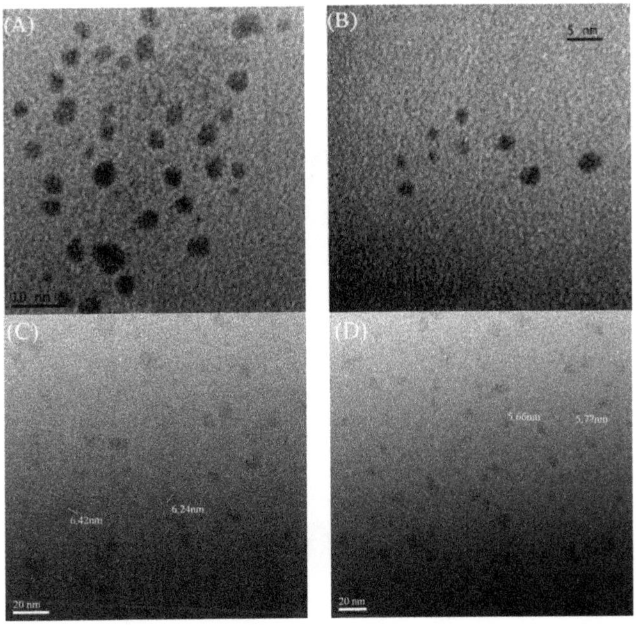

Figure 6. TEM image of Au NCs. (**A**) DTT-Au NCs synthesized by QS-AgNPs. The sizes of the particles are under 5 nm. (**B**) DTT-Au NCs synthesized by AgNPts. The sizes of the particles are under 5 nm. (**C**) GSSG-Au NCs synthesized by S-AgNPs. Most of the particles are approximately 6 nm in size. (**D**) GSSG-Au NCs synthesized by T-AgNPts. Most of the particles are approximately 5–6 nm in size.

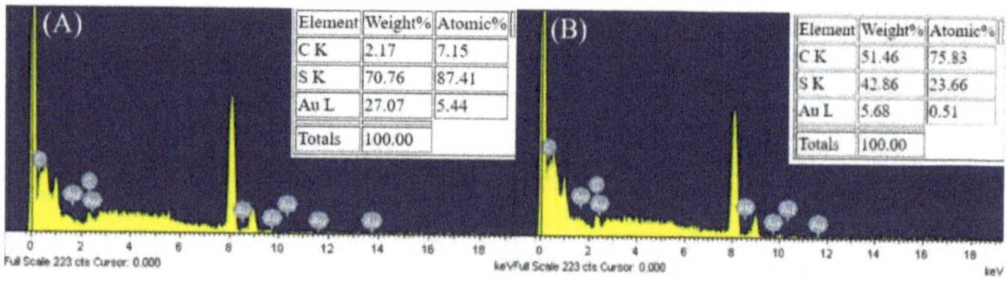

Figure 7. Energy dispersive X-ray spectrometer (EDS). (**A**) EDS analysis of DTT-Au NCs synthesized by QS-AgNPs. (**B**) EDS analysis of GSSG-Au NCs synthesized by S-AgNPs. In both cases, the analyzed particles consist of C, S, and Au elements (upper panel in **A**,**B**). No Ag elements were detected.

A previous report suggested that limited ions have to be entrapped in a sequestered space during the synthesis of nanoclusters [12]. In this study, AgNPs were employed as a reaction platform, providing surfaces that play dual roles in the synthesis, as reducing agents and attracting capping ligands on its surfaces to form small spaces to accommodate limited Au atoms. Figure 8A,C show the TEM images of DTTox-Au NCs, which were synthesized in the early stage of the synthesis (5 h only). Figure 8A demonstrates that the newly synthesized DTTox-Au NCs, which were reduced by QS-AgNPs, were arranged into a round shape, similar to that of the QS-AgNPs. Figure 8C demonstrates that the newly synthesized DTTox-Au NCs, which were reduced by AgNPts (mixtures of round and triangular nanoplates), were arranged into circular and triangular shapes, similar to the outline of AgNPts. These data indicate that Au NCs were indeed synthesized on the surface of AgNPs. The EDS data show that the major metal elements were Au and Cu, and very little Ag was detected (Figure 8B,D). These results reveal that the particles were Au NCs. It has to be noted that hollow spaces without Au NCs in the center can be observed in Figure 8C, indicating that capping ligands cannot easily reach the center of triangular or circular AgNPts. On the contrary, DTTox-Au NCs filled in the whole space when QS-AgNPs were used.

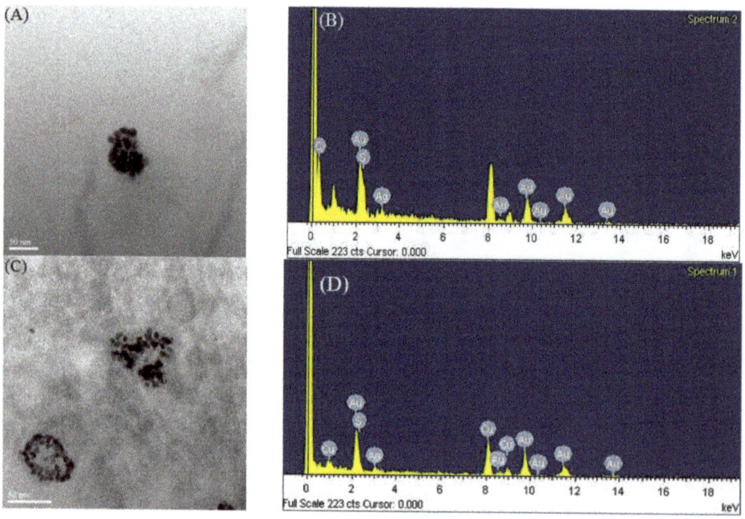

Figure 8. TEM image of DTTox-Au NCs synthesized for 5 h. (**A**) DTTox-Au NCs synthesized by QS-AgNPs. The particles were less than 5 nm in size. (**B**) The EDS analysis shows very little Ag. (**C**) DTTox-Au NCs synthesized by AgNPts. The sizes of the particles are under 5 nm. (**D**) EDS analysis.

To evaluate the properties of these Au NCs, some of the Au NC products were measured to determine their fluorescent wavelengths of excitation and emission (Table 2). Fluorescent quantum yields were also calculated (Table 3). As the data revealed, the synthesized GSSG-Au NCs showed the same emission wavelength at 623 nm, and DTT-Au NCs emitted the highest fluorescent intensity at 640 nm. These results reveal that the fluorescent emission wavelength is related to the capping ligands. The fluorescent quantum yields showed a slight difference, which could have been due to the standard error.

Table 2. Excitation and emission wavelengths of the synthesized Au NCs.

	QS-AgNPs		AgNPts	
Capping Ligands	**Excited Wavelength**	**Emission Wavelength**	**Excited Wavelength**	**Emission Wavelength**
GSSG	354 nm	623 nm	354 nm	623 nm
DTT	354 nm	640 nm	354 nm	643 nm
DTTox	360 nm	676 nm	357 nm	668 nm

Table 3. Quantum yields of the synthesized Au NCs.

Capping Ligands	**DTT**	**DTT**	**GSSG**	**GSSG**	**GSSG**	**GSSG**
AgNPs	QS-AgNP	AgNPt	QS-AgNP	T-AgNPt	S-AgNP	AgNPt
QY(%)	7.2	5.4	6.8	6.3	7.3	7.8

4. Conclusions

We previously reported a facile synthesis method for thiolated Au NCs, mediated by quasi-spherical AgNPs. It was suggested that QS-AgNPs provide the surface of its particles and play two important roles in the synthesis of Au NCs. One is to reduce gold ions, and the other is to create a suitable space with small capping ligands for the synthesis of clusters. Au ions distributed on the surface of AgNPs are trapped and reduced by the AgNPs while the capping ligands are attracted to the surface of the AgNPs, resulting in an association with the reduced Au atoms. Therefore, it is conceivable that the shape and size distributions of AgNPs could be critical for the binding of capping ligands and dictate the successfulness of the synthesis. In this study, AgNPs with four different morphologies (QS-AgNPs, S-AgNPs, T-AgNPts, and AgNPts) were prepared and used to examine whether this approach is dependent on the morphology of AgNPs. Based on the data shown, the results are distinguishable, and QS-AgNPs and AgNPts are the best choices to mediate the synthesis of Au NCs for all three tested small molecules as the capping ligands. The TEM images showed that Au NCs are synthesized on the surface of silver nanoparticles (Figure 8). Differently shaped AgNPs serve as platforms to synthesize Au NCs with small capping ligands (Figure 9). It was demonstrated that Au NCs were transiently assembled on the surface of silver nanoparticles in the method. The formation of Au NCs was observed on the whole surface of the QS-AgNPs if the synthesis of DTTox-Au NCs was mediated by QS-AgNPs (Figure 9A). In contrast, the formation of Au NCs was only observed on the edges and corners of AgNPts if the synthesis of DTTox-Au NCs was mediated by AgNPts. Our data imply that DTT and/or gold ions cannot bind to or reach the center of T-AgNPts (Figure 9B). The synthesized Au NCs display similar fluorescent properties, including quantum yields and excitation and emission wavelengths. However, the size distributions of silver nanoparticles were not discussed in this study, because the synthesis of uniformly sized AgNPs is beyond our capability.

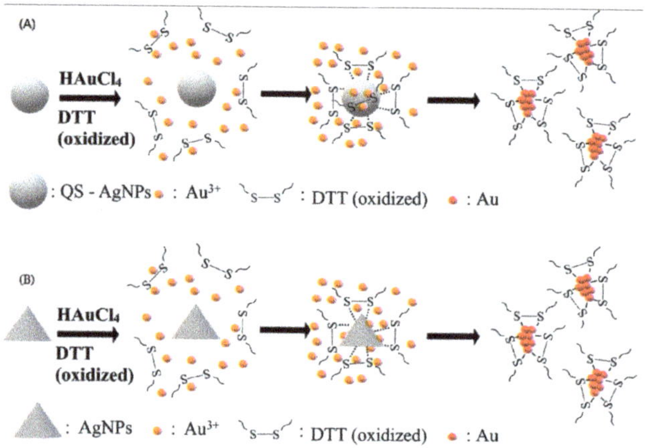

Figure 9. Two differently shaped AgNPs serve as a platform to synthesize Au NCs with small capping ligands. (**A**) The synthesis of DTTox-Au NCs mediated by QS-AgNPs. The formation of Au NCs was observed on the whole surface of the QS-AgNPs. (**B**) The synthesis of DTTox-Au NCs mediated by AgNPts. The formation of Au NCs was only observed on the edges and corners of the AgNPts.

Supplementary Materials: The following supporting information can be downloaded at: https://www.mdpi.com/article/10.3390/nano13162338/s1, Figure S1: The set-up for LED irradiation to generate AgNPs. The cabinet, containing 24 LEDs with wavelengths of 405 nm (violet LED, left panel), 525 nm (green LED, center panel), and 630 nm (red LED, right panel) was used to produce a uniform light intensity (0.01 W/cm^2), surrounding a 50 mL reaction vial; Figure S2: The synthesis of DTT-Au NCs using 0.4 mM $HAuCl_4$ and differently shaped 4.8×10^{-2} mM AgNPs. The concentrations of DTT are indicated on the bottom of each figure. The images of aqueous solutions of the as-prepared DTT-Au NCs were taken during UV-box detection. (A) Odd numbers (1, 3, 5, and 7) indicate that QS-AgNPs were used as the reductants and even numbers (2, 4, 6, and 8) indicate that T-AgNPs were used as the reductants. (B) Odd numbers indicate S-AgNPs were used as the reductants and even numbers indicate that AgNPts were used as the reductants.

Author Contributions: Conceptualization and methodology, J.-C.C.; QS-AgNPs, C.-Y.C., T.-H.T. and Y.-R.W.; synthesis of gold nanoclusters, C.-Y.C., Y.-R.W., J.-H.S. and T.-H.T.; fluorescence analysis, C.-Y.C., Y.-R.W. and T.-H.T.; TEM images and EDS measurements, B.-R.C.; synthesis of S-AgNPs, T-AgNPts, and AgNPts, C.-Y.C., Y.-R.W., Y.-S.W. and F.-Y.J.; writing, review and editing, C.-L.H. and J.-C.C.; project administration, C.-L.H. All authors have read and agreed to the published version of the manuscript.

Funding: This work was supported by the Ministry of Science and Technology (Taiwan) for C-L.H. (MOST 111-2113M-415-005) and the Department of Applied Chemistry, National Chiayi University, Taiwan.

Institutional Review Board Statement: Ethical review and approval were waived for this study due to all the equipment, reagents, and experiments are not involved in animals or humans.

Data Availability Statement: All data generated and analyzed during this study are included in this paper and the attached Supplementary Materials. The data presented in this study are available on request from the corresponding author.

Conflicts of Interest: The authors declare no conflict of interest.

References

1. Cui, H.; Shao, Z.-S.; Song, Z.; Wang, Y.-B.; Wang, H.-S. Development of gold nanoclusters: From preparation to applications in the field of biomedicine. *J. Mater. Chem. C* **2020**, *8*, 14312–14333.
2. Chen, L.-Y.; Wang, C.-W.; Yuan, Z.; Chang, H.-T. Fluorescent gold nanoclusters: Recent advances in sensing and imaging. *Anal. Chem.* **2015**, *87*, 216–229.

3. Li, G.; Jin, R. Atomically precise gold nanoclusters as new model catalysts. *Acc. Chem. Res.* **2013**, *46*, 1749–1758. [CrossRef] [PubMed]
4. Liang, A.; Liu, Q.; Wen, G.; Jiang, Z. The surface-plasmon-resonance effect of nanogold/silver and its analytical applications. *TrAC Trends Anal. Chem.* **2012**, *37*, 32–47.
5. Berends, A.C.; de Mello Donega, C. Ultrathin one-and two-dimensional colloidal semiconductor nanocrystals: Pushing quantum confinement to the limit. *J. Phys. Chem. Lett.* **2017**, *8*, 4077–4090. [PubMed]
6. Qu, X.; Li, Y.; Li, L.; Wang, Y.; Liang, J.; Liang, J. Fluorescent gold nanoclusters: Synthesis and recent biological application. *J. Nanomater.* **2015**, *2015*, 4.
7. Pichugina, D.Y.A.; Kuz, N.E.; Shestakov, A.F. Ligand-protected gold clusters: The structure, synthesis and applications. *Russ. Chem. Rev.* **2015**, *84*, 1114.
8. Rao, T.U.B.; Pradeep, T. Luminescent Ag7 and Ag8 clusters by interfacial synthesis. *Angew. Chem. Int. Ed.* **2010**, *49*, 3925–3929.
9. Duan, H.; Nie, S. Etching colloidal gold nanocrystals with hyperbranched and multivalent polymers: A new route to fluorescent and water-soluble atomic clusters. *J. Am. Chem. Soc.* **2007**, *129*, 2412–2413.
10. Zheng, J.; Dickson, R.M. Individual water-soluble dendrimer-encapsulated silver nanodot fluorescence. *J. Am. Chem. Soc.* **2002**, *124*, 13982–13983. [CrossRef]
11. Jao, Y.-C.; Chen, M.-K.; Lin, S.-Y. Enhanced quantum yield of dendrimer-entrapped gold nanodots by a specific ion-pair association and microwave irradiation for bioimaging. *Chem. Commun.* **2010**, *46*, 2626–2628.
12. Xie, J.; Zheng, Y.; Ying, J.Y. Protein-directed synthesis of highly fluorescent gold nanoclusters. *J. Am. Chem. Soc.* **2009**, *131*, 888–889. [CrossRef]
13. Liu, C.L.; Wu, H.T.; Hsiao, Y.H.; Lai, C.W.; Shih, C.W.; Peng, Y.K.; Tang, K.C.; Chang, H.W.; Chien, Y.C.; Hsiao, J.K. Insulin-directed synthesis of fluorescent gold nanoclusters: Preservation of insulin bioactivity and versatility in cell imaging. *Angew. Chem. Int. Ed.* **2011**, *50*, 7056–7060.
14. Wu, X.; Xu, H.; Luo, F.; Wang, J.; Zhao, L.; Zhou, X.; Yang, Y.; Cai, H.; Sun, P.; Zhou, H. Sizes and ligands tuned gold nanocluster acting as a new type of monoamine oxidase B inhibitor. *Biosens. Bioelectron.* **2021**, *189*, 113377.
15. Siddiqui, M.F.; Jeon, S.; Kim, M.-M. Rapid and sensitive detection of melanin using glutathione conjugated gold nanocluster based fluorescence quenching assay. *Spectrochim. Acta Part A Mol. Biomol. Spectrosc.* **2021**, *247*, 119086.
16. Zhang, Q.; Wang, J.; Meng, Z.; Ling, R.; Ren, H.; Qin, W.; Wu, Z.; Shao, N. Glutathione disulfide as a reducing, capping, and mass-separating agent for the synthesis and enrichment of gold nanoclusters. *Nanomaterials* **2021**, *11*, 2258. [CrossRef] [PubMed]
17. Qi, S.; Al-mashriqi, H.S.; Salah, A.; Zhai, H. Glutathione capped gold nanoclusters-based fluorescence probe for highly sensitive and selective detection of transferrin in serum. *Microchem. J.* **2022**, *175*, 107163.
18. Chen, Y.-S.; Kamat, P.V. Glutathione-capped gold nanoclusters as photosensitizers. Visible light-induced hydrogen generation in neutral water. *J. Am. Chem. Soc.* **2014**, *136*, 6075–6082.
19. Zhang, C.; Zhou, Z.; Qian, Q.; Gao, G.; Li, C.; Feng, L.; Wang, Q.; Cui, D. Glutathione-capped fluorescent gold nanoclusters for dual-modal fluorescence/X-ray computed tomography imaging. *J. Mater. Chem. B* **2013**, *1*, 5045–5053. [CrossRef]
20. Chang, C.-Y.; Tseng, T.-H.; Chen, B.-R.; Wu, Y.-R.; Huang, C.-L.; Chen, J.-C. Silver Nanoparticle-Mediated Synthesis of Fluorescent Thiolated Gold Nanoclusters. *Nanomaterials* **2021**, *11*, 2835. [CrossRef]
21. Xie, Z.X.; Tzeng, W.C.; Huang, C.L. One-pot synthesis of icosahedral silver nanoparticles by using a photoassisted tartrate reduction method under UV light with a wavelength of 310 nm. *ChemPhysChem* **2016**, *17*, 2551–2557. [PubMed]
22. Huang, C.-C.; Chen, H.-J.; Leong, Q.L.; Lai, W.K.; Hsu, C.-Y.; Chen, J.-C.; Huang, C.-L. Synthesis of silver nanoplates with a narrow LSPR band for chemical sensing through a plasmon-mediated process using photochemical seeds. *Materialia* **2022**, *21*, 101279.

nanomaterials MDPI

Article

Electric Field-Induced Nano-Assembly Formation: First Evidence of Silicon Superclusters with a Giant Permanent Dipole Moment

Fatme Jardali [1], Jacqueline Tran [1], Frédéric Liège [1,2], Ileana Florea [1,3], Mohamed E. Leulmi [4] and Holger Vach [1,5,*]

1 Laboratoire de Physique des Interfaces et des Couches Minces, CNRS, École Polytechnique, IP Paris, 91128 Palaiseau, France; fjardali0@gmail.com (F.J.); jacqueline.tran@polytechnique.edu (J.T.); frederic.liege@universite-paris-saclay.fr (F.L.); if@crhea.cnrs.fr (I.F.)
2 LMF, École Normale Supérieure, Paris-Saclay, 91190 Gif-sur-Yvette, France
3 CRHEA, CNRS, Université Côte d'Azur, 06903 Sophia-Antipolis, France
4 Center for SiNC Applications, 75000 Paris, France; mohamed.leulmi@siliconnanoclusters.com
5 Centre for Research in Molecular Modeling, Concordia University, Montreal, QC H4B 1R6, Canada
* Correspondence: holger.vach@polytechnique.edu

Abstract: The outstanding properties of silicon nanoparticles have been extensively investigated during the last few decades. Experimental evidence and applications of their theoretically predicted permanent electric dipole moment, however, have only been reported for silicon nanoclusters (SiNCs) for a size of about one to two nanometers. Here, we have explored the question of whether suitable plasma conditions could lead to much larger silicon clusters with significantly stronger permanent electric dipole moments. A pulsed plasma approach was used for SiNC production and surface deposition. The absorption spectra of the deposited SiNCs were recorded using enhanced dark-field hyperspectral microscopy and compared to time-dependent DFT calculations. Atomic force microscopy and transmission electron microscopy observations completed our study, showing that one-to-two-nanometer SiNCs can, indeed, be used to assemble much larger "superclusters" with a size of tens of nanometers. These superclusters possess extremely high permanent electric dipole moments that can be exploited to orient and guide these clusters with external electric fields, opening the path to the controlled architecture of silicon nanostructures.

Keywords: hydrogenated silicon nanoclusters; permanent electric dipole moment; self-assembled nanostructures; surface deposition; superclusters

1. Introduction

The existence of hydrogenated silicon nanoclusters (SiNCs) with a permanent electric dipole moment and size of about one nanometer was theoretically predicted as early as 2005 [1]. Their outstanding stability has been traced back to extensive electron delocalization caused by the significant over-coordination of one or several of their constituting silicon atoms [2–4]. Ab initio simulations of the growth dynamics in a plasma reactor demonstrated the self-organized assembly of such SiNCs [5]. Possible applications were predicted in the field of photovoltaics and terahertz communication [6,7]. Recently, their permanent electric dipole moment was directly measured in a plasma reactor to be of the order of 2.2 Debye, in excellent agreement to prior theoretical predictions. Clusters with this relatively large dipole moment have consequently been used to reduce the work function of LaB_6 cathodes employed, for example, in satellite thrusters. The simple presence of these SiNCs on a LaB_6 cathode actually increased the thermionic electron density emission by a factor of up to 30. To this end, it was necessary, however, to deposit the polar SiNCs in a well-oriented manner, which was simply accomplished with the application of an electric field during their surface deposition on the cathode [8]. Interestingly, Schäfer et al. have very recently confirmed the existence of silicon clusters with a permanent electric dipole moment by means of electric molecular beam deflection experiments at cryogenic temperatures for clusters with 30 to 90 atoms [9], following the theoretical work of Jackson and Jellinek [10]. In this work, Schäfer et al. concluded that the dipole moment per atom is almost constant, indicating that the dipole moment should increase with cluster size. Our previous investigations, both theoretical and experimental, concerned quasi-monodisperse cluster size distributions with cluster sizes between 1 and 2 nm. Therefore, the work by Schaefer et al. immediately triggered our interest in the permanent electric dipole moments of much larger silicon clusters.

A silicon nanoparticle with a perfect crystalline structure (i.e., with all atoms being four-times coordinated) cannot have a permanent electric dipole moment. Therefore, we must restrict our search for large polar clusters to those that are not perfectly crystalline. However, it is not trivial to predict whether a cluster of a certain size will be crystalline or amorphous. There is a general consensus that growing silicon clusters undergo an auto-organized phase transition from that which is ill-defined to that of a crystalline morphology at a certain size, but it has not been possible to clarify at which size this transition takes place. On the one hand, it is well known that amorphous dusty clusters might be as large as hundreds of micrometers, while on the other hand, clusters as small as 2 nm have been observed to be crystalline. In this latter work, Kortshagen et al. chose plasma conditions in such a way that nanocrystal agglomeration was essentially suppressed, and the crystallization of even the smallest clusters was attributed to unconventional heating through electron–ion recombination processes [11].

To make sure that our nanoclusters do not have a crystalline structure, we thus adopted a scheme that is exactly contrary to the one of Kortshagen et al. Specifically, we created the same plasma environment as the one that already permitted us to produce one-to-two-nanometer amorphous-like SiNCs with a relatively strong permanent dipole moment [8]. Furthermore, we increased the on-time of our pulsed plasma from about 2 s to 8 s, which was sufficient time for the tiny SiNCs to agglomerate before the end of the plasma pulse. We will show that the resulting self-organized "superclusters" are, indeed, non-crystalline, and possess an unexpectedly large absorption spectrum as well as an extremely strong permanent electric dipole moment. This allows us to present the first experimental evidence of electric field-induced assembly formation with silicon superclusters.

2. Materials and Methods

2.1. Cluster Preparation

Since the experimental details have been published elsewhere [8], we will only concentrate on some major features here. Inside a 0.02 m^3 vacuum chamber, an RF (13.56 MHz) CCP discharge was generated (Cesar 133 RF power supply) at low pressure (10 to 15 Pascals) between a grounded and a powered electrode. The circular electrode surfaces had a diameter of 7 cm and were separated by 4 cm. The shower head shape of the powered electrode allowed the injection of a gas mixture of silane (2%) diluted in argon (98%) directly between the electrodes, to produce hydrogenated silicon nanoclusters in a pulsed discharge. The flow rates of both the SiH_4/Ar gas mixture and pure argon were controlled by mass flow controllers (MFC-Bronkhorst F-201DV, Low-ΔP-Flow), which were operated at typical flow rates of 10 sccm and 40 sccm, respectively. To also control the polar SiNCs in a horizontal direction (i.e., parallel to the substrate) before their deposition, a second set of vertical electrodes of 25 mm diameter was installed on both sides of the standard horizontal reactor electrodes.

The particle charging in low-pressure cold plasmas, like the one used in this work, has been extensively studied (see, for example, Fridman et al. [12]). It has been demonstrated, for instance, that all silicon particles undergo charge fluctuations in the plasma due to ion and electron bombardment occurring on their surfaces. For the decharging of dust nanoparticles in the plasma afterglow, it was shown that some particles keep a residual charge that has been evaluated to be 2% of the charge they acquired in the plasma [13,14]. For 3 nm SiNCs, it was proven that they never have more than one elementary charge in the plasma [12]. This means that all SiNCs considered here are neutral in the afterglow. We underline that we have already given experimental proof for this conclusion regarding our specific nanoclusters [8]: After their creation, our SiNCs had to travel a certain distance before laser detection. Even if they had been charged with only one single elementary charge, they should have traveled this distance within 9 μs because of the applied electric DC field. In the experiment, however, the majority of SiNCs only arrived after about 150 ms at the detecting laser beam. This travel time roughly corresponds to the time imposed by the gas flow. We can, thus, be sure that our SiNCs are neutral in the afterglow. Consequently, the electric DC field applied in the afterglow only acts on their permanent electric dipole moment.

As concerns the role of the SiNC charge in the plasma for the creation of a dipole moment, we have used ab initio methods to evaluate the influence of one elementary charge on their growth mechanism. It turns out that our 1 nm SiNCs only suffer minor deformations, but acquire a dipole moment that is considerably larger (about 4.0 Debye) than the one of their neutral counterparts (about 2.3 Debye). Therefore, the temporary charging of the SiNCs in the plasma actually facilitates the agglomeration for supercluster formation. We expect this mechanism to be operational until a supercluster size of about 50 nm in diameter. Thereafter, the electrostatic Coulomb repulsion is expected to overcome the kinetic energy of the colliding SiNCs [15].

We have previously demonstrated that we can totally neglect collisions between individual SiNCs in the time frame between the end of the plasma pulse and the moment of their surface deposition [8]. However, during the duration of the plasma on-time, the SiNCs are trapped in the plasma bulk and their probability of collision increases with the duration of the discharge. In our previous work, we have shown that a plasma on-time of 6 s or longer will strongly favor SiNC agglomeration. Therefore, we have chosen a discharge duration of 8 s for the present study. At 50 ms after the end of the discharge pulse, a DC bias voltage of 120 V or 50 V was applied either to the horizontal electrodes or to the vertical ones, respectively, to orient and guide the SiNCs in space before their deposition.

2.2. Enhanced Darkfield Hyperspectral Microscopy

CytoViva's patented enhanced darkfield illumination performs two functions very differently from standard darkfield microscopes: First, it couples the source illumination directly to the condenser system, minimizing light loss. Second, it collimates the source illumination and focuses it precisely onto the condenser annulus. This maximizes the number of oblique angle photons focused on a very shallow focal plane. As a result, a significant increase in the signal-to-noise ratio detection capability of the darkfield is observed (up to 10-fold). In addition, a diffraction grating spectrophotometer is installed onto a microscope camera mount. It captures the unique reflectance spectra of objects from the microscope field of view in the wavelength region of about 400 nm to 1700 nm. The complete spectra for each pixel of the CCD detector are captured. Spectral data are reported in high resolution (down to 2.0 nm). The development of enhanced darkfield hyperspectral microscopy provides the ability to identify nanoscale materials in complex environments in a semi-quantitative manner. Hyperspectral images captured with the CytoViva system look similar to optical images. However, each pixel of a hyperspectral image contains the spectral response for the spatial area of that pixel. Using integrated hyperspectral image analysis software, the unique spectral responses of nanomaterials can be identified and easily mapped throughout the sample [16].

2.3. Atomic Force Microscopy (AFM)

For the acquisition of the AFM images, a commercial Bruker Dimension Icon in the PeakForce Tapping (PFT) mode was used. PFT is a Bruker-exclusive mode optimized to simultaneously achieve high-resolution images and quantitative mechanical measurements based on the deflection of the cantilever. During operation, the cantilever is brought in and out of contact with the surface in a non-resonant cycle, where the PFT algorithms directly control the instantaneous force interaction. Because the entire cycle is monitored and recorded, FT provides force spectroscopy information at each pixel. In the feedback mode, the instantaneous peak force is limited, potentially down to the piconewton level [17].

2.4. TEM, EDX, and Focused Ion Beam Analysis

Transmission electron microscopy (TEM), high-angle annular darkfield (HAADF), and energy-dispersive X-ray (EDX) spectroscopy analyses were performed on a Titan Themis transmission electron microscope operating at 200 kV equipped with a Cs aberration probe corrector and a Super X detector that allows chemical analyses of light and heavy elements with a spatial resolution within the nm range. In this case, a specific specimen focused ion beam (FIB) lamella preparation was required. Thus, cross-section lamellae were prepared using a standard lift-out procedure within a FIB dual beam microscope (FIB, FEI-Scios Dual Beam). For the chemical analyses, we chose, as main elements of interest, silicon (Si) with the K_α = 1.74 KeV ionization edge and oxygen (O) with the K_α = 0.523 KeV ionization edge. Carbon (C) and platinum (Pt) protective layers were deposited on top of the sample prior to the FIB milling process to prevent possible gallium (Ga) ion implantation during the milling process.

2.5. First Principles Calculations

In our previous works, we have demonstrated the reliability of the B3LYP functional combined with Grimme's DFT-D2 dispersion correction scheme for our SiNCs in comparison to MP2 calculations [3,6] and experimental measurements [8]. Therefore, both structural optimizations and time-dependent density functional theory (TD-DFT) calculations were accomplished at a B3LYP + DFT-D2 level of theory using Gaussian G16 [18]. After all structural optimizations, frequency analyses were performed to assure that the optimized geometries really corresponded to minima on the corresponding potential energy surfaces. For the TD-DFT calculations, we always calculated 600 excited states leading to the shown absorption spectra. Since all computational details have already been published elsewhere [19], we only focus on some specifications here: To treat the relatively large cluster aggregates on the same level of theory as the individual clusters, we limited ourselves to the 6–31G(d,p) basis set, including dispersive interactions, as described by Grimme's DFT-D2 method [20]. We underline that this approach led to a perfect agreement between the dipole moment for one SiNC, calculated here to be 2.28 Debye, and one previously measured to be 2.25 $\pm$ 0.25 Debye [8]. Nevertheless, we tested a possible influence of diffuse functions and a more recent Grimme correction scheme, but no major changes could be observed. For an individual SiNC, for instance, the inclusion of diffuse functions for both Si and H atoms only slightly decreases the dipole moment by 0.14 Debye. Replacing Grimme's DFT-D2 method by his DFT-D3 method essentially gives the same results for the dipole moments within 1%, and reduces the total energy by less than 0.001%.

UV–Visible spectra are plotted as molar absorption coefficient (ϵ) vs. wavelength (λ). We have assumed that all absorption peaks possess a Gaussian band shape. The overall spectrum, $\epsilon(\nu)$, as a function of the excitation energy, ν, then comes from the sum of all the individual bands, $\epsilon(\nu_i)$:

$$\epsilon(\nu) = \sum_{i=1}^{n}\left(B \times \frac{f_i}{\sigma} \times exp\left[-\left(\frac{\nu - \nu_i}{\sigma}\right)^2\right]\right) \quad (1)$$

where i runs from the first to the nth electronic excitation. In the present work, we have calculated the first 600 transitions for each absorption curve and assumed a value of 0.4 eV for the standard deviation σ. The constant B has a value of $1.3062974 \times 10^8 \times \mathrm{L} \times \mathrm{eV} \times \mathrm{mol}^{-1} \times \mathrm{cm}^{-1}$. The (dimensionless) oscillator strength is f_i corresponding to the ith electronic excitation of interest with an excitation energy ν_i.

3. Results

In our search for larger SiNCs with a permanent dipole moment, we started out with essentially the same experimental conditions as in our last work, but we increased the plasma on-time from 2 s to 8 s, which significantly favors the agglomeration of small one-to-two-nanometer SiNCs [8]. This agglomeration takes place under the influence of the ambipolar electric field naturally present in the plasma discharge [12]. This electric field assures that the orientation of the small polar SiNCs is the same as that of the growing supercluster during their agglomeration.

3.1. Bare-Eye Observation

For our first deposition, we used simple Corning glass substrates with sizes of approximately 50 mm × 50 mm. We deposited our plasma-born SiNCs during two succeeding runs with two orthogonal electric field directions. We achieved this by using our above-mentioned two sets of electrodes, under otherwise identical deposition conditions. We used different times of deposition to control the layer thickness. Using ellipsometry, we determined that the deposited layer thicknesses of the samples shown in Figure 1 were about 50 nm, which excludes optical interference effects, in order to explain the difference in colors.

Instead, we propose that the deposition of SiNCs with two different, orthogonal dipole directions leads to thin films with different optical properties. To explore this hypothesis, we then employed a state-of-the-art procedure to measure the absorption spectra resulting from horizontally and vertically oriented SiNCs.

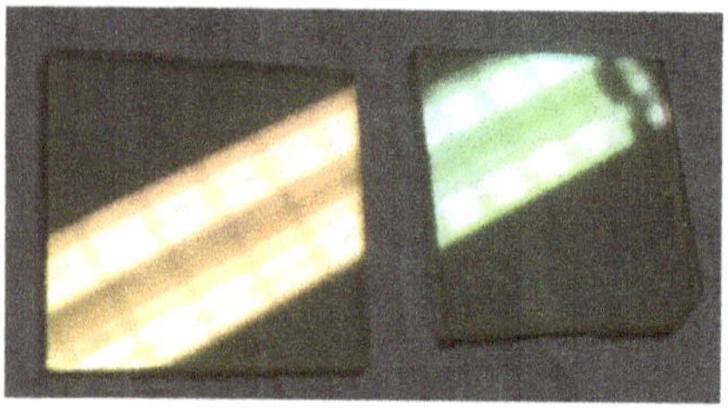

Figure 1. Photo taken after the deposition of two samples obtained with two orthogonal bias voltages (**left** side, horizontal E-field; **right** side, vertical E-field) with otherwise identical plasma conditions. Looking at the reflection of the room illumination, we can clearly distinguish two different colors.

3.2. Hyperspectral Darkfield Absorption Measurements

For these measurements, we deposited our SiNCs onto 100 µm thick microscope glass slides (50 mm × 15 mm). To perform the planned enhanced darkfield hyperspectral microscopy measurements, we had to send our samples to Auburn, AL, USA. To protect our samples against possible oxidation during transport, we added a transparent 15 nm thick Teflon-like layer to cover them in situ using octafluorocyclobutane (C_4F_8) plasma directly after the SiNC deposition, before venting the reactor to the atmosphere. On site, our samples were then prepared by adding a drop of immersion oil onto a slide and placing the glass with the SiNCs directly on top. Hyperspectral images were acquired with transmitted enhanced darkfield illumination at 40× magnification. Data were recorded in the VNIR (400–1000 nm) range.

As can be seen in the hyperspectral images in the two top panels of Figure 2, there is a remarkable difference in the observed density of SiNCs for the two electric field orientations. This difference, however, has already been predicted based on the inhomogeneity of the applied electric fields in our setup. In fact, we have previously shown that the vertical electric field between our two horizontal electrodes does not only provide a torque that orients the molecular dipole moments, but that it also exerts a force on the polar clusters to displace them because of its inhomogeneity. It actually turns out that this field "steers" a considerable fraction of our SiNCs toward a "magic ring" that has a diameter of about

6 cm [8]. Our samples, however, were placed in the very center of the cathode during depositions, explaining the relatively low SiNC density for the vertical E-field polarization. This striking difference in SiNC density for the two orthogonal E-fields already manifests convincing evidence that SiNCs produced under the present plasma conditions possess a significant permanent electric dipole moment—plasma conditions that are prone to lead to significantly larger agglomerated SiNCs than those in previous works [8,9].

Despite the relatively small number of SiNCs appearing for vertical polarization, there were enough of such clusters to pursue our optical absorption measurements. We have actually been able to perform two distinctively different nanocluster analyses: one for relatively small SiNCs (see inset of Figure 2) and one for relatively large SiNCs (see main bottom panel of Figure 2). Note, that we will address the question of absolute cluster size in Section 5. While the absorption spectra recorded for the small SiNCs are reasonably comparable for both E-field polarizations, there is an unexpectedly important difference for the large ones: the vertical E-field (V) leads to a much broader absorption spectrum, and the maximum absorption wavelength is shifted across the entire visible wavelength region by more than 200 nm relative to the one corresponding to the horizontal field (H). To assure that the H sample does not exhibit peaks as broad as those of the V sample, a Peak Location Classifier algorithm [21] was run to find any particles with peaks of 700 $\pm$ 100 nm, but none were found. It appears as though H may have activity in the UV region, whereas V is fairly active throughout the entire visible region until the near-infrared region.

To address the question of how SiNCs deposited with a vertical E-field can have such a broad absorption profile, and specifically, how they can so efficiently absorb light in the red and even in the near-infrared spectral region, we realized a series of ab initio simulations. As we have seen in Figure 2, the absorption spectra not only show a crucial dependence toward the applied E-field direction, but also on the size of the deposited SiNCs. Therefore, we calculated the absorption spectra for SiNCs with four different sizes. To this end, we placed between two to four SiNCs in close proximity and optimized the geometries of the resulting ensembles by minimizing their energy. In all three cases, the lowest-energy structures showed one covalent bond between neighbor clusters. With time-dependent density functional theory (TD-DFT), we then calculated the first 600 transitions for each aggregate, and determined the corresponding molar absorption coefficient ϵ as a function of wavelength, as described in detail above. The results are shown in the main panel of Figure 3.

As expected, the overall absorption capacity increases with the number of aggregated SiNCs. It appears that the spectra are composed of two components: one that is dominant in the UV region and one that shifts toward longer wavelengths and increases its relative importance as the aggregate size becomes larger. It is obvious, however, that we need much larger aggregates to reproduce the experimentally observed intensity ratio between these two components, as displayed in Figure 2 for the vertical E-field. To obtain at least an order of magnitude estimate of how large such an aggregate should be, we have traced, in the upper inset of Figure 3, the intensity ratio as calculated at the wavelengths roughly corresponding to the two experimental absorption maxima. Unfortunately, we have too few points for a reliable extrapolation. As a highly conservative estimate, we first calculated a linear regression curve, which suggests that we need at least 12 SiNCs (i.e., about 228 silicon atoms) to come to the experimentally observed ratio of about 0.92. However, a logarithmic regression on the same four points (see the cyan line) shows a much better fit, and suggests a minimum aggregate size of about 2400 SiNCs (i.e., 46,000 silicon atoms). While the latter fit agrees much more with our four data points, we can only qualitatively conclude at this point that we need much larger aggregates than those accessible for our present ab initio calculations to explain the measured absorption curve for the vertical polarization. In Section 5, we will actually measure the aggregate size that is necessary for the observed spectra, and will compare it to the present estimates.

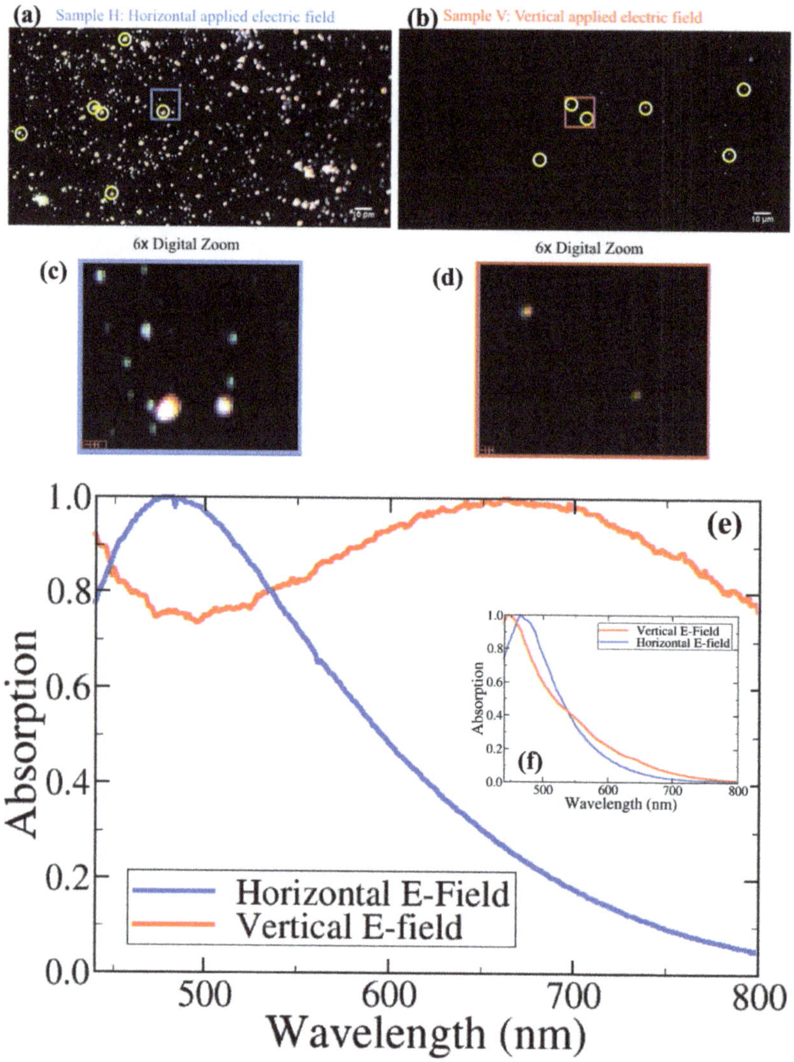

Figure 2. Darkfield hyperspectral imaging of dipole-oriented SiNCs deposited onto a glass substrate exhibiting unique spectral profiles. **Top panels (a,b)**: hyperspectral images (taken at 40× magnification) of SiNCs deposited with a horizontal electric field (**a**) and a vertical electric field (**b**) applied during their deposition; six representative "regions of interest" (ROIs) are drawn around the chosen SiNCs (yellow circles). **Middle panels (c,d)**: the selected SiNCs are shown at a higher magnification. **Bottom panel (e,f)**: optical spectra measured from the ROIs defined above and averaged for mean comparison for both samples. Inset (**f**): normalized absorption spectra measured for the smallest observed SiNCs. **Main panel (e)**: normalized absorption spectra measured for the largest SiNCs.

In the lower inset of Figure 3, we have displayed the permanent dipole moment calculated for the four investigated structures. As recently suggested [9], we also found a nearly linear increase in the dipole moment with aggregate size (i.e., number of silicon atoms). Different to reference [9], however, we are not here dealing with pure silicon clusters, but with hydrogenated silicon clusters; i.e., instead of having a dipole moment of about 0.02 Debye/Si atom, typical of pure silicon clusters, we report ~0.08 Debye/Si atom.

In the present calculations, we always had one covalent bond between all clusters. As we will see below, this dipole moment per silicon atom will become even more important for SiNCs that are held together only by non-covalent bonds.

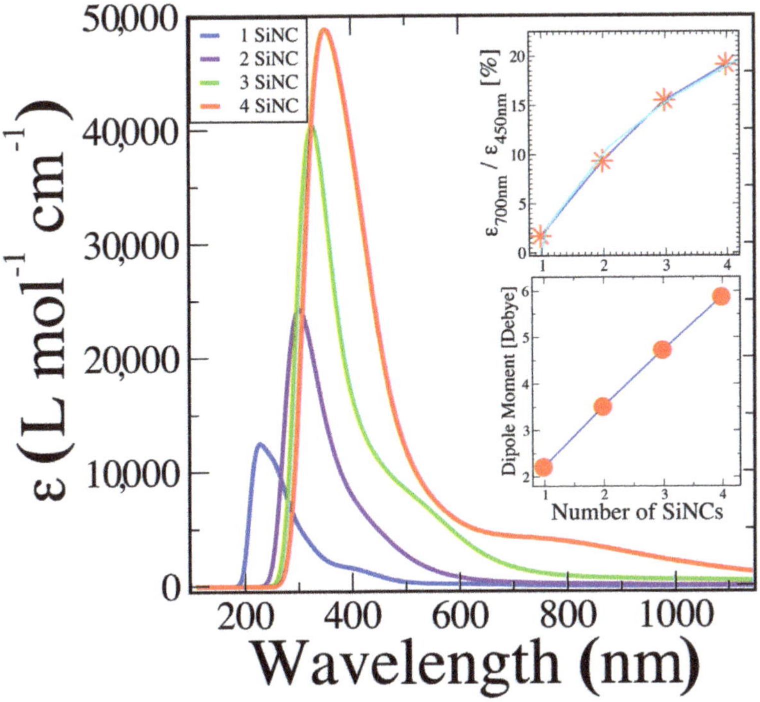

Figure 3. Theoretical molar absorption coefficient ϵ as a function of wavelength for clusters self-assembled from one to four of the most stable, presently known, individual 1 nm SiNCs (see text). **Upper inset**: absorption intensity ratio at the wavelengths roughly corresponding to the experimentally observed absorption maxima as a function of cluster size; the cyan curve corresponds to a logarithmic regression. **Lower inset**: calculated permanent electric dipole moments as a function of cluster size.

3.3. Atomic Force Microscopy Observation

After having explored some of the optical properties that depend on the orientation of the permanent electric dipole moments of our deposited SiNCs, we investigated how this E-field programmable SiNC orientation can influence the mechanical properties of a substrate. In Figure 4, we show the AFM images resulting from two subsequent depositions of our SiNCs on intrinsic silicon substrates with two orthogonal E-fields applied during depositions with otherwise identical plasma conditions.

While the average peak height for the horizontal E-field polarization is obviously much lower than for the vertical one, some higher peaks do appear at the right edge of the sample. We tentatively attribute the appearance of these peaks to the breakdown of the homogeneity of the horizontal E-field. Nevertheless, even including these relatively high edge peaks, the sample deposited with V-polarization exhibits vertical structures that are, on the average, six times higher than the structures of the sample deposited with H-polarization.

As we have previously shown, the inhomogeneity becomes strongest at the edges of the electrodes [8]. For the horizontal polarization, this effect becomes further amplified by the fact that the electrodes are much smaller and further apart than for the vertical polarization. In addition, for the V-polarization, SiNCs that start their journey in the very center of the electrode will experience a perfectly homogeneous E-field until their surface deposition. For the H-polarization, however, the same SiNCs must cross the most inhomogeneous part of the H-field twice on their way to the substrate, potentially causing some misalignment among them. Therefore, the H-polarization cannot align the dipole moments as precisely as the vertical one can, in our present setup. As a result, the steering and orienting capacity are less efficient for the H-polarization, leading to smaller ensembles and no apparent horizontal structures lying on the substrate within the limited lateral resolution power of the AFM. As a consequence, these much smaller structures cannot efficiently absorb in the red region, as we can see in Figure 2.

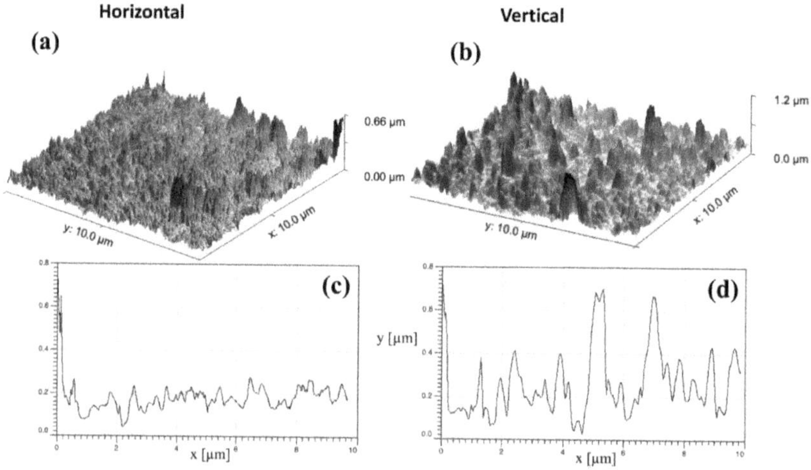

Figure 4. Top panels (a,b): three-dimensional representation of two AFM scans of two typical samples deposited with two orthogonal polarizations of the DC bias during deposition with otherwise identical plasma conditions. **Lower panels (c,d)**: profile scans of above samples in two randomly chosen directions. Note that the overall roughness is about six times higher for the vertically than for the horizontally applied electric field deposition.

3.4. TEM, STEM-HAADF, and EDX Observations

Finally, we address the question of how large the deposited clusters are, which shape they possess, and whether they are oxidized or not. To this end, we simultaneously deposited our SiNCs onto an ITO substrate and a TEM grid ("holey" carbon Cu 300 mesh). To visualize the SiNCs deposited onto the ITO substrate, we first tried to "scratch" off some material with a scalpel for the TEM analysis. However, we did not succeed in doing so because the deposited layer was too resistant. Therefore, we employed the *focused ion beam* (FIB) technique, as described above. In Figure 5, we can distinguish an impressive quantity of SiNCs ranging in size from about 4 nm to 20 nm, with an average size of about 14 nm. There seem to be some even larger clusters, but careful inspection suggests that they are the result of agglomeration of smaller ones. Note that a spherical cluster with a 14 nm diameter could result from the agglomeration of about 2700 SiNCs, each ~1 nm in diameter. This estimate agrees surprisingly well with our logarithmic regression, shown in the upper inset of Figure 3, that suggests an aggregate size of 2400 SiNCs. A qualitative EDX analysis of the same sample suggests that the level of silicon cluster oxidation is reasonably marginal, and seems to be limited to the outer surface of the aggregates. Performing electron diffraction, we did not observe any diffraction spots or rings underlining the non-crystalline structure of our superclusters.

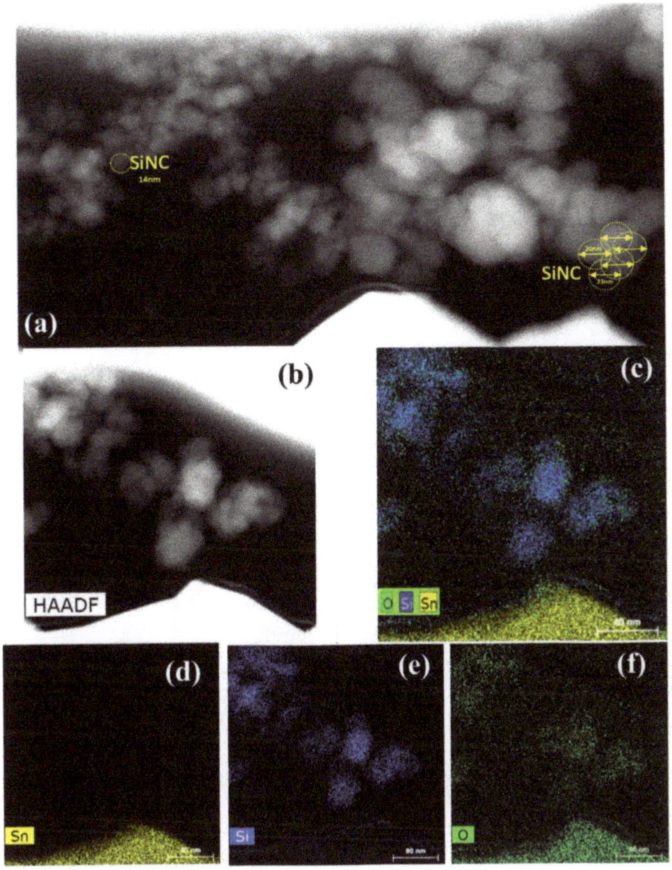

Figure 5. Top panel (**a**): STEM-HAADF image of SiNCs deposited onto an indium tin oxide (ITO) substrate together with an approximate diameter estimate. On top of the image, the carbon and platinum protective layers can be seen. **Middle panel** (**b**,**c**): STEM-HAADF images of the same silicon clusters and their corresponding EDX mapping with silicon (blue), oxygen (green), and tin (yellow). **Bottom panel** (**d**–**f**): the elemental EDX maps for the selected elements.

In Figure 6, we present a high-resolution image of a 20 nm SiNC deposited onto a TEM grid resulting from the same deposition run. Its shape appears rather spherical, and it seems that it is, indeed, composed of many smaller SiNCs.

Figure 7 shows the results of an ab initio simulation conducted in order to have a rough idea of what a cluster composed of several individual SiNCs looks like, and which kind of dipole moment we could expect. To this end, we placed 12 SiNCs in close proximity and let the electronic structure program optimize the ensemble to obtain a supercluster. The SiNCs we used in the present simulations were not fully passivated [22] which, fortunately, seems to describe our experimental SiNCs rather well. In the present case, two covalent bonds formed between three neighboring SiNCs during the structural optimization process, as we had already seen in our four-SiNC study described above. Nevertheless, there are still nine individual SiNCs that remain strongly bonded to the supercluster by only non-covalent bonds. Despite the two covalent bonds, the shown supercluster possesses a permanent electric dipole moment of 34.6 Debye. This roughly corresponds to 0.150 Debye per silicon atom, which is more than we found above for the four SiNCs, simply because covalent bonds were formed between all four SiNCs. Actually, this dipole moment per silicon atom turned out to be higher than for an individual SiNC (i.e., 0.118 Debye per silicon atom), attributable to induced dipole moments that have to be added to the initial permanent electric dipole moment. We can thus confirm the suggested quasi-linear dependence of the dipole moment value on the number of silicon atoms at least up to 228 atoms.

Figure 6. High-resolution TEM image of a typical nearly spherical silicon supercluster deposited onto a TEM grid.

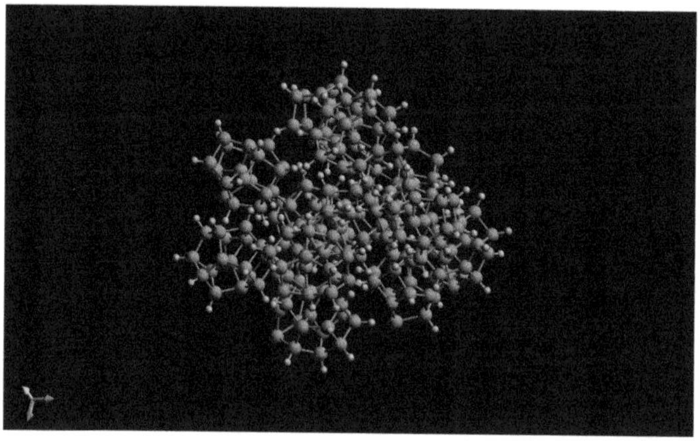

Figure 7. Ab initio simulation of a supercluster composed of 12 SiNCs with only two covalent bonds between all individual SiNCs, resulting in a theoretical dipole moment of 35 Debye.

4. Discussion

Molecular entities with giant permanent electric dipole moments have an uneven distribution of electric charge, which gives them unique properties. For example, they tend to have high polarity, which makes them useful in solvents, as they can dissolve polar substances. They can also be used as building blocks for materials with specific properties, such as high dielectric constants, which are important in electronic devices. In addition, entities with large dipole moments can have important implications for biochemistry, as they can interact with biological molecules such as proteins and nucleic acids, affecting their structure and function. Other potential applications include material work function tailoring, nonlinear optics, ferroelectrics, and organic photovoltaics.

The search for molecules with giant permanent dipole moments is, consequently, a very active field of research. Recently, 5,6-diaminobenzene-1,2,3,4-tetracarbonitrile was unveiled because, with a dipole moment of 14 Debye, it has the highest known dipole moment of any neutral molecule [23]. A yet higher dipole moment was shown by a rather exotic species: a trilobite Rydberg dimer, a highly excited type of two-atom molecule Cs_2

with its atoms 100 nm apart; it was found to have a dipole moment of thousands of Debye, albeit under a high vacuum at just 40 μK [24].

In the present work, we used plasma conditions that are reasonably comparable to our previous work, in which we slowed down the reaction dynamics in a plasma reactor in such a way that polar SiNCs, with sizes of roughly one to two nanometers and permanent dipole moments of about 2.25 Debye, were formed [8]. In contrast to our previous work, however, we lengthened the duration of the plasma discharge here from 2 s to 8 s, which allows the rather-well-passivated 1–2 nm SiNCs to agglomerate. As a result, we are working here with aggregates that have sizes of roughly 20 nm.

During the supercluster growth by agglomeration, the ambipolar electric field plays a crucial role. In general, this field points from the center of the plasma to the electrodes or the walls. Specifically, the field points toward the upper electrode for the upper half of the plasma and toward the bottom electrode for the lower half. However, wherever the growing supercluster "picks up" another tiny polar SiNC, the ambipolar electric field naturally occurring in the discharge [12] assures that the two dipole moments of the two agglomerating entities possess the same orientation, causing the permanent electric dipole moment of the supercluster to increase with each newly attached SiNC. After the plasma is switched off, the ambipolar field disappears immediately and the strongly polar supercluster is oriented (and guided) by the then switched-on applied DC bias, either between the set of vertical or horizontal electrodes.

In this sense, we define a "supercluster" as a nanoscale entity that has been constructed by the agglomeration of one-to-two-nanometer polar silicon clusters with their permanent dipole moments directed in the same direction due to the ambipolar electric field of the plasma discharge during its creation. The precise plasma conditions determine to which degree the small SiNC building blocks are passivated with hydrogen atoms. Perfectly passivated SiNCs will lead to the highest possible permanent dipole moment for the resulting supercluster, while non-passivated silicon atoms might form covalent bonds between the small building blocks, increasing the overall stability of the ensemble. We underline, however, the importance of non-covalent forces between our SiNCs. For the above example of 12 SiNCs, the binding energy resulting for perfectly passivated SiNCs (i.e., without any inter-cluster covalent bonds) can readily be calculated to be approximately 6.78 eV. This means that (even for the ultimate case of absolutely no covalent bonds between the individual building blocks) the proposed superclusters are expected to remain stable even at very high temperatures because of the strong dipole–dipole interaction between all SiNCs due to their very close proximity.

Based on the measured absorption spectra shown in Figure 2, we can readily understand the colors observed in Figure 1. For the horizontal polarization, the absorption peak appears around 480 nm, which leads to a complementary orange/red color for the reflected light. For the vertical polarization, however, the main absorption peak is located around 700 nm, which leads to a reflection in the green [25].

There is a very important conclusion we can draw from the measured absorption spectra shown in Figure 2. Comparing the field dependence of absorption for the smallest and the largest SiNCs, it is obvious that the absorption spectra of large SiNCs depend much more strongly on the E-field polarization than those of smaller ones. Consequently, we can qualitatively conclude that larger SiNCs have stronger dipole moments than smaller ones. This, however, can only be true if our large clusters do not possess a crystalline structure. Straightforward experimental evidence for this conclusion is given by the absence of any diffraction spots or rings during our TEM analysis. The non-crystalline nature, even for our largest clusters, can become possible if they are assembled from non-crystalline 1–2 nm SiNCs, such as those of our previous study [8]. The cauliflower-like structure of a 20 nm supercluster, as shown in our TEM image in Figure 6, supports such a hypothesis.

For the absorption spectra shown in Figure 2, we observed the appearance of a second spectral component that increases its importance and progressively shifts toward the infrared region as the number of agglomerated SiNCs increases. As we point out elsewhere, we must remember that delocalized electrons are less tightly bonded to their chemical structures than localized ones. Therefore, it is well known that the absorption spectra of chemical compounds shift to longer wavelengths as their degree of aromaticity increases [6,25], indicating that the degree of electron delocalization might increase with the size of the supercluster. Such an increase, however, could, in principle, decrease the expected permanent dipole moment because the electron density becomes more evenly distributed throughout the cluster, resulting in a more balanced distribution of charge and a smaller overall dipole moment. In our present case, however, we believe that this effect is relatively negligible because the position of the maximum absorption, corresponding to this spectral component, appears at about 700 nm for both the four-SiNC ensemble (as calculated in Figure 3) and for a supercluster of about 2400 SiNCs (as measured for the vertical polarization in Figure 2).

From the TEM images in Figure 5, we can estimate that the average aggregate size is about 14 nm (i.e., there are roughly 51,000 silicon atoms or the equivalent of 2700 SiNCs 1 nm in size). This observation agrees remarkably well with our estimate of 2400 SiNCs based on the logarithmic regression shown in the upper inset of Figure 3. We point out that the samples analyzed in Figure 5 were exposed to ambient air after their deposition for 6 years prior to the shown measurements. Individual one-to-two-nanometer SiNCs deposited at the same time, also exposed to air, totally disappeared due to oxidation in the same time interval. This observation can clearly be understood by the fact that 1–2 nm clusters do not have any volume, and nearly all silicon atoms are located on the surface of the cluster (i.e., once these surface atoms are oxidized, there is no silicon cluster left). For the 20 nm superclusters, however, we have shown that the oxidation apparently only took place on the outer surfaces, leaving the majority of the inner 1–2 nm aggregated and aligned SiNCs intact. We can safely assume that this surface oxidation corresponds to the well-known formation of a native oxide layer with a typical thickness of about 2 nm. It is this oxide layer that protects the inner 1–2 nm SiNCs, which should largely conserve their individual properties if they are sufficiently well passivated.

5. Conclusions

To date, there have only been two experimental reports confirming the theoretically predicted permanent electric dipole moments of small silicon clusters [8,9]. While the technique of laser evaporation is very powerful [9], we tentatively believe that our pulsed plasma approach could be more efficient in creating clusters with large dipole moments; this is because the naturally present ambipolar electric field in the plasma aligns the dipole moments of the small SiNCs during agglomeration. Wherever the growing supercluster "picks up" another small SiNC in the plasma, the naturally occurring ambipolar electric field assures that the two dipole moments possess the same orientation. This causes the permanent electric dipole moment of the supercluster to increase with each newly attached SiNC. For the laser evaporation technique, however, there seems to be a risk that the randomly oriented electric dipole moments created between under- and over-coordinated silicon atoms might cancel each other out once a certain cluster size is exceeded because of the absence of any atomic alignment of the randomly oriented dipole moments.

With our method, it is the ambipolar electric field in the plasma discharge that aligns the dipole moments and enables the creation of superclusters with giant permanent electric dipole moments, as we have seen in Figure 2. Thereafter, it is the electric field that we only switch on after the supercluster creation (i.e., when the pulsed plasma discharge is switched off) that orients the supercluster on its way to the substrate. In the future, more homogeneous electric fields will help us to align the permanent dipole moments of the superclusters even more precisely before surface deposition to assure that they are arriving perpendicularly or at a well-defined angle to the substrate. With more sophisticated control of the inhomogeneities of the employed electric fields during deposition, one can realize the building of ingenious nanostructures in the future based on the here-demonstrated electric field-induced assembly formation. In this sense, the externally applied electric fields will assure the nano-architecture on a sub-micrometer scale, while the inherent dipole–dipole interaction between the superclusters assures the building block alignment on an atomic scale. The precise sizes of our supercluster building blocks are readily controlled by the duration of the plasma discharge, while the dimensions of our assembled nanostructures are controlled by the duration of the surface deposition.

At this point, however, the dipole moments of our superclusters remain to be measured, as we have only presented qualitative experimental evidence that the dipole moments of the superclusters should be significantly higher than the ones of individual SiNCs, which have previously been measured to be 2.25 Debye. This experimental challenge, however, should be motivated by our present computational prediction that a small supercluster of only 12 SiNCs should already have a dipole moment of 35 Debye. The experimental feasibility of such superclusters with PECVD plasma is straightforward, as can be seen in Figures 5 and 6. Their dipole moments before surface deposition can either be measured by analyzing their trajectories in electric fields [8] or with time-of-flight techniques [9]. After their surface deposition, scanning tunneling microscopy (STM) can, for instance, be used to measure their dipole moments by analyzing the changes in the tunneling current as the tip is moved over the cluster. There is no theoretical reason why the addition of dipole moments should saturate above a certain value. However, if the dipole moments of individual SiNCs continue to add up, then the experimentally observed 14 nm clusters would be expected to have dipole moments greater than 6000 Debye, which would even eclipse the present world record of the exotic cesium dimer excited to a Rydberg state in a vacuum at 40 µK [24]. It all depends on which length scale the atomic-scale alignment precision of the electric dipole moments can be preserved, which would open an exciting path to more in-depth studies of our demonstrated proof of concept. Ultimately, the giant permanent electric dipole moments of our silicon superclusters can be exploited in many practical applications, for instance, as a new ferroelectric material, in nonlinear optics, or even for certain catalytic reactions.

We conclude with an extension inspired by our present work: It is well known that hydrogenated amorphous silicon (a-Si:H) grown in a plasma-enhanced chemical vapor deposition (PECVD) reactor can contain over-coordinated and under-coordinated silicon atoms, which contribute to the disorder and structural heterogeneity of the material. Similar to the small silicon clusters created by Schäfer et al. [9], we propose that dipole moments are created between these over- and under-coordinated silicon atoms. However, due to the random nature of amorphous silicon, these randomly oriented dipole moments cancel each other out over a sufficiently large piece of material. Therefore, we highlight the need to explore the possible creation of a well-oriented permanent electric dipole moment for an amorphous silicon film, grown under the influence of an homogeneous DC electric field. Once again, in this instance, the possible success depends on how well the individual dipole moments can be aligned with atomic precision.

Author Contributions: Conceptualization, F.J., M.E.L. and H.V.; methodology, F.J., M.E.L. and H.V.; software, F.L. and H.V.; validation, F.J., M.E.L. and H.V.; formal analysis, H.V.; investigation, F.J., J.T., I.F., M.E.L. and H.V.; resources, F.J., M.E.L. and H.V.; data curation, F.J., M.E.L. and H.V.; writing—original draft preparation, H.V.; writing—review and editing, F.J., J.T., F.L., I.F., M.E.L. and H.V.; visualization, F.J., J.T., F.L., I.F., M.E.L. and H.V.; supervision, H.V.; project administration, F.J., M.E.L. and H.V.; funding acquisition, H.V. All authors have read and agreed to the published version of the manuscript.

Funding: This research was funded by the CNRS and Ecole Polytechnique. In particular, we gratefully acknowledge the efficient help of CNRS in the framework of their Innovation Prematuration 2020 program. The HPC centers of IDRIS (Grant A0080900642) and CERMM are acknowledged for computational resources, and the Hariri Foundation for Sustainable Human Development is acknowledged for the scholarship awarded to F.J.; I. F. would like to give thanks for the funding from ANR under the contract number ANR-10-EQPX-50 (pole NanoTEM) for the FIB sample preparation and STEM-HAADF observations.

Data Availability Statement: Additional data and information are available upon request to the authors.

Acknowledgments: Our most sincere thanks go out to Elyse V. Johnson of CytoViva, Inc., Auburn, AL, USA, for her skillful data acquisition using the enhanced darkfield hyperspectral microscope, as well as for her professional comments and suggestions, which were crucial for the success of our work. The Centre Interdisciplinaire de Microscopie électronique de l'X (CIMEX) is gratefully acknowledged. We would like to acknowledge James R. Creel for his proof reading of our manuscript.

Conflicts of Interest: The authors declare no conflict of interest.

Abbreviations

The following abbreviations are used in this manuscript:

SiNC	Silicon Nanocluster
AFM	Atomic Force Microscopy
TEM	Transmission Electron Microscopy
STEM-HAADF	Scanning Transmission Electron Microscopy—High-Angle Annular Darkfield
EDX	Energy-Dispersive X-ray spectroscopy
FIB	Focused Ion Beam
ROI	Region Of Interest
TD-DFT	Time-Dependent Density Functional Theory

References

1. Vach, H.; Brulin, Q. Controlled growth of silicon nanocrystals in a plasma reactor. *Phys. Rev. Lett.* **2005**, *95*, 165502. [CrossRef] [PubMed]
2. Vach, H. Ultrastable silicon nanocrystals due to electron delocalization. *Nano Lett.* **2011**, *11*, 5477–5481. [CrossRef] [PubMed]
3. Vach, H. Electron-deficiency aromaticity in silicon nanoclusters. *J. Chem. Theory Comput.* **2012**, *8*, 2088–2094. [CrossRef]
4. Vach, H. Symmetric and irregular aromatic silicon nanoclusters. *Chem. Phys. Lett.* **2014**, *614*, 199–203. [CrossRef]
5. Vach, H.; Brulin, Q.; Chaabane, N.; Novikova, T.; i Cabarroca, P.R.; Kalache, B.; Hassouni, K.; Botti, S.; Reining, L. Growth dynamics of hydrogenated silicon nanoparticles under realistic conditions of a plasma reactor. *Comput. Mater. Sci.* **2006**, *35*, 216–222. [CrossRef]
6. Vach, H.; Ivanova, L.V.; Timerghazin, Q.K.; Jardali, F.; Le, H.L.T. Metallic-like bonding in plasma-born silicon nanocrystals for nanoscale bandgap engineering. *Nanoscale* **2016**, *8*, 18062–18069. [CrossRef]
7. Vach, H. Terahertz and Gigahertz emission from an all-silicon nanocrystal. *Phys. Rev. Lett.* **2014**, *112*, 197401. [CrossRef]
8. Jardali, F.; Keary, B.P.; Perrotin, T.; Silva, F.; Vanel, J.C.; Bonnassieux, Y.; Mazouffre, S.; Ruth, A.A.; Leulmi, M.E.; Vach, H. Hydrogenated Silicon Nanoclusters with a Permanent Electric Dipole Moment for the Controlled Assembly of Silicon-Based Nanostructures. *ACS Appl. Nano Mater.* **2021**, *4*, 12250–12260. [CrossRef]
9. Rivic, F.; Lehr, A.; Schäfer, R. Scaling of the permanent electric dipole moment in isolated silicon clusters with near-spherical shape. *Phys. Chem. Chem. Phys.* **2023**, *25*, 13376–13382. [CrossRef]
10. Jackson, K.; Jellinek, J. Si clusters are more metallic than bulk Si. *J. Chem. Phys.* **2016**, *145*, 244302. [CrossRef]
11. Mangolini, L.; Thimsen, E.; Kortshagen, U. High-Yield Plasma Synthesis of Luminescent Silicon Nanocrystals. *Nano Lett.* **2005**, *5*, 655–659. [CrossRef]
12. Fridman, A.A.; Boufendi, L.; Hbid, T.; Potapkin, B.V.; Bouchoule, A. Dusty plasma formation: Physics and critical phenomena. Theoretical approach. *J. Appl. Phys.* **1996**, *79*, 1303–1313. [CrossRef]
13. Ivlev, A.V.; Kretschmer, M.; Zuzic, M.; Morfill, G.E.; Rothermel, H.; Thomas, H.M.; Fortov, V.; Molotkov, V.I.; Nefedov, A.P.; Lipaev, A.M.; et al. Decharging of complex plasmas: First kinetic observations. *Phys. Rev. Lett.* **2003**, *90*, 055003. [CrossRef]
14. Couëdel, L.; Mikikian, M.; Boufendi, L.; Samarian, A.A. Residual dust charges in discharge afterglow. *Phys. Rev. E* **2006**, *74*, 026403. [CrossRef] [PubMed]
15. Boufendi, L.; Bouchoule, A. Particle nucleation and growth in a low-temperature argon-silane discharge. *Plasma Sources Sci. Technol.* **1994**, *3*, 262–267. [CrossRef]
16. Misra, S.K.; Ostadhossein, F.; Daza, E.; Johnson, E.V.; Pan, D. Hyperspectral Imaging Offers Visual and Quantitative Evidence of Drug Release from Zwitterionic-Phospholipid-Nanocarbon When Concurrently Tracked in 3D Intracellular Space. *Adv. Funct. Mater.* **2016**, *26*, 8031–8041. [CrossRef]
17. Xu, K.; Sun, W.; Shao, Y.; Wei, F.; Zhang, X.; Wang, W.; Li, P. Recent development of PeakForce Tapping mode atomic force microscopy and its applications on nanoscience. *Nanotechnol. Rev.* **2018**, *7*, 605–621. [CrossRef]
18. Frisch, M.J.; Trucks, G.W.; Schlegel, H.B.; Scuseria, G.E.; Robb, M.A.; Cheeseman, J.R.; Scalmani, G.; Barone, V.; Petersson, G.A.; Nakatsuji, H.; et al. *Gaussian 16 Revision C.01*; Gaussian Inc.: Wallingford, CT, USA, 2016.
19. Manzetti, S.; Lu, T.; Behzadi, H.; Esrafili, M.; Le, H.L.; Vach, H. Intriguing properties of unusual silicon nanocrystals. *RSC Adv.* **2015**, *5*, 78192–78208. [CrossRef]

20. Grimme, S. Semiempirical GGA-Type Density Functional Constructed with a Long-Range Dispersion Correction. *J. Comput. Chem.* **2006**, *27*, 1787–1799. [CrossRef]
21. Zamora-Perez, P.; Tsoutsi, D.; Xu, R.; Rivera-Gil, P. Hyperspectral-Enhanced Dark Field Microscopy for Single and Collective Nanoparticle Characterization in Biological Environments. *Materials* **2018**, *11*, 243. [CrossRef]
22. Gordeychuk, M.V.; Katin, K.P.; Grishakov, K.S.; Maslov, M.M. Silicon buckyballs versus prismanes: Influence of spatial confinement on the structural properties and optical spectra of the $Si_{18}H_{12}$ and $Si_{19}H_{12}$ clusters. *Int. J. Quantum Chem.* **2018**, *118*, e25609. [CrossRef]
23. Wudarczyk, J.; Papamokos, G.; Margaritis, V.; Schollmeyer, D.; Hinkel, F.; Baumgarten, M.; Floudas, G.; Müllen, K. Hexasubstituted Benzenes with Ultrastrong Dipole Moments. *Angew. Chem. Int. Ed.* **2016**, *55*, 3220–3223. [CrossRef] [PubMed]
24. Booth, D.; Rittenhouse, S.T.; Yang, J.; Sadeghpour, H.R.; Shaffer, J.P. Production of trilobite Rydberg molecule dimers with kilo-Debye permanent electric dipole moments. *Science* **2015**, *348*, 99–102. [CrossRef] [PubMed]
25. Juster, N.J. Color and chemical constitution. *J. Chem. Educ.* **1962**, *39*, 596. [CrossRef]

nanomaterials

MDPI

Article

The Kinetics of Aragonite Formation from Solution via Amorphous Calcium Carbonate

Simon M. Clark [1,2,*], Vili Grigorova [1], Bruno Colas [1,2], Tamim A. Darwish [3], Kathleen Wood [2], Joerg Neuefeind [4] and Dorrit E. Jacob [5]

1 School of Engineering, Faculty of Science and Engineering, Macquarie University, North Macquarie Park, Shellharbour, NSW 2109, Australia
2 Australian Centre for Neutron Scattering, Australian Nuclear Science and Technology Organisation, Locked Bag 2001, Kirrawee DC, Sydney, NSW 2232, Australia
3 National Deuteration Facility, Australian Nuclear Science and Technology Organisation, Kirrawee DC, Sydney, NSW 2232, Australia
4 Spallation Neutron Source, Oak Ridge National Laboratory, Oak Ridge, TN 37831, USA
5 Research School of Earth Sciences, The Australian National University, Canberra, ACT 2600, Australia
* Correspondence: simon.clark@mq.edu.au

Abstract: Magnesium doped Amorphous Calcium Carbonate was synthesised from precursor solutions containing varying amounts of calcium, magnesium, H_2O and D_2O. The Mg/Ca ratio in the resultant Amorphous Calcium Carbonate was found to vary linearly with the Mg/Ca ratio in the precursor solution. All samples crystallised as aragonite. No Mg was found in the final aragonite crystals. Changes in the Mg to Ca ratio were found to only marginally effect nucleation rates but strongly effect crystal growth rates. These results are consistent with a dissolution-reprecipitation model for aragonite formation via an Amorphous Calcium Carbonate intermediate.

Keywords: neutron scattering; X-ray diffraction; Small Angle Neutron Scattering; Laser-Ablation Inductively Coupled Plasma Mass Spectrometry; themo gravimetric analysis; NOMAD; QUOKKA; ACC; aragonite; Mg/Ca

1. Introduction

Amorphous calcium carbonate (ACC) plays a key role in the early stages of biomineralisation as a precursor to crystalline calcium carbonate phases (e.g., [1]) and is a model system in the development of biomimetic materials. Many organisms take advantage of the mouldable character of ACC in the formation of their intricate shells and skeletons. ACC also seems to lower the activation energy for nucleation of subsequent crystalline carbonates due to lower surface energies [2] enabling an ambient temperature non-classical crystallisation pathway [3]. It is known that ACC is polyamorphous with different water content, local order and mode of formation [4,5], and it was recently shown that ACC is best described as a nano-fluid [6]. Understanding the mechanism by which these nanoparticles self-assemble is essential if we are to understand the effect of climate change on marine life and how well we are presently using the fossil record to constrain climate change models.

In the laboratory ACC converts to the crystalline carbonate phases in a matter of minutes although, depending on the aqueous fluid composition and any additives [7], its stability can be increased allowing ACC to be preserved for weeks to years [8,9]. The additives most commonly used in the laboratory to prolong ACC lifetime are organic macromolecules, such as polyacrylic acid [10,11] and magnesium [11–14]. In natural biominerals, in addition to Mg, a large range of organic macromolecules can exert a stabilising effect on ACC, which can extend the stability of ACC almost indefinitely [15]. Mg-ACC usually transforms into calcite at low Mg content [16–19], but at high Mg content transforms into aragonite [17,18].

Non-classical nucleation and growth models often provide a better description of biomineralization kinetic data than classical crystallisation models [3]. In this context, two models have been used to describe the transformation mechanism from ACC to aragonite or calcite in non-classical crystallisation: a solid-state transformation, where ACC dehydrates and crystallises into a calcium carbonate polymorph directly in the solid phase [1,20]; or a dissolution-reprecipitation process [20–22], where ACC dissolves and crystalline calcium carbonate precipitates from this solution. In the laboratory, ACC usually transforms via dissolution precipitation which is typically associated with significant morphological changes to the texture of the solid phase [20,23,24]. In vivo, preservation of intricate hierarchical structural details [25] during stepwise crystallisation seems to contradict the findings in the laboratory, but supports either solid-state transformation, or very localised dissolution-reprecipitation processes or perhaps a combination of both [23,26].

To try to distinguish between these two models, we measured the rate of formation of crystalline aragonite from ACC in the laboratory for a range of Mg/Ca ratio and H_2O/D_2O compositions.

2. Methodology

2.1. Sample Preparation

A set of ACC samples with a range of Mg/Ca ratios and water contents were synthesised with Mg/Ca ratios expected to give measurable formation and crystallisation rates. A commonly used synthesis procedure was followed [27] and modified with or without an additional freeze-drying step to allow variation of water content. Stock 0.02 M solutions of Na_2CO_3 and 0.02 M solutions made by dissolving various mixtures of solid $CaCl_2$ and $MgCl_2$ in water or heavy water to give solutions with a range of Mg/Ca content, were prepared and stored at 4 °C. About 100 mL of both the stock Na_2CO_3 and $CaCl_2$/$MgCl_2$ solutions were then quickly mixed together in a 250 mL bottle held at 4C while stirring using with a magnetic stirrer. Almost instantly the solutions were mixed a white precipitate formed. The white precipitate was filtered and washed with acetone and then either stored under nitrogen at room temperature or freeze-dried for one hour before being stored under nitrogen at room temperature. Stock solutions were prepared using H_2O, D_2O and mixtures of H_2O and D_2O as the solvent in order to vary the solvent chemistry and probe the effect of solvent on sample structure, stability, and crystallisation products.

2.2. Analysis Techniques

Each ACC sample synthesised was characterised using Thermal Gravimetric Analysis (TGA), to determine the amount of water present, Laser-Ablation Inductively Coupled Plasma Mass Spectrometry (LA-ICPMS), to determine the Mg/Ca ratio, Neutron scattering to confirm the LA-ICPMS measurements, and X-ray diffraction (XRD) to confirm that there were no crystalline phases present in the sample. Each of the carbonates crystallised from our ACC samples were characterised using Raman Spectrometry and powder X-ray Diffraction to determine which calcium carbonate polymorph was present, Scanning Electron Microscopy to provide visual evidence of sample composition, and a second powder X-ray diffraction measurement incorporating an internal standard to provide accurate crystalline phase lattice parameters. Finally, time resolved Small Angle Neutron Scattering (SANS) was used to measure the rate of formation of ACC from solution, and time resolved powder X-ray diffraction was used to determine the rate of formation of crystalline carbonate phases upon transformation of the ACC samples.

The TGA measurements utilised a TA Instruments TGA 2050. The sample temperature was increased from 20 °C to 800 °C, at a rate of 10 °C/min in a N_2 atmosphere while measuring the weight of the sample. For the LA-ICPMS we used a Photon Machines G2 Excimer laser system (wavelength 193 nm) coupled with an Agilent 7700 quadrupole ICP-MS. ACC samples were dried at 100 °C for 36 h, pressed into pellets and measured with laser spot size of 50 μm and pulse rates of 5 Hz (energy density 5.42 J/cm^2). Backgrounds were measured for 30 s followed by 40 s of sample ablation. NIST SRM (US National Institute of Standard and Technology Standard Reference Material) 610 glass was used as the external standard with ^{43}Ca as the internal standard. The mg for NIST SRM 610 reported in the Geo- ReM database was used as the "true" concentration in this reference glass [28]. Neutron scattering data were collected using the NOMAD instrument at the Spallation Neutron Source, Oak Ridge National Laboratory, USA. Samples synthesized using 100% D_2O to reduce background noise from inelastic scattering of neutrons from protons, were loaded into 5 mm diameter quartz capillary tubes inside a glove box under a nitrogen atmosphere, sealed and quickly transported to the instrument. Scattering data were collected in 30 min frames at room temperature in an argon atmosphere for a total of 1.5 h. The standard instrument data reduction software was used to normalize the data and for background subtraction. XRD patterns were collected using a PANalytical Expert Pro MPD (Multi-Purpose Diffractometer) with a Copper tube (45 kV, 40 mA), Soller slits (0.04 rad.) and fixed incident beam mask (15 mm). The instrument zero point was calibrated using the NIST 640c Si standard. The samples were placed on zero background sample holders, and the wavelength used was CuKα (λ = 1.54 Å). For the accurate lattice parameter measurements appropriate amounts of the NIST, Silicon Powder, Standard Reference Material® 640c was mixed with each sample to increase the accuracy of the measurement. Unit cell parameters were then determined using the Profex software package [29]. Series of powder XRD patterns were also collected from ACC samples using the same system at ambient temperature while they transformed into crystalline carbonate phases. Diffraction patterns were collected over the 10° to 50° 2θ range every 30 min for 24 to 48 h for a range of initial conditions. Peak intensities were determined from these spectra as a function of time and converted into a measure of the degree of reaction for subsequent analysis. Raman spectra were collected using a Horiba LABRAM HR Evolution confocal laser Raman spectrometer. The spectrometer used a red excitation laser (λ = 633 nm) and a 1800 gr/mm grating. Before collecting data from each sample, the spectrometer was calibrated using the Rayleigh line and a Si-wafer with a sharp peak at 520 cm^{-1}. A JEOL JSM- 6480 LA Scanning Electron Microscope (SEM) was used to obtain SEM micrographs from gold coated samples. SANS data were collected using the QUOKKA instrument [30] at the Australian Centre for Neutron Scattering, Australian Nuclear Science and Technology Organisation, Lucas Heights, Australia. Samples were placed in quartz Hellma cells mounted in a thermally controlled sample changer. The neutron path length through the sample was 2 mm and the cell volume 600 µL. For each measurement, the solutions made from Na_2CO_3, $CaCl_2$ and $MgCl_2$ were mixed in the cells minimising any lag between mixing and the start of data collection. All solutions were made with only D_2O, because samples with H_2O were found to give too large a background signal due to incoherent scattering from protons. The source was set to a wavelength of λ = 5 Å, and the detector placed 8 m from the sample. Samples with Mg/Ca ratios of 0.03, 0.2 and 0.26 were studied at ambient temperature to simulate in vivo conditions as closely as possible.

3. Results

3.1. Characterisation of the Amorphous Phase

The main aims of the ACC characterisation were to check that the samples had not transformed to a crystalline carbonate phase, to determine the Mg/Ca ratios and the amount of water.

3.1.1. Thermal Gravimetric Analysis

A typical TGA pattern is shown in Supplementary Materials Figure S1. The weight loss at around 100 °C was attributed to loss of surface water, the small weight loss around 350 °C was attributed to the loss of included water and the weight loss around 700 °C was attributed to the loss of carbon dioxide in keeping with previous studies [16,19]. The stoichiometric water loss was calculated from this weight loss assuming a sample formula of $(Ca_{1-x}Mg_x)CO_3 \cdot nH_2O$. The results are summarised in Supplementary Material Table S1.

Freeze-dried samples were found to contain about 1.18 molecules of water per formula unit, and filtered samples were found to contain about 2.69 molecules of water. This was independent of the Mg/Ca ratio or whether the solvent contained H_2O, D_2O or a mixture of the two (Supplementary Materials Table S1). These values are comparable with those found previously, such as n = 1.42–1.63 [19], and n = 0.98–1.4 [16]. Freeze drying can be seen to have the effect of reducing the amount of water percent in an ACC sample.

3.1.2. X-ray Diffraction

XRD patterns of the freshly synthesised ACC samples were found to contain no Bragg peaks. We observed only broad humps indicative of an amorphous or nano-crystalline phase with two diffuse maxima at 32° (d = 2.8 Å) and 46° (d = 2.0 Å) 2θ (Supplementary Figure S2) consistent with those previously measured from ACC samples [16].

3.1.3. Laser-Ablation Inductively Coupled Plasma Mass Spectrometry

Representative measured Mg/Ca values are contained in Supplementary Materials Table S1. In Figure 1 the Mg/Ca ratios determined using LA-ICPMS are plotted against the Mg/Ca ratios of the solutions from which the ACC samples formed. Comparing data from this study (blue spots) with data from other studies [23,30–33] (Figure 1) shows a linear correlation with the Mg/Ca ratio in solution, with an apparent distribution coefficient ($K_d = [(Mg/Ca)_{ACC}/(Mg/Ca)_{solution}]$) of 0.12. The exact relationship seems to be highly dependent upon the reported study with our data in the same range as data from two previous measurements [23,34]. Further reference to Mg/Ca ratios in this manuscript will refer to the Mg/Ca ratio of the solid sample and not the nascent solution. Potentially, the measured Mg/Ca ratio in the solid sample could be due to variations in the washing and drying protocols with some measurement signal coming from fluid still adhering to solid particles. To investigate this possibility, we also used neutron scattering to determine Mg/Ca ratios.

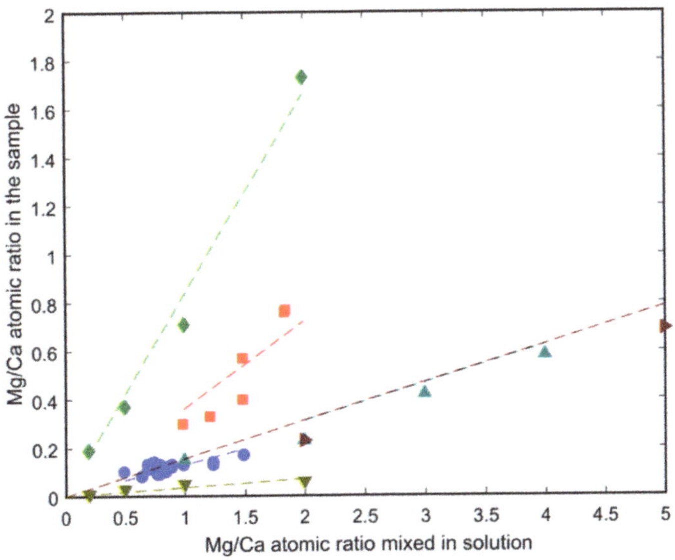

Figure 1. Comparison of Mg/Ca ratios measured in ACC samples to Mg/Ca ratios of the solutions from which they formed for samples: ● from this study, ◆ mixed with Formamide [33], ■ [31], ▲ [32], ▶ [23], ▼ with water [33]; the dashed lines are linear fits to the data points of corresponding colour.

3.1.4. Neutron Scattering

A previous study showed that scattering from an ACC sample containing only Ca has a peak in the real space scattering intensity at G(r) $\approx$ 2.3 Å, due to Ca-O correlations. In comparison, an ACC sample containing only Mg has a peak in the real space scattering intensity at G(r) $\approx$ 2.1 Å. ACC samples with mixtures of Ca and Mg have both peaks with the positions of these peaks varying linearly between the two solid solution end members (Supplementary Material Figure S3). This provides a methodology for the measurement of Mg/Ca ratio in an ACC sample free from any remnant liquid phase contamination. Neutron scattering patterns of six samples containing different Mg/Ca proportions, x, given by the general formula: $(Ca_{1-x}Mg_x)CO_3 \cdot nH_2O$, are shown in Supplementary Material Figure S4. The sample with Ca only contains one peak at ~2.4 Å due to Ca-O correlations, while the samples made from mixtures of Mg and Ca contain peaks at ~2.2 Å and ~2.4 Å due to Mg-O and Ca-O correlations, respectively. This observation demonstrates that Mg is indeed incorporated in the ACC atomic structure and is not a measurement artefact.

The positions of these two peaks are presented in Figure 2 as a function of Mg/Ca ratio together with data from a previous study [31]. The two datasets are seen to be well correlated. The peak due to Ca-O correlations at ~2.4 Å does not change with Mg/Ca ratio while the peak due to Mg-O correlations is seen to vary from ~2.2 Å at low Mg content to ~2.1 Å at 100% Mg content. These data confirm that the LA-ICPMS measurements provide an accurate determination of the Mg/Ca ratios in our samples. The data also demonstrate that comparison of the positions of the Ca-O and Mg-O correlation peaks from neutron scattering data provides an accurate method for the determination of Mg/Ca ratios with no need for any calibrations using standards.

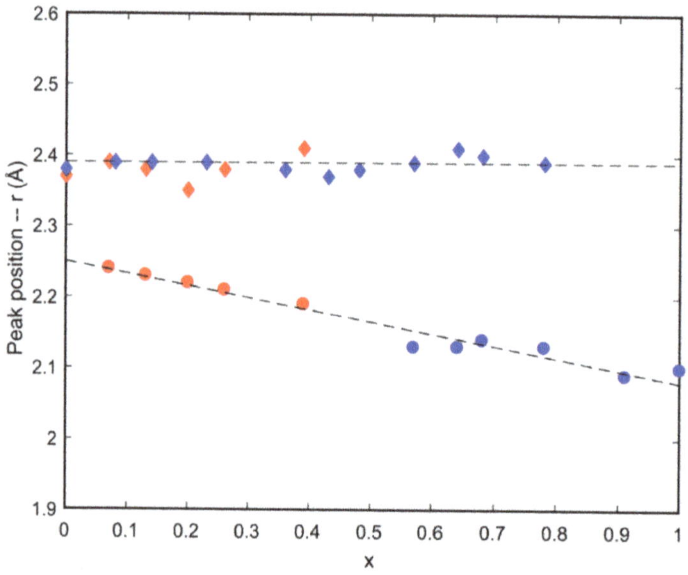

Figure 2. Peak positions of the first Mg-O and Ca-O correlations, taken from Supplementary Materials Figures S3 and S4, Where x is the proportion of magnesium in the sample as given by: $(Ca_{1-x}\,Mg_x)CO_3$. The data shown as red markers are from this study (Figure S4), and the blue markers from [31] (Figure S3). Filled diamond symbols denote the first Ca-O correlation peak around 2.4 Å; Filled circle symbols denote the first Mg-O correlation peak around 2.1–2.2 Å.

The background intensity of the neutron scattering patterns can be used to estimate the H_2O/D_2O ratios in our samples by use of a calibration curve made from known pure H_2O/D_2O mixtures (Supplementary Materials Figure S5). This leads us to conclude that there is less than 10% H_2O in our fully deuterated samples and, thus, more than 90% D_2O. This provides an error estimate for our determined H_2O/D_2O ratios of about 5% due to contamination related to the highly hygroscopic nature of D_2O. It translates into an uncertainty in the number of water molecules per formula unit of about ±0.06 for the freeze-dried samples and ±0.13 for the filtered samples. Comparing these uncertainties with the range of measured values using TGA of 0.9 to 1.37 for the freeze-dried samples and 2.12 to 3.46 for the filtered samples, it implies that the effect of freeze-drying vs. filtering is negligible.

3.2. *Characterisation of the Crystalline Phases*

The main aim of the characterisation of the crystalline phases was to determine which carbonate polytypes formed and their Mg/Ca ratios.

3.2.1. Raman Spectroscopy

Raman spectra collected from our samples were compared to spectra we collected from reference samples of pure aragonite and calcite (Supplementary Materials Figure S6) Comparison of these spectra showed that the crystallised sample consisted solely of aragonite. This was confirmed by comparison with reference data from [35,36] and the RRUFF™ database [37]. The spectra were found to contain three peak groupings: the ν_1 symmetric stretching of the carbonate groups at 1085 cm^{-1}; the ν_4 in-plane bending of calcite at 711 cm^{-1}, and the ν_4 in-plane bending of aragonite at 706 cm^{-1} and 701 cm^{-1} and the lattice modes of calcite at 155 cm^{-1} and 281 cm^{-1}, and the lattice modes of aragonite at 206 cm^{-1} and 155 cm^{-1}. All crystallised samples measured consisted solely of the aragonite phase.

3.2.2. X-ray Diffraction

XRD patterns from the crystalline phases produced from our ACC samples were fitted using the *Profex* (Solothurn, Switzerland) software [29] and were found to contain only reflections due to aragonite (Supplementary Materials Figure S7) showing that only aragonite crystals formed in agreement with our Raman data.

3.2.3. Scanning Electron Microscopy

A representative SEM image is contained in the Supplement Materials Figure S5. Clusters of tablet-shaped aragonite crystals can be seen. These are surrounded by xenomorphic, rounded grains that we interpret as ACC particles. The aragonite tablets are of similar size and shape as ones observed in nacre, 0.5 µm thick, 5–15 µm in diameter polygons [38]. This supports the conclusions of both Raman and XRD measurements that only aragonite is present in the crystallised samples.

3.2.4. High-Precision X-ray Diffraction

Accurate determination of the Mg/Ca ratio in carbonates crystallised from ACC is difficult due to the presence of ACC particles on crystal faces (Supplementary Figure S8) and in between the carbonate platelets making up the crystalline phase. The unit cell volume of carbonate phases is highly sensitive to any incorporation of other ions such as magnesium and should prove an accurate method for determining the Mg/Ca ratio. The unit cell volume of aragonite crystallised from our ACC as a function of Mg/Ca ratio in the precursor ACC is shown in Figure 3. The volume is seen to show only a small change from the volume of pure calcium containing aragonite with Mg/Ca ratio. How significant this small change is in the absence of literature aragonite unit cell volume data for varying Mg/Ca ratios can be estimated by comparison of volume change with substitution of other similar atoms. Volumes calculated for substitution of Ca by Sr in the aragonite

($CaCO_3$) to strontianite ($SrCO_3$) solid solution series [39] are also plotted in Figure 3. It was observed that there is an increase in unit cell volume with increasing amounts of Sr which is probably due to the Sr atom being bigger (in a crystal 1.32 Å) than a Ca atom (1.14 Å). In a similar manner, the unit cell volume of aragonite on substitution of calcium (1.14 Å) with magnesium (0.86 Å) was calculated the result of which is also plotted in Figure 3. Our conclusion is that there is no significant substitution of calcium by magnesium in the crystalline aragonite samples although it was previously incorporated in the ACC before transformation to aragonite.

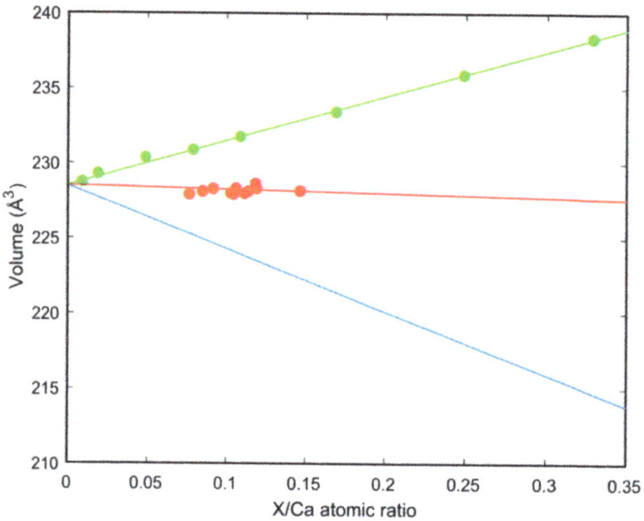

Figure 3. Data from this study compared to the unit cell volume change of the lattice estimated for substitution of Sr and Mg to Ca. X represents either Sr or Mg. ● Data from this study with the linear fit of the data points; ● Strontianite (SrCO3)-aragonite (CaCO3) solid solutions from [39] with the linear fit of the data points; —— Estimate for Mg substituting Ca in the aragonite.

3.3. *Kinetics of Particle Growth and Crystallisation of Calcium Carbonate*

SANS is the most suitable method for measuring the kinetics of ACC formation given the previously observed length scales of ACC particles. For determining the kinetics of aragonite formation time resolved powder X-ray diffraction is more appropriate given the crystalline nature of the product phase.

3.3.1. Small-Angle Neutron Scattering

A time resolved SANS measurement determines scattering intensity as a function of momentum transfer for particles of nanometre to micro size in a series of time slices. The shape and slope of each of these scattering curves depends on the size and shape of the particles in the sample. A typical series of SANS spectra collected during the formation of one of our ACC samples is shown in Supplementary Materials Figure S9 together with the extracted degree of formation curve and determined rate values. The data were fitted with two models; the first one was a simple spherical particle model, that considers all particles to be spheres and fits the radius to the data; and a Porod-Guinier model, that has a shape parameter as well as a size parameter using the SasView software package [40]. Both models gave similar results consistent with a spherical particle shape. ACC particles were found to start with an average radius of around 35 nm which grew to around 60 nm after 4–5 h for the samples with Mg/Ca ratios of 0.03 and 0.2. ACC particles with a Mg/Ca ratio of 0.26 also started with a radius of about 35 nm which remained constant for the duration of the measurement. A summary of particle size evolution under varying conditions is shown in Figure 4. The height of the peak in the SANS patterns was used to estimate the amount of ACC particles of each size as a function of time. These data were then fit with the Avrami equation [5]. This gave an estimate of the rate of formation of the initial 35 nm particles (rate = 0.17 h^{-1} for a Mg/Ca ratio of 0.26, rate = 0.27 h^{-1} for a Mg/Ca ratio of 0.2 and rate = 0.3 h^{-1} for a Mg/Ca ratio of 0.03) and the rate of transformation from 35 to 60 nm particles (rate = 2.8 h^{-1} for a Mg/Ca ratio of 0.03 and rate = 0.9 h^{-1} for a Mg/Ca ratio of 0.2 (Figure 5). The Mg/Ca ratio of 0.26 sample did not show any growth during the measurement period.

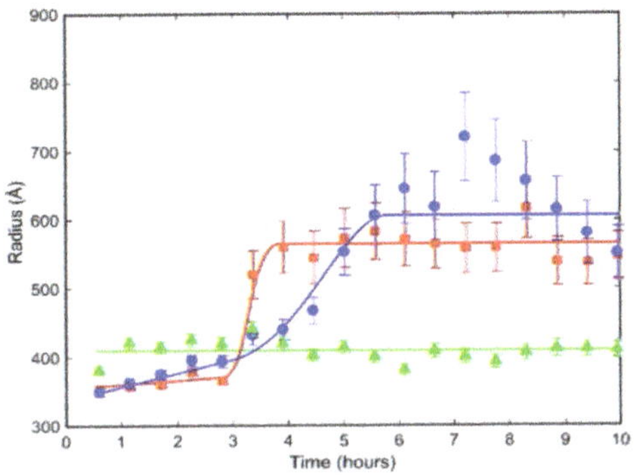

Mg/Ca ratio	k (hr^{-1})
0.26	0.17
0.20	0.27
0.03	0.3

Figure 4. Evolution of the particle size in the samples analysed at 20 °C (ambient) with time. ▲ is for the sample with Mg/Ca = 0.26; ● is for the sample with Mg/Ca = 0.20; and ■ is for the sample with Mg/Ca = 0.03. The corresponding lines are guides for the eye. The table contains estimates of the rate of transformation between the 35 nm and nm particles for three Mg/Ca ratios.

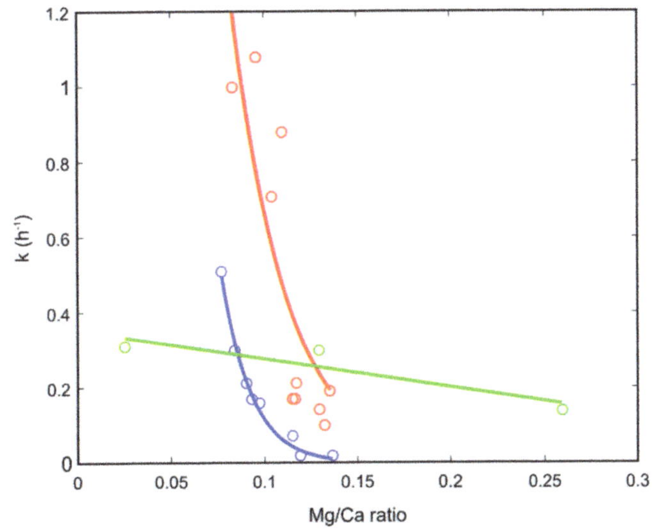

Figure 5. Reaction rates for the formation of initial ACC particles (green symbols and line) and subsequent formation of aragonite crystals (red symbols and line for formation in H_2O and blue symbols and lines formation in D_2O). The red and blue lines are exponential fits to the corresponding symbols, their equations are $y = 22.87\ e^{-35.11\ x}$ for the red line and $y = 76.38\ e^{-64.57\ x}$ for the blue line. The green line is a linear fit to the corresponding symbols giving: $y = -0.75\ x + 0.35$.

3.3.2. X-ray Diffraction

A representative series of powder diffraction patterns are presented in Supplementary Materials Figure S10. The aragonite (111) and (221) peaks are seen to grow out of the background due to nano-crystalline calcium carbonate. The intensity of the (111) peaks were used to determine the degree of reaction for each series of XRD patterns which was then used with the Avrami equation [5] to give the corresponding reaction rate Figure 6). Broad amorphous scattering was observed but no sharp diffraction signal from a crystalline product within the time frame of our measurements when the ACC precursor had an Mg/Ca ration of above about 0.14. Some samples with very high Mg/Ca ratios were found to be still amorphous many months after they were synthesised.

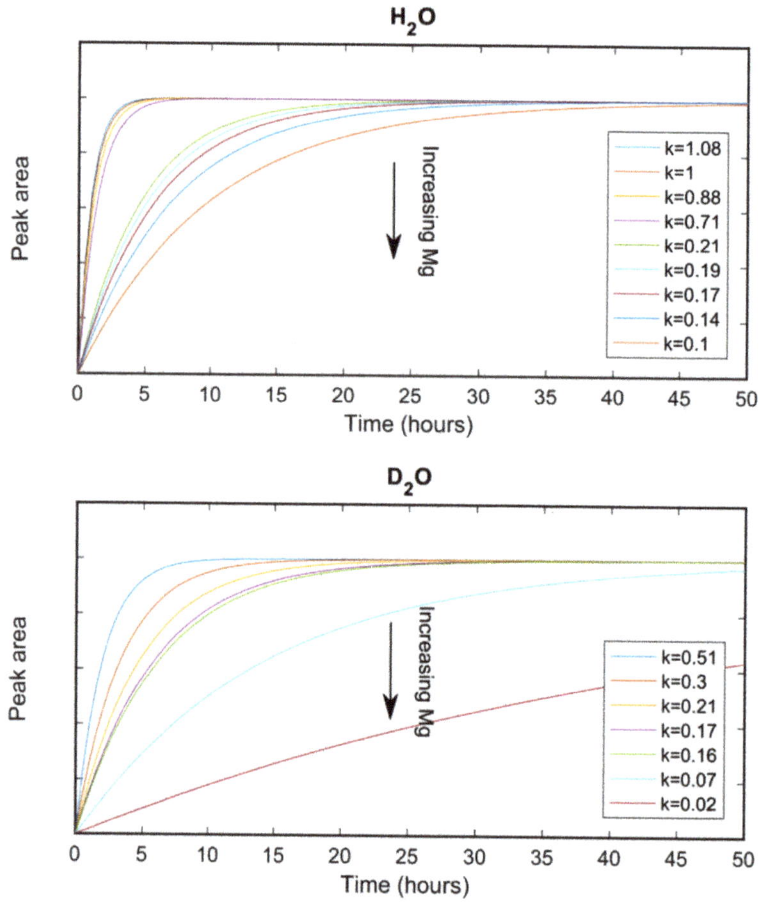

Figure 6. Compilation of the degree of aragonite formation as a function of magnesium content for samples formed from ACC with varying Mg/Ca ratios using H_2O and D_2O as the solvent.

4. Discussion and Conclusions

All samples of ACC synthesised during this study crystallised into aragonite as confirmed by XRD and Raman data. After the completion of crystallisation, a small amount of remnant ACC particles were observed on aragonite crystals, consistent with observations in natural systems such as mollusc shells [41,42]. Mg was found to be present in the nano-crystalline ACC atomic structure but not in the crystalline aragonite atomic structure. This suggests that Mg was lost during the transformation of ACC into aragonite. The lost Mg atoms most likely were incorporated into the remnant ACC particles observed on the surface of the aragonite crystals in our SEM images, stabilising this portion of the ACC to allow it to exist for extended periods of time as observed in ACC cortices around crystalline nacre tablets in mollusc shells [43,44].

The redistribution of Mg upon transformation of ACC into aragonite observed here requires transport of ions from the ACC nano-particles forming the crystalline phase, and thus, is consistent with a dissolution-reprecipitation process and discounts a pathway of solid-state transformation [3]. A schematic representation of the transformation process from solution through ACC to crystalline aragonite consistent with the data collected in this work is contained in Figure 7. We observe three steps:

1. Precipitation of ACC from solution forming 35 nm particles: reaction R_1.
2. The 35 nm particles grow into 70 nm particles: reaction R_2.
3. Formation of aragonite crystals by localised dissolution and reprecipitation: reaction R_3.

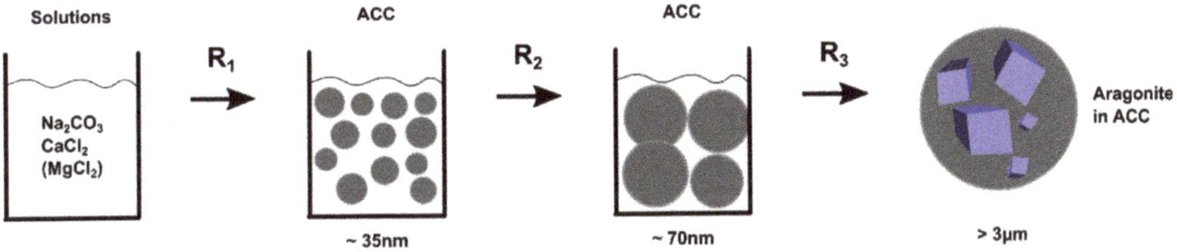

Figure 7. Schematic diagram summarizing the average particle pathway from solution to crystal for calcium carbonate, as observed in this work.

This is broadly in line with previous observations, for example: the precipitation of ACC particles of 35–40 nm [5], the observation of 100 nm particles in amorphous and crystal phases [42,45], 50–60 nm crystals [46], and 100 nm silica-coated ACC particles with 20% water [47]. Electron microscopy [48] as revealed the evolution of particles prior to step 1 above, with ACC particles growing from ~1 nm to ~20 nm in a range of a few seconds. Prenucleation clusters of size ~1 nm are thought to precede the nanoparticles of ~30 nm [49].

Values of the rates for step R_1 were found to be largely independent of Mg/Ca ratio while the Mg/Ca ratio had a very large effect on the rates for steps R_2 and R_3. The change in rate for steps R_2 and R_3 with Mg/Ca ratio were found to be very similar. From a classical nucleation and growth perspective this might imply that the addition of magnesium mainly effects particle growth and has little effect on nucleation. This might explain why biomineralization predominantly involves nanoparticles since the nucleation rate is largely unregulated while the growth rate can be controlled.

All ACC samples were found to contain magnesium, but all the crystalline aragonite products were found not to contain any magnesium. The high rate for crystallisation observed suggests that this process does not involve diffusion in the solid state. This suggests that our results would favour a dissolution and reprecipitation model for ACC transformation to aragonite over a solid-state transformation model.

In this case, the nanoparticles were found to not simply orientate and attach to form larger structures but to dissolve and recrystallize as part of the self-assembly process. The bigger question we then ask is to what extent is this mechanism employed in self-assembly for both biological and synthetic systems.

Supplementary Materials: The following supporting information can be downloaded at: https://www.mdpi.com/article/10.3390/nano12234151/s1. Figure S1. Typical TGA pattern of an ACC sample. Figure S2. X-ray powder diffraction pattern of our Mg-ACC sample with two diffuse maxima. Figure S3. X-ray scattering of solid-solution mixtures of $(Ca_{1-x}\,Mg_x)CO_3$. Figure S4. Neutron scattering patterns of samples prepared with different Mg/Ca ratios. Figure S5. Neutron scattering data of H_2O/D_2O mixtures. Figure S6. Representative Raman spectrum. Figure S7. X-ray powder diffraction pattern of an ACC converted into aragonite. Figure S8. SEM micrograph of a synthesised sample from this study showing tablets of aragonite and xenomorphic aggregates believed to be ACC. Figure S9. SANS patterns evolving with time. Figure S10. XRD patterns during crystallization of aragonite from ACC. Table S1. Water content and Mg/Ca ratios determined for ACC.

Author Contributions: S.M.C. devised the project, assisted with data collection, processing and interpretation and wrote the manuscript. V.G. contributed to preparation of the manuscript. B.C. synthesised the samples, carried out sample characterisation, measured the neutron and X-ray data, analysed the bulk of the data and prepared the first draft of this manuscript. T.A.D. helped devise the synthesis protocols and oversaw the synthesis procedures. K.W. oversaw the SANS data collection and processing. J.N. oversaw the neutron scattering data collection and processing on NOMAD. D.E.J. helped devise the project and contributed to data interpretation and manuscript preparation. All authors have read and agreed to the published version of the manuscript.

Funding: We would like to thank the Australian Centre for Neutron Scattering for the award of neutron beamtime on QUOKKA under proposals number NDF3219 and NDF2769. Thanks also to the Oak Ridge National Laboratory for provision of beam time on the NOMAD instrument under proposal number IPTS-17259.1.

Data Availability Statement: All data is available on request from the corresponding author.

Acknowledgments: We thank colleagues at the Department of Earth and Environmental Sciences at Macquarie University for technical support and fruitful discussions, specifically we are grateful to Russell Field, Will Powell and Norman Pearson. LA-ICPMS data were obtained using instrumentation funded by DEST Systemic Infrastructure Grants, ARC LIEF, NCRIS/AuScope, industry partners and Macquarie University. Aragonite and calcite reference samples were made available from the mineral collection of the Department of Earth and Planetary Sciences at Macquarie University. We thank the Australian Centre for Neutron Scattering for provision of beamtime on QUOKKA and Oak Ridge National Laboratory for provision of beam time on NOMAD. SMC and BC would like to acknowledge partial financial support from the Australian Nuclear Science and Technology Organization. DEJ acknowledges support from the Australian Research Council (DP210101268). We thank Andrew Whitten for assistance with measurements on QUOKKA. The National Deuteration Facility is partly supported by the National Collaborative Research Infrastructure Strategy an initiative of the Australian Government.

Conflicts of Interest: The authors declare no conflict of interest.

References

1. Politi, Y.; Metzler, R.A.; Abrecht, M.; Gilbert, B.; Wilt, F.H.; Sagi, I.; Addadi, L.; Weiner, S.; Gilbert, P.U.P.A. Transformation mechanism of amorphous calcium carbonate into calcite in the sea urchin larval spicule. *Proc. Natl. Acad. Sci. USA* **2008**, *105*, 17362–17366. [CrossRef]
2. Navrotsky, A. Energetic clues to pathways to biomineralization: Precursors, clusters, and nanoparticles. *Proc. Natl. Acad. Sci. USA* **2004**, *101*, 12096–12101. [CrossRef]
3. De Yoreo, J.J.; Gilbert, P.U.P.A.; Sommerdijk, N.A.; Penn, R.L.; Whitelam, S.; Joester, D.; Zhang, H.; Rimer, J.D.; Navrotsky, A.; Banfield, J.F. Crystallisation by particle attachment in synthetic, biogenic, and geologic environments. *Science* **2015**, *349*, aaa6760. [CrossRef]
4. Cartwright, J.H.E.; Checa, A.G.; Gale, J.D.; Gebauer, D.; Sainz-Días, C.I. Calcium carbonate polyamorphism and its role in biomineralisation: How many amorphous calcium carbonates are there? *Angew. Chem. Int. Ed.* **2012**, *51*, 11960–11970. [CrossRef] [PubMed]
5. Bots, P.; Benning, L.G.; Rodriguez-Blanco, J.-D.; Roncal-Herrero, T.; Shaw, S. Mechanistic Insights into the Crystallization of Amorphous Calcium Carbonate (ACC). *Cryst. Growth Des.* **2012**, *12*, 3806–3814. [CrossRef]

6. Clark, S.M.; Colas, B.; Jacob, D.E.; Neuefeind, J.C.; Wang, H.-W.; Page, K.L.; Soper, A.K.; Schodder, P.I.; Duchstein, P.; Zubiri, B.A.; et al. The nano- and meso-scale structure of amorphous calcium carbonate indicate nanoscale assembly processes. *Sci. Rep.* **2022**, *12*, 6870. [CrossRef]
7. Rodriguez-Blanco, J.D.; Shaw, S.; Benning, L.G. The kinetics and mechanisms of amorphous calcium carbonate (ACC) crystallization to calcite, via vaterite. *Nanoscale* **2011**, *3*, 265–271. [CrossRef] [PubMed]
8. Stephens, C.J.; Ladden, S.F.; Meldrum, F.C.; Christenson, H.K. Amorphous Calcium Carbonate is Stabilized in Confinement. *Adv. Funct. Mater.* **2010**, *20*, 2108–2115. [CrossRef]
9. Addadi, L.; Raz, S.; Weiner, S. Taking Advantage of Disorder: Amorphous Calcium Carbonate and Its Roles in Biomineralization. *Adv. Mater.* **2003**, *15*, 959–970. [CrossRef]
10. Ouhenia, S.; Chateigner, D.; Belkhir, M.; Guilmeau, E.; Krauss, C. Synthesis of calcium carbonate polymorphs in the presence of polyacrylic acid. *J. Cryst. Growth* **2008**, *310*, 2832–2841. [CrossRef]
11. Huang, S.-C.; Naka, K.; Chujo, Y. A Carbonate Controlled-Addition Method for Amorphous Calcium Carbonate Spheres Stabilized by Poly(acrylic acid)s. *Langmuir* **2007**, *23*, 12086–12095. [CrossRef] [PubMed]
12. Lam, R.S.K.; Charnock, J.M.; Lennie, A.; Meldrum, F.C. Synthesis-dependant structural variations in amorphous calcium carbonate. *CrystEngComm* **2007**, *9*, 1226–1236. [CrossRef]
13. Loste, E.; Wilson, R.M.; Seshadri, R.; Meldrum, F.C. The role of magnesium in stabilising amorphous calcium carbonate and controlling calcite morphologies. *J. Cryst. Growth* **2003**, *254*, 206–218. [CrossRef]
14. Rodriguez-Blanco, J.; Shaw, S.; Bots, P.; Roncal-Herrero, T.; Benning, L. The role of pH and Mg on the stability and crystallization of amorphous calcium carbonate. *J. Alloys Compd.* **2011**, *536*, S477–S479. [CrossRef]
15. Foran, E.; Weiner, S.; Fine, M. Biogenic Fish-gut Calcium Carbonate is a Stable Amorphous Phase in the Gilt-head Seabream, *Sparus aurata*. *Sci. Rep.* **2013**, *3*, 1700–1705. [CrossRef]
16. Pai, R.K.; Pillai, S. Nanoparticles of amorphous calcium carbonate by miniemulsion: Synthesis and mechanism. *CrystEngComm* **2008**, *10*, 865–872. [CrossRef]
17. Sun, W.; Jayaraman, S.; Chen, W.; Persson, K.A.; Ceder, G. Nucleation of metastable aragonite $CaCO_3$ in seawater. *Proc. Natl. Acad. Sci. USA* **2015**, *112*, 3199–3204. [CrossRef]
18. Zhang, Z.; Xie, Y.; Xu, X.; Pan, H.; Tang, R. Transformation of amorphous calcium carbonate into aragonite. *J. Cryst. Growth* **2012**, *343*, 62–67. [CrossRef]
19. Blue, C.; Dove, P. Chemical controls on the magnesium content of amorphous calcium carbonate. *Geochim. Cosmochim. Acta* **2015**, *148*, 23–33. [CrossRef]
20. Gal, A.; Habraken, W.; Gur, D.; Fratzl, P.; Weiner, S.; Addadi, L. Calcite Crystal Growth by a Solid-State Transformation of Stabilized Amorphous Calcium Carbonate Nanospheres in a Hydrogel. *Angew. Chem. Int. Ed.* **2013**, *52*, 4867–4870. [CrossRef]
21. Ihli, J.; Kim, Y.-Y.; Noel, E.H.; Meldrum, F.C. The Effect of Additives on Amorphous Calcium Carbonate (ACC): Janus Behavior in Solution and the Solid State. *Adv. Funct. Mater.* **2012**, *23*, 1575–1585. [CrossRef]
22. Pontoni, D.; Bolze, J.; Dingenouts, N.; Narayanan, T.; Ballauff, M. Crystallization of Calcium Carbonate Observed In-situ by Combined Small- and Wide-angle X-ray Scattering. *J. Phys. Chem. B* **2003**, *107*, 5123–5125. [CrossRef]
23. Giuffre, A.J.; Gagnon, A.C.; De Yoreo, J.J.; Dove, P.M. Isotopic tracer evidence for the amorphous calcium carbonate to calcite transformation by dissolution–reprecipitation. *Geochim. Cosmochim. Acta* **2015**, *165*, 407–417. [CrossRef]
24. Liu, Z.; Zhang, Z.; Wang, Z.; Jin, B.; Li, D.; Tao, J.; Tang, R.; De Yoreo, J.J. Shape-preserving amorphous-to-crystalline transformation of $CaCO_3$ revealed by in situ TEM. *Proc. Natl. Acad. Sci. USA* **2020**, *117*, 3397–3404. [CrossRef] [PubMed]
25. Jacob, D.E.; Wirth, R.; Agbaje, O.B.A.; Branson, O.; Eggins, S.M. Planktic foraminifera form their shells via metastable carbonate phases. *Nat. Commun.* **2017**, *8*, 1265. [CrossRef]
26. Nielsen, M.H.; Aloni, S.; De Yoreo, J.J. In situ TEM imaging of $CaCO_3$ nucleation reveals coexistence of direct and indirect pathways. *Science* **2014**, *345*, 1158–1162. [CrossRef] [PubMed]
27. Fernandez-Martinez, A.; Kalkan, B.; Clark, S.M.; Waychunas, G.A. Pressure-Induced Polyamorphism and Formation of 'Aragonitic' Amorphous Calcium Carbonate. *Angew. Chem. Int. Ed.* **2013**, *52*, 8354–8357. [CrossRef]
28. Jochum, K.P.; Nohl, U.; Herwig, K.; Lammel, E.; Stoll, B.; Hofmann, A.W. GeoReM: A new geochemical database for reference materials and standards. *Geostand. Geoanal. Res.* **2006**, *29*, 333–338. [CrossRef]
29. Doebelin, N.; Kleeberg, R. Profex: A graphical user interface for the Rietveld refinement program BGMN. *J. Appl. Crystallogr.* **2015**, *48*, 1573–1580. [CrossRef]
30. Radha, A.; Fernandez-Martinez, A.; Hu, Y.; Jun, Y.-S.; Waychunas, G.A.; Navrotsky, A. Energetic and structural studies of amorphous $Ca1-xMgxCO_3{\cdot}nH_2O$ ($0 \leqslant x \leqslant 1$). *Geochim. Cosmochim. Acta* **2012**, *90*, 83–95. [CrossRef]
31. Wood, K.; Mata, J.; Garvey, C.J.; Wu, C.-M.; Hamilton, W.A.; Abbeywick, P.; Bartlett, D.; Bartsch, F.; Baxter, P.; Booth, N.; et al. QUOKKA, the pinhole small-angle neutron scattering instrument at the OPAL Research Reactor, Australia: Design, performance, operation and scientific highlights. *J. Appl. Crystallogr.* **2018**, *51*, 294–314. [CrossRef]
32. Wang, D.; Hamm, L.M.; Giuffre, A.J.; Echigo, T.; Rimstidt, J.D.; De Yoreo, J.J.; Grotzinger, J.; Dove, P.M. Revisiting geochemical controls on patterns of carbonate deposition through the lens of multiple pathways to mineralization. *Faraday Discuss.* **2012**, *159*, 371–386. [CrossRef]
33. Xu, J.; Yan, C.; Zhang, F.; Konishi, H.; Xu, H.; Teng, H.H. Testing the cation-hydration effect on the crystallization of $Ca–Mg–CO_3$ systems. *Proc. Natl. Acad. Sci. USA* **2013**, *110*, 17750–17755. [CrossRef]

34. Wang, Y.-W.; Kim, Y.-Y.; Stephens, C.J.; Meldrum, F.C.; Christenson, H.K. In Situ Study of the Precipitation and Crystallization of Amorphous Calcium Carbonate (ACC). *Cryst. Growth Des.* **2012**, *12*, 1212–1217. [CrossRef]
35. Parker, J.E.; Thompson, S.P.; Lennie, A.R.; Potter, J.; Tang, C.C. A study of the aragonite-calcite transformation using Raman spectroscopy, synchrotron powder diffraction and scanning electron microscopy. *CrystEngComm* **2010**, *12*, 1590–1599. [CrossRef]
36. Wehrmeister, U.; Jacob, D.E.; Soldati, A.L.; Loges, N.; Häger, T.; Hofmeister, W. Amorphous, nanocrystalline and crystalline calcium carbonates in biological materials. *J. Raman Spectrosc.* **2010**, *42*, 926–935. [CrossRef]
37. Lafuente, B.; Downs, R.T.; Yang, H.; Stone, N. 1. The Power of Databases: The RRUFF Project. In *Highlights in Mineralogical Crystallography*; Armbruster, T., Danisi, R.M., Eds.; De Gruyter: Berlin, Germany, 2016; pp. 1–30. ISBN 9783110417104.
38. Nudelman, F. Nacre biomineralisation: A review on the mechanisms of crystal nucleation. *Semin. Cell Dev. Biol.* **2015**, *46*, 2–10. [CrossRef] [PubMed]
39. Dudnikova, V.B.; Eremin, N.N. Simulation of the local structure, properties of mixing, and stability of solid solutions Ba x Sr1−x CO3 by the interatomic potential method. *Phys. Solid State* **2016**, *58*, 1101–1107. [CrossRef]
40. Stuhrmann, H. Small-angle scattering of X-rays. *Prog. Cryst. Growth Charact.* **1989**, *18*, 1–19. [CrossRef]
41. Jacob, D.; Wirth, R.; Soldati, A.; Wehrmeister, U.; Schreiber, A. Amorphous calcium carbonate in the shells of adult Unionoida. *J. Struct. Biol.* **2011**, *173*, 241–249. [CrossRef]
42. Wolf, S.E.; Böhm, C.; Harris, J.; Hajir, M.; Mondeshki, M.; Marin, F. Single Nanogranules Preserve Intracrystalline Amorphicity in Biominerals. *Key Eng. Mater.* **2016**, *672*, 47–59. [CrossRef]
43. Nassif, N.; Pinna, N.; Gehrke, N.; Antonietti, M.; Jäger, C.; Cölfen, H. Amorphous layer around aragonite platelets in nacre. *Proc. Natl. Acad. Sci. USA* **2005**, *102*, 12653–12655. [CrossRef] [PubMed]
44. Macías-Sánchez, E.; Willinger, M.G.; Pina, C.M.; Checa, A.G. Transformation of ACC into aragonite and the origin of the nanogranular structure of nacre. *Sci. Rep.* **2017**, *7*, 12728. [CrossRef] [PubMed]
45. Jacob, D.; Soldati, A.; Wirth, R.; Huth, J.; Wehrmeister, U.; Hofmeister, W. Nanostructure, composition and mechanisms of bivalve shell growth. *Geochim. Cosmochim. Acta* **2008**, *72*, 5401–5415. [CrossRef]
46. Hovden, R.; Wolf, S.E.; Holtz, M.E.; Marin, F.; Muller, D.A.; Estroff, L.A. Nanoscale assembly processes revealed in the nacroprismatic transition zone of Pinna nobilis mollusc shells. *Nat. Commun.* **2015**, *6*, 10097. [CrossRef]
47. Ihli, J.; Wong, W.C.; Noel, E.H.; Kim, Y.-Y.; Kulak, A.N.; Christenson, H.K.; Duer, M.J.; Meldrum, F.C. Dehydration and crystallization of amorphous calcium carbonate in solution and in air. *Nat. Commun.* **2014**, *5*, 3169. [CrossRef] [PubMed]
48. Smeets, P.J.M.; Cho, K.R.; Kempen, R.G.E.; Sommerdijk, N.A.J.M.; De Yoreo, J.J. Calcium carbonate nucleation driven by ion binding in a biomimetic matrix revealed by in situ electron microscopy. *Nat. Mater.* **2015**, *14*, 394–399. [CrossRef]
49. Pouget, E.M.; Bomans, P.H.H.; Goos, J.A.C.M.; Frederik, P.M.; de With, G.; Sommerdijk, N.A.J.M. The Initial Stages of Template-Controlled $CaCO_3$ Formation Revealed by Cryo-TEM. *Science* **2009**, *323*, 1455–1458. [CrossRef]

nanomaterials

Article

Supramolecular Self-Assembly of Atomically Precise Silver Nanoclusters with Chiral Peptide for Temperature Sensing and Detection of Arginine

Wenjuan Wang, Zhi Wang, Di Sun, Shulin Li, Quanhua Deng and Xia Xin *

National Engineering Research Center for Colloidal Materials, Key Laboratory of Colloid and Interface Chemistry (Ministry of Education), School of Chemistry and Chemical Engineering, Shandong University, Jinan 250100, China; wenjuanwang@mail.sdu.edu.cn (W.W.); zwang@mail.sdu.edu.cn (Z.W.); dsun@sdu.edu.cn (D.S.); 202032370@mail.sdu.edu.cn (S.L.); qhdeng@sdu.edu.cn (Q.D.)

* Correspondence: xinx@sdu.edu.cn; Tel.: +86-531-8836-3597

Abstract: Metal nanoclusters (NCs) as a new type of fluorescent material have attracted great interest due to their good biocompatibilities and outstanding optical properties. However, most of the studies on metal NCs focus on the synthesis, atomic or molecular assembly, whereas metal NCs ability to self-assemble to higher-level hierarchical nanomaterials through supramolecular interactions has rarely been reported. Herein, we investigate atomic precise silver NCs (Ag_9-NCs, $[Ag_9(mba)_9]$, where H_2mba = 2-mercaptobenzoic acid) and peptide DD-5 were used to induce self-assembly, which can trigger an aggregation-induced luminescence (AIE) effect of Ag_9-NCs through non-covalent forces (H-bond, π–π stacking) and argentophilic interactions [Ag(I)–Ag(I)]. The large Stokes shift (~140 nm) and the microsecond fluorescence lifetime (6.1 μs) indicate that Ag_9-NCs/DD-5 hydrogel is phosphor. At the same time, the chirality of the peptide was successfully transferred to the achiral Ag_9-NCs because of the supramolecular self-assembly, and the Ag_9-NCs/DD-5 hydrogel also has good circularly polarized luminescence (CPL) properties. In addition, Ag_9-NCs/DD-5 luminescent hydrogel is selective and sensitive to the detection of small biological molecule arginine. This work shows that DD-5 successfully induces the self-assembly of Ag_9-NCs to obtain high luminescent gel, which maybe become a candidate material in the fields of sensors and biological sciences.

Keywords: supramolecular self-assembly; metal nanoclusters; peptide; aggregation-induced emission; chirality

1. Introduction

As a bridge between atom and nanoparticle, metal nanoclusters (metal NCs, mainly Au, Ag and Cu) which are composed of a few to hundreds of mental atoms and covered with organic ligands on the surface of the metal core, have been receiving extensive attention due to their important roles in the fields of catalysis, sensing, electrochemistry, energy transfer and biomedicine [1–5]. Among various applications, metal nanoclusters are widely used as sensors because of their sensitive responses to temperature, pH value, metal ions and small biological molecules [6–10]. For example, Hu et al. utilized Cu-MOF as a precursor to prepare highly stable Cu NCs, which can be used to construct a pH ratiometric fluorescence sensor to monitor pH of microorganism [8]. Kailasa et al. found that the addition of La^{3+} can significantly enhance the fluorescence emission of BSA-AuNCs, and constructed an La^{3+} ion-BSA-AuNCs fluorescence sensor, thus realizing the detection of four divalent metal ions (Hg^{2+}, Cy^{2+}, Pb^{2+} and Cd^{2+}) [9].

In particular, the biocompatibility and fluorescent properties are potentially useful in biological systems of Ag NCs, making them a research hotspot [11]. However, the current research on Ag NCs is mainly focused on the synthesis and atomic or molecular assembly, while the nanostructures obtained by the means of supramolecular self-assembly are rarely reported [12,13]. This may be due to the small size of the Ag NCs, and to the higher surface energy inducing further growth of Ag NCs into larger Ag nanoparticles, which is not conducive to the progress of self-assembly. Therefore, in order to solve this problem, the subtle non-covalent interactions (H-bond, π–π stacking, van der Waals forces, electrostatic interactions and hydrophobic interaction) should be controlled between Ag NCs by adding small molecules through supramolecular self-assembly to obtain ordered aggregates.

As a natural biomolecule, peptides are usually composed of less than 50 amino acids, are easy to manipulate and synthesize, and can self-assemble into various ordered aggregates, such as nanotubes, nanofibers, nanovesicles, nanobelts and hydrogel [14–18]. However, in order to overcome the inherent limitations of single-component materials and make multi-component materials more widely used, peptides can be co-assembled with a variety of functional molecules to obtain supramolecular materials [19–21]. For example, Gazit et al., through the co-assembly between Fmoc-RGD and chitosan, obtained a hydrogel with stronger mechanical properties and stronger durability than Fmoc-RGD self-assembled hydrogel, which has a wide range of applications in the field of cell culture scaffolds [22]. Xu et al. obtained the hydrogel of nanofibers through coordination, H-bond and hydrophobic interaction between hairpin peptides and copper ions, and the nanofibers can be utilized as templates for the synthesis of long, ultrathin CuS nanowires which have near-infrared (NIR) laser-induced thermal effect [23]. Therefore, using peptides to induce the self-assembly of Ag NCs can not only improve the biocompatibility of multi-component materials, but also expand the range of application for metal NCs-based materials [6,7,24].

In this work, we used water-soluble, atomically precise Ag NCs (Ag_9-NCs, $[Ag_9(mba)_9]$, where H_2mba = 2-mercaptobenzoic acid) to interact with the peptide DD-5 to construct luminescent hydrogel. The Ag_9-NCs/DD-5 hydrogel was realized through non-covalent forces (H-bond, π–π stacking) and argentophilic interactions [Ag(I)–Ag(I)], and phosphorescent emission was obtained through aggregation-induced luminescence (AIE). The Ag_9-NCs/DD-5 xerogel has good application in temperature sensing and the orange-red emission of the hydrogel can be used to detect arginine (Scheme 1). This work provides a new example for the construction of metal NC-peptide complexes through non-covalent bonds used in the fields of temperature fluorescence sensing and biological detection.

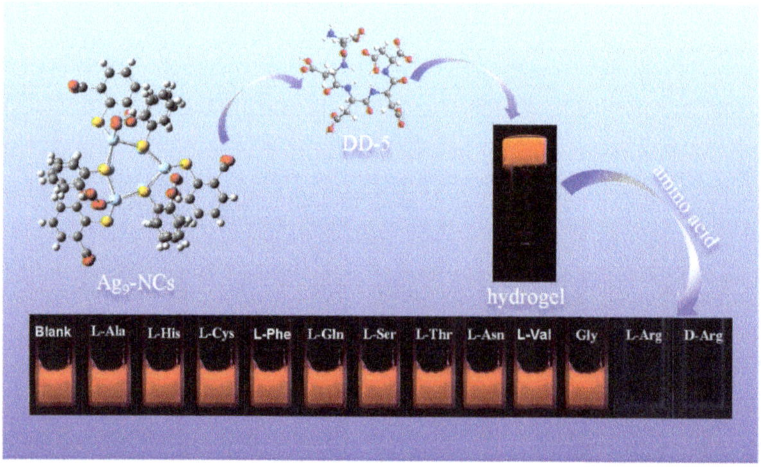

Scheme 1. Schematic illustration of the self-assembly of the Ag_9-NCs driven by DD-5 and their application in the detect of Arginine.

2. Materials and Methods

2.1. Materials

The synthesis, purification and detailed characterization of Ag_9-NCs can be found in literature [6,25]. DD-5 was purchased from GL Biochem Ltd. (Shanghai, China) and used without further purification. It was polymerized by five aspartic acids and its relative molecular mass was 593.46 g mol^{-1}. The molecular structure of DD-5 is shown in Figure 1a. L-Arginine (L-Arg), D-Arginine (D-Arg), L-alanine (L-Ala), L-histidine (L-His), L-cysteine (L-Cys), L-phenylalanine (L-Phe), L-tyrosine (L-Tyr), L-asparagine (L-Asn), L-valine (L-Val) and glycine (Gly) were purchased from Sinopharm Chemical Reagent Co., Ltd, (Shanghai, China). L-Glutamine (L-Glu) and L-serine (L-Ser) were purchased from Sigma-Aldrich (St. Louis, MO, USA). Ultrapure water used in the experiments with a resistivity of 18.25 MΩ cm^{-1} was obtained using a UPH-IV ultrapure water purifier (Shanghai Youpu Industry Company Ltd., Shanghai, China).

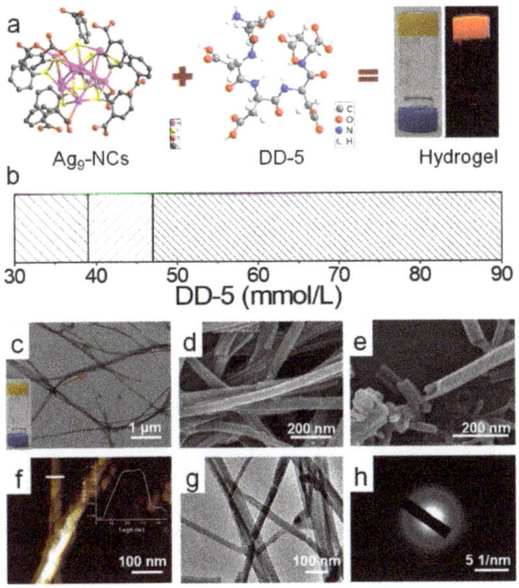

Figure 1. Self-assembly phase behavior and detailed characterization of 5 mM Ag_9-NCs/70 mM DD-5 hydrogel. (**a**) The structure of Ag_9-NCs and DD-5, the schematic diagram of the formation of the core sample hydrogel; (**b**) the concentration of Ag_9-NCs was fixed at 5 mM, and the concentration of DD-5 was changed to obtain a phase diagram; (**c**) TEM, the inset is a visual diagram of the sample; (**d**,**e**) SEM; (**f**) AFM, inset: a plan view of the height extending from the indicator line; (**g**) HR-TEM; (**h**) SAED pattern of HR-TEM.

2.2. Self-Assembly of Ag9-NCs/DD-5 Hydrogel

In this typical experiment, we added 20.0 mg DD-5 to 355 μL ultrapure water and stirred, then added 145 μL Ag_9-NCs solution (15.87 mM) and stirred. The hydrogel was successfully prepared after 8 h of constant temperature (20 °C) in a thermostat. The hydrogel was lyophilized in a vacuum extractor at −60 °C for 5 days to collect the orange-yellow powder.

2.3. The Detection of L-Arg and D-Arg

Amino acids were added to the solution of DD-5 and Ag_9-NCs to ensure that the final concentration of amino acids was 100 mM. After incubating in 20 °C incubator for 8 h, we performed a fluorescence test to study the selectivity of the hydrogel toward L-Arg and D-Arg, because both D-Arg and L-Arg can completely quench the fluorescence of nanotubes. We took L-Arg as an example to investigate in detail: we added different concentrations of L-Arg to the formed 100 μL of Ag_9-NCs/DD-5 hydrogel, and placed it in a thermostat at 20 °C for 8 h. We used a triangular cuvette with a capacity of 2 mL for fluorescence detection to obtain concentration change spectra.

2.4. Characterizations

A copper mesh was inserted into the gel to obtain a sample, and after drying under an IR lamp for 45 min, transmission electron microscopy (TEM) images were observed under a JCR-100CX II (JEOL, Tokyo, Japan) microscope. Field emission scanning electron microscopy (FE-SEM) observations were carried out on a Hitachi SU8010 (Hitachi, Tokyo, Japan) under 10 kV. High-resolution TEM (HRTEM) images and selected area electron diffraction (SAED) were recorded by HR-JEOL 2100 (JEOL, Tokyo, Japan) system with an accelerating voltage of 200 kV. Atomic force microscope (AFM) tapping mode measurements were performed on Bruke Bioscope Resolve. UV-vis data were recorded on a Shimadzu UV-2600 spectrophotometer (Shimadzu, Kyoto, Japan). Fluorescence data were tested on a LS-55 spectrofluorometer (PerkinElmer, Waltham, MA, USA) and an Edinburgh Instruments FLS920 luminescence spectrometer (xenon lamp, 450 W) (Edinburgh Instruments Ltd., Livingston, UK). Fourier transform infrared (FT-IR) spectra in KBr wafer were recorded on a VERTEX-70/70v spectrophotometer (Bruker, Billerica, MA, USA). CD spectra were taken on J-810 Spectra Manager system (ChirascanV 100). Small-angle X-ray spectroscopy (SAXS) measurements were performed using an Anton-Paar SAX Sess mc^2 system (Anton Paar, Graz, Austria) with nickel-filtered Cu Kα radiation (1.54 Å) operating at 50 kV and 40 mA. X-ray diffraction (XRD) patterns were taken on a D8 ADVANCE (Bruker, Germany) diffractometer equipped with Cu Kα radiation and a graphite monochromator. X-ray photoelectron spectroscope (XPS) date were collected by an X-ray photoelectron spectrometer (ESCALAB 250, Thermo Fisher Scientific, Waltham, MA, USA) with a monochromatized AI Kα X-ray source (1486.71 eV). The rheological measurements were carried out on an MARS60 rheometer (Thermo Fisher Scientific, Waltham, MA, USA) with a cone−plate system. Before the frequency sweep, an amplitude sweep at a fixed frequency of 1 Hz was carried out to ensure that the selected stress was in the linear viscoelastic region. The frequency sweep was carried out from 0.01 to 100 Hz at a fixed stress of 10 Pa. The variable temperature spectrum is recorded in the UV-Vis-microspectrophotometer (20/30 PV^{TM}, Craic Technologies, San Dimas, CA, United States). Thermogravimetric analyses (TGA) were performed under a nitrogen atmosphere at 25–1000 °C with a heating speed of 10 °C min^{-1} on a TA SDT Q600 thermal analyzer (TA Instruments, New Castle, DE, USA).

3. Results

3.1. Self-Assembly of Ag_9-NCs/DD-5 Hydrogel

The average diameter of Ag_9-NCs in the aqueous solution was about 1.4 ± 0.5 nm, and they were in a non-fluorescent state (Figure S1a). We wonder whether DD-5 peptides, which are non-fluorescent at room temperature, can be used to induce the AIE effect of Ag_9-NCs through a self-assembly strategy (Figure S1b). The schematic diagram of the core sample self-assembly is shown in Figure 1a. The concentration of Ag_9-NCs was fixed at 5 mM, the concentration of DD-5 was changed, and after incubation for 8 h, the phase diagram was obtained (Figure 1b). Under 365 nm UV light, the solution state (30~38 mM DD-5) did not emit fluorescence, while the precipitated state (38~47 mM DD-5) and the hydrogel state (47~90 mM DD-5) both emitted orange-red fluorescence (Figure S1c), which implied that as the concentration of DD-5 increases, Ag_9-NCs gelation restricted the rotation or vibration of the Ag_9-NCs ligand on the spatial scale, enhancing AIE and emitted fluorescence.

From the TEM and SEM images, it was found that the precipitate was a solid nanorod structure, and the hydrogel was hollow nanotube with a little spiral (Figure S2). When $c_{DD\text{-}5}$ = 70 mM, it is located in the center of the hydrogel and its morphology is better (Figure S2), so it is defined as the core sample for follow-up research. The microstructures within the hydrogel were characterized in detail by imaging studies. TEM shows that the fibers formed in 5 mM Ag_9-NCs/70 mM DD-5 hydrogel were about 30–50 nm in width and 5–20 μm micrometers in length (Figure 1c,d) and the inside of the fiber had lower contrast than the edge (Figure 1c,g). The fracture surface observed by SEM (Figure 1d,e) indicated that the fibers were hollow nanotube structures. AFM image shows that the thickness of the nanotube was 40 nm, which was basically the same as the width of nanotubes (Figure 1f). In addition, SAED results in Figure 1h indicate that the nanotubes were a polycrystalline structure.

3.2. Fluorescence and Chirality of Ag_9-NCs/DD-5 Hydrogel

As metal NCs are known as potential AIE molecules, we next explored the fluorescent properties of Ag_9-NCs/DD-5 hydrogel. In the solution state of Ag_9-NCs, due to the free vibration or rotation of the ligand mba^{2-}, the non-radiative inactivation channel is opened and the fluorescence disappears. Therefore, Ag_9-NCs have a shorter fluorescence lifetime (~3.277 ns) (Figure S3a, Table S1). However, under the excitation wavelength of 490 nm, the quantum yield of the Ag_9-NCs/DD-5 hydrogel is 8.11% with a microsecond fluorescence lifetime (~6.10 μs), and the emission wavelength is located at 630 nm with a large Stokes shift (~140 nm) (Figures 2a and S3b, Table S2), indicating that it is essentially a phosphor.

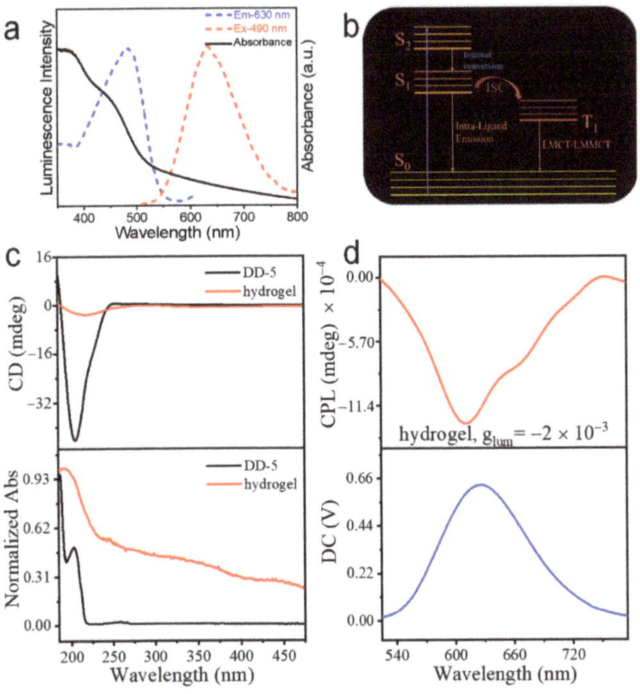

Figure 2. Study on the optical and chiral behavior of Ag_9-NCs/DD-5 hydrogel. (**a**) Excitation (blue dotted line, λ_{em} = 630 nm), emission (red dotted line, λ_{ex} = 490 nm) spectra and UV-vis absorption (black line); (**b**) schematic diagram of the excite-state relaxation dynamics of the nanotubes; (**c**) CD spectrum of pure DD-5 and hydrogel, the bottom are the UV absorption spectra of pure DD-5 and hydrogel; (**d**) CPL spectrum of hydrogel, the DC value in the bottom spectrum stands for fluorescence intensity of hydrogel.

There are several reasons to prove that the system is phosphorescence produced by triple transitions: (i) the optimal excitation peak for Ag_9-NCs/DD-5 hydrogel is located at approximately 490 nm, overlapping with the absorption peak of the charge transfer from the ligand to the metal of Ag_9-NCs caused by the addition of DD-5 (Figure 2a), inducing a phosphor with long lifetime emission [26]; (ii) the aromatic carboxyl group at room temperature is phosphorescence (RTP) molecule, the energy levels of singlet and triplet of H_2mba are very similar, and after the introduction of DD-5 through non-covalent interactions the amount of charge transfer from ligand-to-metal charge transfer (LMCT) increases, which makes intersystem crossing (ISC) prone to occur [27]; (iii) charge transfer can be caused by argentophilic interactions, which can also be attributed to the triplet state of argentophilic interactions with ligand-metal-mental charge transfer (LMMCT) [28,29]. Thus, it can be concluded that after DD-5 is added, DD-5 combines with Ag_9-NCs through non-covalent interactions, which limits the intramolecular vibration and rotation of the Ag_9-NCs ligand mba^{2-}, reducing the non-radiative loss of triplet excitons, and promotes phosphorescence emission (Figure 2b).

Based on the chirality of the peptide DD-5, we next examined the chirality of Ag_9-NCs/DD-5 hydrogel. From the CD spectrum (Figure 2c), it is found that DD-5 has a strong negative Cotton effect at 204 nm, while Ag_9-NCs/DD-5 hydrogel has a relatively weak negative Cotton effect at 217 nm, which indicates that the configuration after assembly has changed [30]. At the same time, the red shift 13 nm indicates that the chirality of DD-5 was successfully transferred to the assembly, and supramolecular chirality is obtained. Due to Ag_9-NCs/DD-5 hydrogel having good luminescence performance and chirality, the circularly polarized luminescence (CPL) performance of the assembly is investigated. Generally speaking, the strength of CPL can be evaluated by the luminescence dissymmetry factor (g_{lum}), which is defined as $g_{lum} = 2 \times (I_L - I_R)/(I_L + I_R)$, where I_L and I_R refer to the intensity of left- and right-hand CPL, respectively. From the CPL spectrum (Figure 2d), the assembly has good CPL performance and asymmetry factor glum is -2.0×10^{-3}, indicating that Ag_9-NCs/DD-5 hydrogel with a negative Cotton effect displayed left-handed CPL.

3.3. Structure and Mechanism Analysis of the Hydrogel

To dissect the self-assembly mechanism of Ag_9-NCs/DD-5 hydrogels, it is essential to analyze the composition of supramolecular assembly. The peak splitting results of C_{1s}, N_{1s}, O_{1s}, S_{2p} and Ag_{3d} for Ag_9-NCs/DD-5 xerogel in XPS indicate that the hydrogel is formed by Ag_9-NCs and DD-5 (Figures 3a and S4). Moreover, Ag_9-NCs/DD-5 xerogel and Ag_9-NCs powder have the similar signals in Ag $3d_{5/2}$ and Ag $3d_{3/2}$, indicating that Ag_9-NCs do not undergo chemical or structural conversions during gelation (Figure 3b). In order to understand the deposition pattern and spatial structure of the hydrogel, SAXS and XRD were carried out. For the SAXS result of Ag_9-NCs/DD-5 xerogel, four diffraction peaks are found at q = 10.59, 9.48, 6.64 and 4.66 nm^{-1} with a scattering factor q ratio of $1{:}\sqrt{2}{:}2{:}\sqrt{5}$, which is a typical tetragonal phase stack (Figure 3c) [31]. In addition, the smallest repeating unit of its aggregate d = 1.35 nm, which is equivalent to the size of Ag_9-NCs. Considering the length of the Ag_9-NCs ligand mba^{2-} ligand (~6.5 Å) and the length of DD-5 molecule, the smallest repeating unit of 1.35 nm indicates a strong crossover between adjacent Ag_9-NCs and DD-5, supporting the π–π stacking form of Ag_9-NCs.

As shown in the XRD spectrum (Figure 3d), the Ag_9-NCs/DD-5 xerogel has several relatively significant peaks were recorded in the range of 20–45°. The diffraction peaks at $2\theta = 22.14°, 28.62°$ and $34.10°$ correspond to π–π stacking (peak at ▲ 4.0 Å), Ag-Ag (peak at • 3.1 Å), and Ag-S (peak at ▪ 2.6 Å) and other possible interplanar distances [32,33], which indicate that the presence of π-π stacking and Ag–Ag interactions contribute to an ordered arrangement in the assembled hydrogel. In contrast, the XRD of lyophilized Ag_9-NCs solution and pure DD-5 only showed a diffuse reflection peak, indicating that they are of an amorphous nature. Based on the above data, we can conclude that the self-assembly process is closely related to the non-covalent interactions (H-bond, π–π stacking) between Ag_9-NCs and DD-5 and argentophilic interactions [Ag(I)–Ag(I)] between Ag_9-NCs, and finally highly ordered fluorescent nanotubes are obtained.

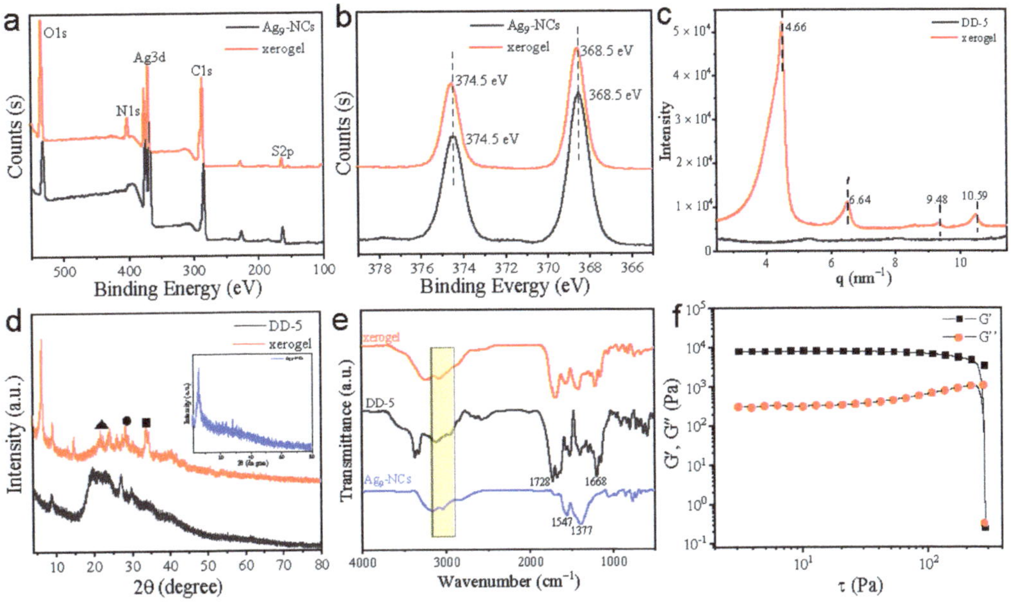

Figure 3. The structure characterization results of Ag_9-NCs/DD-5 xerogel (or hydrogel). (**a**) XPS spectra of lyophilized Ag_9-NCs and xerogel; (**b**) High-resolution XPS spectra of Ag element; (**c**) SAXS results of DD-5 and xerogel; (**d**) XRD results of DD-5 and xerogel, the inset shows the result of lyophilized Ag_9-NCs; (**e**) FT-IR spectra of a lyophilized Ag_9-NCs, DD-5 and xerogel; (**f**) stress sweep test at $f = 1$ Hz for the hydrogel Study on the optical and chiral behavior of Ag_9-NCs/DD-5 hydrogel.

FT-IR was performed to explore the non-covalent forces between Ag_9-NCs and DD-5 in the self-assembly (Figure 3d). For pure Ag_9-NCs powder, 1537 and 1377 cm^{-1} are assigned to the antisymmetric and symmetric stretching vibrations of C=O in the ligand mba^{2-} (Figure 3e) [34]. For pure DD-5 powder, the peak at 1410–1260 cm^{-1} is assigned to the in-plane curvature of the free carboxyl group, the peaks at 1668 and 1728 cm^{-1} belong to the amide I band, which is attributed to the stretching vibration of the peptide backbone, and the peak at 3370–3320 cm^{-1} is attributed to the stretching vibration of N-H. In Ag_9-NCs/DD-5 xerogel, it was found that the C=O of the Ag_9-NCs ligand showed a significant red shift, the DD-5 amide I band peak disappeared and the free carboxyl group in-plane curvature peak weakened or even disappeared, indicating that there are hydrogen bonds between Ag_9-NCs and DD-5. Moreover, the widening of the stretching vibration absorption band of 3200–3000 cm^{-1} belongs to –OH, which also proves the formation of hydrogen bonds in this system [35].

The rheological characteristics are of great significance to supramolecular materials in the gel state, so we next tested the rheological properties of Ag_9-NCs/DD-5 hydrogel to evaluate the mechanical properties of the hydrogel [36]. In the stress scan, the storage modulus (G′) is much larger than the loss modulus (G″), indicating that the hydrogel exhibits solid-like nature (Figure 3f). The yield stress corresponds to the transition from gel to fluid, that is, the network structure of the gel is destroyed [37], and Ag_9-NCs/DD-5 hydrogel basically remains unchanged before reaching the yield stress of 285 Pa, indicating that the formed hydrogel has high rigidity and resistance to damage ability. In the frequency sweep experiment, the storage modulus (G′) and loss modulus (G″) of the hydrogel remained basically unchanged, and G′ (2300 Pa) was much larger than G″ (115 Pa), indicating that Ag_9-NCs/DD-5 hydrogel has a good mechanical strength (Figure S5a).

TGA was performed to explore the thermal stability of the hydrogel (Figure S5b). The weight loss of the xerogel sample before 100 °C can be attributed to water loss. The weight loss at 200–285 °C is attributed to the vaporization of DD-5 carbon and the removal of oxidized functional groups (the first stage decomposition temperature of pure DD-5 at 174 °C), together with the decomposition of H_2mba (the decomposition temperature of lyophilized Ag_9-NCs ligand H_2mba is 181 °C). Compared with the original DD-5 and Ag_9-NCs, the decomposition temperature of Ag_9-NCs/DD-5 xerogel increased slightly (200 °C), indicating that the thermal stability of the xerogel was further improved.

Based on the above results, it can be concluded that DD-5 and Ag_9-NCs were successfully co-assembled to obtain highly stable hydrogel. The hydrogel consists of highly ordered hollow nanotubes with a little spiral forming a stable three-dimensional network. We can propose the formation mechanism of nanotubes: (i) After the addition of DD-5, the rotational vibration of the mba^{2-} is restricted by H-bond formed which was surmised as being further divided in several parts, such as $-COO^-$ of mba^{2-} with –OH (or –NH–) of DD-5, and/or –OH of protonated mba^{2-} (H_2mba) with $-COO^-$ of DD-5 and so on. (ii) The adjacent components with Ag(I)-rich surface led to the strong argentophilic interactions between Ag_9-NCs which induces the formation of highly oriented nanotubes. The π–π stacking of adjacent ligands also contributes to the formation of spiral nanotubes. Therefore, the assembly process of hollow spiral nanotubes is controlled by inter-ligand non-covalent interactions (e.g., H-bond, π–π stacking) and argentophilic interactions between Ag_9-NCs.

3.4. *Kinetic Tracing of the Formation of Ag_9-NCs/DD-5 Hydrogel*

In order to explore the relationship between the gelation process and fluorescent properties of Ag_9-NCs/DD-5 hydrogel, we followed the kinetic tracking of the self-assembly (Figure 4d). TEM showed the evolution of the morphology during the gelation (Figure 4a). At 1 min, the self-assembled body is a small particle of Ag_9-NCs combined with DD-5, there is no fluorescence emission (Figure 4a,c); The small particles self-assemble into a hollow short rod structure with a length of about 500 nm in 5 min, and weak fluorescence appears at this time; at 30 min, the hollow short rod structure further grows into nanotubes with a length of about 1 μm, and the fluorescence of the nanotubes is obviously enhanced; nanotubes with a little spiral structure are about 2 μm in length at 3 h, and the fluorescence is continually increased compared with that at 30 min, but it is not much different from the final state fluorescence. Therefore, we believe that fluorescence emission is inseparable from the gelation process, that is, the increase in fluorescence intensity is caused by the transition from loose assembly to tightly packed spiral nanotubes. Based on the above results, we believe that in the presence of argentophilic interactions and non-covalent forces (H-bond, π–π stacking), a tightly ordered assembly structure is formed which spatially limits the vibration and rotation of Ag_9-NCs ligands, thus triggers the AIE effect of the Ag_9-NCs and induces fluorescence. That is, as the order of the assembled structure of Ag_9-NCs/DD-5 hydrogel increases, the fluorescence gradually increases.

3.5. *Temperature Sensing*

Optical sensors have the advantages of wireless operation and imaging in harsh environments. Moreover, it is very important to realize temperature monitoring in scientific production, which also promotes the development of temperature sensing. The ultra-small size, good biocompatibility and colloidal stability of metal NCs are considered to be good materials for the development of high-intensity fluorescence thermometers. However, temperature sensors based on metal NCs are mainly concentrated in solution and gel states, while solid-state temperature sensors are rarely explored. In addition, most solid-state phosphorescent temperature sensing materials have poor photoluminescence capabilities at high temperatures, and the fluorescence generally disappears when the temperature is higher than 100 °C [38–40]. Therefore, it is particularly important to develop a solid material with good photoluminescence ability at high temperature. Based on this, we freeze-dried the hydrogel to obtain phosphorescent xerogel, and explored the luminescence of the xerogel with temperature changes in detail.

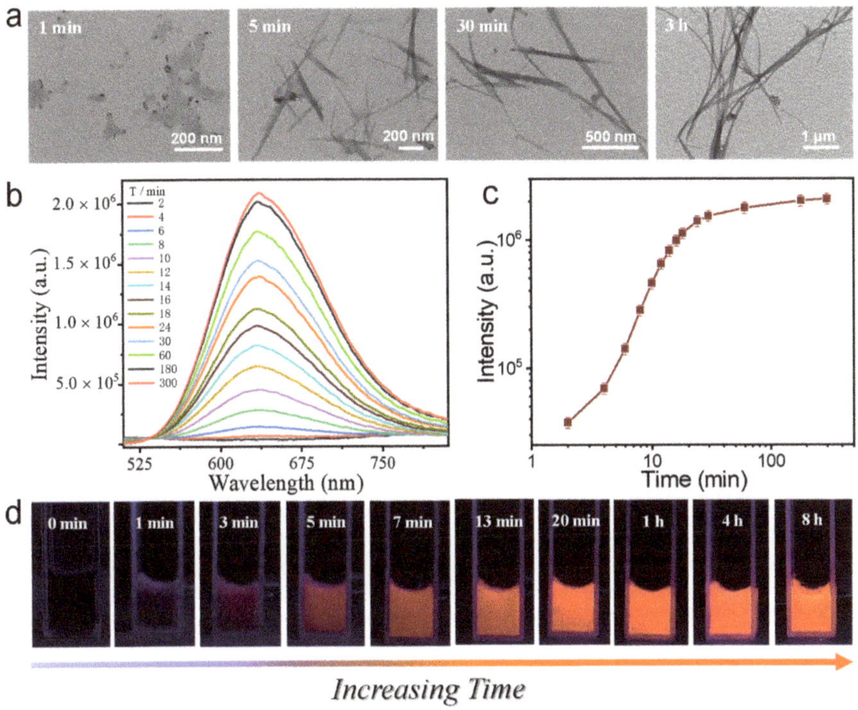

Figure 4. Ag_9-NCs/DD-5 hydrogel kinetic tacking. (**a**) TEM images of hydrogel changing over time; (**b**) Time-varying PL spectrum of hydrogel under excitation at 490 nm; (**c**) PL intensity graph at λ = 630 nm (λ_{ex} = 490 nm); (**d**) Photographs of the formation process of the hydrogel irradiated with 365 nm UV light.

As the temperature rises from −160 °C to 260 °C, the PL intensity of Ag_9-NCs/DD-5 xerogel continues to decrease and the fluorescence peak position also appears slightly blue-shifted (Figure 5a–c). The reasons for this change are as follows: (i) as the temperature increases, the oxygen molecules collide with their luminescent center; (ii) the high temperature weakens the Ag-S bond between the H_2mba and the sliver core, resulting in a decrease in the charge from the ligand to the sliver core of Ag_9-NCs; (iii) as the temperature rises, non-covalent interactions (H-bond, π–π stacking) gradually weaken, resulting in a compact network being stretched, the rotation limit of the mba^{2-} ligand is reduced, non-radiative relaxation is prone to occur; and (iv) as the temperature increases, the conformation of DD-5 changes, causing Ag_9-NCs to be directly exposed to the air and oxidized to larger silver nanoparticles at high temperatures. All these factors lead to the decrease of fluorescence intensity and blue shift of fluorescence peak. Figure 5d shows the change in PL intensity over the entire temperature range, which can be described by two linear parts with an inflection point at 80 °C (Figure 5d). The xerogel satisfies a good linear relationship in the low temperature range from −160 °C to 80 °C and the high temperature range from 80 °C to 260 °C, which will have a wide range of applications in the field of both low and high temperature fluorescence sensing.

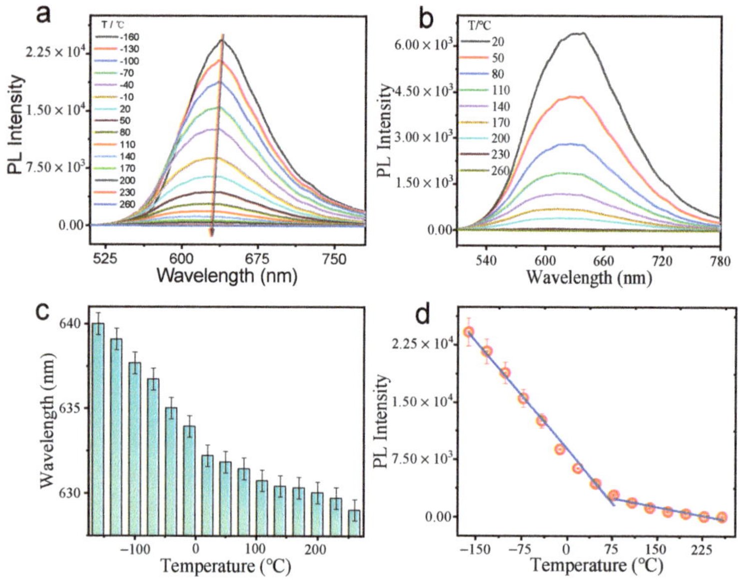

Figure 5. Fluorescence change of xerogel with temperature. (**a**,**b**) Fluorescence spectra with increasing temperature; (**c**) PL peak position changes with increasing temperature; (**d**) The corresponding PL intensity under different temperature.

3.6. The Detection of L-Arg and D-Arg

It is well-known that amino acids play an important role in the human body. Among various amino acids, Arg can stimulate the secretion of hormones such as insulin, growth hormone, glucagon and prolactin. At the same time, Arg content is also one of the key parameters for evaluating the pathophysiology of hyperammonemia and astrocytes cultured with aggregated nerve cells [41–43]. Therefore, it is very important for human health to detect Arg with high selectivity and sensitivity. Fortunately, Ag_9-NCs/DD-5 hydrogel with good biocompatibility can be used as a biosensor to detect L-Arg and D-Arg. It can be found that the addition of L-Arg and D-Arg can completely quench the fluorescence of Ag_9-NCs/DD-5 hydrogel, while when the other amino acids (L-Ala, L-His, L-Cys, L-Phe, L-Gln, L-Ser, L-Thr, L-Asn, L-Val and Gly) were added, the fluorescence intensity changed slightly (Figure 6a–c), we then take L-Arg as an example to investigate in detail. With the increase of the concentration of L-Arg, the fluorescence intensity gradually decreases (Figure 6d), I_0/I have a linear relationship with L-Arg at a lower concentration, and the detection line calculated from 3δ/slope is 240 μM (Figure 6e), indicating that Ag_9-NCs/DD-5 hydrogel can detect L-Arg with high sensitivity. Adding L-Arg and other amino acids together to the hydrogel system, it could be found that even in the presence of other amino acids, L-Arg could still quench its fluorescence, indicating that Ag_9-NCs/DD-5 hydrogel also has high selectivity for L-Arg (Figure 6c).

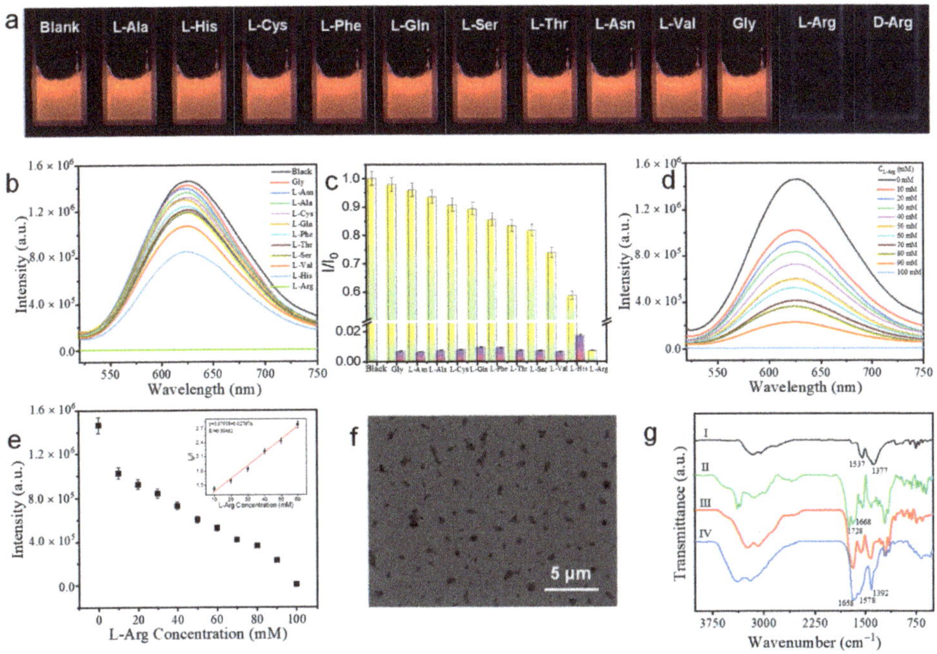

Figure 6. Detection of arginine by hydrogel. (**a**) Photographs of the samples observed under 365 nm UV lamp with the addition of amino acids; (**b**) PL spectra of nanotubes with the addition of amino acids; (**c**) Selectivity of nanotubes toward L-Arg, the lighter color is the fluorescence intensity after adding different kinds of amino acids, the darker color is the fluorescence intensity after adding L-Arg to the above sample; (**d**,**e**) Concentration-dependent fluorescence spectra and fluorescence intensity with the addition of different contents of L-Arg, inset: Stern-Volmer quenching curve of the fluorescence intensity of the hydrogel at 630 nm against L-Arg concentration; (**f**) TEM image of L-Arg added to Ag_9-NCs/DD-5 hydrogel; (**g**) FT-IR spectrum, where I represents lyophilized Ag_9-NCs; II represents DD-5; III represents Ag_9-NCs/DD-5 xerogel; IV represents freeze-dried sample after adding L-Arg to Ag_9-NCs/DD-5 hydrogel.

In addition, it was found from the TEM image that the nanotubes with a little spiral disappeared after the addition of L-Arg, and more particles with superfine nanowires appeared (Figure 6f), indicating that the addition of L-Arg destroyed the ordered structure of the hydrogel. From FT-IR results, as shown in Figure 6g, after the addition of L-Arg, there is a reappearance of peaks at 1658 and 1578 cm^{-1} belonging to the amide I band (its red shift is attributed to the addition of L-Arg) and at 1392 cm^{-1} belonging to the symmetric vibrational absorption peak of the C=O group in mba^{2-}, indicating that the addition of L-Arg destroyed the intermolecular H-bond and reduce the radiation relaxation of the Ag_9-NCs ligand mba^{2-} and made the fluorescence disappear. In the FT-IR, it is found that L-Arg is added to break the H-bond between Ag_9-NCs and DD-5 to achieve the purpose of detection, which has nothing to do with the molecular configuration of Arg. Therefore, we reasonably believe that Ag_9-NCs/DD-5 hydrogel has good detection capabilities for both L-Arg and D-Arg.

4. Conclusions

In this work, an Ag_9-NCs/DD-5 hydrogel with highly ordered hollow nanotubes with a little spiral and phosphorescent emission was successfully obtained by supramolecular self-assembly through inter-ligand non-covalent interactions (H-bond, π–π stacking) and argentophilic interactions [Ag(I)–Ag(I)] between Ag_9-NCs. After self-assembly, the transfer of chirality of DD-5 to supramolecular chirality of Ag_9-NCs/DD-5 hydrogel was successfully realized together with good CPL performance. In addition, Ag_9-NCs/DD-5 xerogel can still emit fluorescence at 200 °C, making it an ideal choice for a new generation of luminous temperature-sensing agents both in low and high temperatures. Moreover, Ag_9-NCs/DD-5 hydrogel has selectivity and sensitivity for the detection of L-Arg and D-Arg. Our work implied that supramolecular self-assembled materials for metal NCs not only have potential applications in the field of fluorescence sensing, but also have huge application prospects in the field of biological detection.

Supplementary Materials: The following supporting information can be downloaded at: https://www.mdpi.com/article/10.3390/nano12030424/s1, Figure S1: Co-assembled primitive photos and co-assembled phase behavior photos, Figure S2: TEM and SEM images at different concentrations of DD-5, Figure S3: PL decay curve of lyophilized Ag_9-NCs solution and Ag_9-NCs/DD-5 hydrogel, Figure S4: High-resolution XPS spectra of the Ag_9-NCs/DD-5 xerogel, Figure S5: Hydrogel rheology and xerogel TGA, Table S1: Lifetime of the powder of lyophilized Ag_9-NCs solution, Table S2. Lifetime of Ag_9-NCs/DD-5 hydrogel.

Author Contributions: W.W.: Conceptualization, methodology, formal analysis, investigation, writing—reviewing and editing; Z.W.: formal analysis, writing—reviewing and editing; D.S.: methodology, writing-reviewing and editing; S.L.: conceptualization, writing-reviewing and editing; Q.D.: resources, funding acquisition; X.X.; writing—reviewing and editing, supervision, funding acquisition. All authors have read and agreed to the published version of the manuscript.

Funding: We gratefully acknowledge the financial support from the National Natural Science Foundation of China (21972077), and Key Technology Research and Development Program of Shandong (2019GGX102019).

Data Availability Statement: Not applicable.

Conflicts of Interest: The authors declare no conflict of interest.

References

1. Kailasa, S.; Borse, S.; Koduru, J.; Muthy, Z. Biomolecules as Promising Ligands in the Synthesis of Metal Nanoclusters: Sensing, Bioimaging and Catalytic Applications. *Trends Environ. Anal. Chem.* **2021**, *32*, e00104. [CrossRef]
2. Han, S.; Zhao, Y.; Zhang, Z.; Xu, G. Recent Advances in Electrochemiluminescence and Chemiluminescence of Metal Nanoclusters. *Molecules* **2020**, *20*, 5208. [CrossRef] [PubMed]
3. Shan, P.; Yang, J.; Zang, Z.; Zhao, Q.; Cheng, Y.; Li, L.; Yang, X.; Yu, X.; Lu, Z.; Zhang, X. Effects of Silver Nanoclusters on the Spectral Properties for Fluorescein Isothiocyanate with Restrained Photobleaching. *Appl. Surf. Sci.* **2021**, *548*, 14928–14938. [CrossRef]
4. Zhang, Y.; Zhang, C.; Xu, C.; Wang, X.; Liu, C.; Waterhouse, G.I.N.; Wang, Y.; Yin, H. Ultrasmall Au Nanoclusters for Biomedical and Biosensing Applications: A Mini-review. *Talanta* **2019**, *200*, 432–442. [CrossRef] [PubMed]
5. Li, J.; Zhu, J.; Xu, K. Fluorescent Metal Nanoclusters: From Synthesis to Applications. *TrAC Trends Anal. Chem.* **2014**, *58*, 90–98. [CrossRef]
6. Xie, Z.; Sun, P.; Wang, Z.; Li, H.; Yu, L.; Sun, D.; Chen, M.; Bi, Y.; Xin, X.; Hao, H. Metal-Organic Gels from Silver Nanoclusters with Aggregation-Induced Emission and Fluorescence-to-Phosphorescence Switching. *Angew. Chem. Int. Ed.* **2020**, *59*, 9922–9927. [CrossRef]
7. Sun, P.; Wang, Z.; Bi, Y.; Sun, D.; Zhao, T.; Zhao, F.; Wang, W.; Xin, X. Self-assembly Driven Aggregation-Induced Emission of Silver Nanoclusters for Light Conversion and Temperature Sensing. *ACS Appl. Nano Mater.* **2020**, *3*, 2038–2046. [CrossRef]
8. Hu, X.; Zhang, X.; Gao, H.; Huang, Y. Cu-Based Metal-Organic Frameworks-Derived Copper Nanoclusters with Tunable Emission for Ratiometric pH Sensing. *Sens. Actuators B Chem.* **2022**, *353*, 131130–131140. [CrossRef]
9. Desai, M.; Basu, H.; Saha, S.; Singhal, R.; Kailasa, S. Fluorescence Enhancement of Bovine Serum Albumin Gold Nanoclusters from La^{3+} Ion: Detection of Four Divalent Metal Ions (Hg^{2+}, Cu^{2+}, Pb^{2+} and Cd^{2+}). *J. Mol. Liq.* **2021**, *336*, 116239–116249. [CrossRef]
10. Borse, S.; Murthy, Z.; Park, T.; Kailasa, S. Lysozyme-Decorated Gold and Molybdenum Bimetallic Nanoclusters for the Selective Detection of Bilirubin as a Jaundice Biomarker. *ACS Appl. Nano Mater.* **2021**, *4*, 11949–11959. [CrossRef]
11. Wen, Q.; Peng, J.; Liu, A.; Hu, Y.; Wang, J.; Ling, J.; Cao, Q. Fluorescent Silver Nanoclusters Stabilized in Guanine-Enhanced DNA Hybridization for Recognizing Different Small Biological Molecules. *J. Lumin.* **2020**, *221*, 117038–117046. [CrossRef]
12. Zhang, S.; Su, H.; Wang, Z.; Wang, L.; Zhao, Q.; Tung, C.; Sun, D.; Zheng, L. Anion-Templated Nanosized Silver Alkynyl Clusters: Cluster Engineering and Solution Behavior. *Chem. Eur. J.* **2017**, *23*, 3432–3437. [CrossRef] [PubMed]
13. Jia, X.; Li, J.; Wang, E. Supramolecular Self-assembly of Morphology-Dependent Luminescent Ag Nanoclusters. *Chem. Commun.* **2014**, *50*, 9565–9568. [CrossRef] [PubMed]

14. Khanra, S.; Abdullah-Al Mamun, M.; Ferreira, F.F.; Ghosh, K.; Guha, S. Functionalized Self-Assembled Peptide Nanotubes with Cobalt Ferrite Nanoparticles for Applications in Organic Electronics. *ACS Appl. Nano Mater.* **2018**, *1*, 1175–1187. [CrossRef]
15. Khan, J.M.; Khan, M.S.; Qadeer, A.; Alsenaidy, M.A.; Ahmed, A.; Al-Shabib, N.A.; Khan, R.H. Cationic Gemini Surfactant (16-4-16) Interact Electrostatically with Anionic Plant Lectin and Facilitates Amyloid Fibril Formation at Neutral pH. *Colloids Surf. A Physicochem. Eng. Asp.* **2017**, *522*, 494–502. [CrossRef]
16. Fatouros, D.G.; Lamprou, D.A.; Urquhart, A.J.; Yannopoulos, S.N.; Viziranakis, I.S.; Zhang, S.; Koutsopoulos, S. Lipid-like Self-Assembling Peptide Nanovesicles for Drug Delivery. *ACS Appl. Mater. Interfaces* **2014**, *6*, 8184–8189. [CrossRef]
17. Xing, R.; Li, S.; Zhang, N.; Shen, G.; Mohwald, H.; Yan, X. Self-Assembled Injectable Peptide Hydrogels Capable of Triggering Antitumor Immune Response. *Biomacromolecules* **2017**, *18*, 3514–3523. [CrossRef]
18. Chen, X.; Zhang, M.; Wang, X.; Chen, Y.; Yan, Y.; Zhang, L.; Zhang, L. Peptide-Modified Chitosan Hydrogels Promote Skin Wound Healing by Enhancing Wound Angiogenesis and Inhibiting Inflammation. *Am. J. Transl. Res.* **2017**, *9*, 2352–2362.
19. Xing, R.; Zou, Q.; Yan, X. Peptide-Based Supramolecular Colloids. *Acta Phys. Chim. Sin.* **2020**, *36*, 1909048–1909064.
20. Shen, J.; Xiao, Q.; Sun, P.; Feng, J.; Xin, X.; Yu, Y.; Qi, W. Self-Assembled Chiral Phosphorescent Microflowers from Au Nanoclusters with Dual-Mode pH Sensing and Information Encryption. *ACS Nano* **2017**, *15*, 4947–4955. [CrossRef]
21. Liu, J.; Yu, Y.; Wang, C.; Shen, J.; Feng, J.; Qi, W. Fabrication of a Chiral Luminescent Hydrogel from Gold Nanoclusters via Molecular Recognition. *Chem. Commun.* **2021**, *57*, 10202–10205. [CrossRef] [PubMed]
22. Chakraborty, P.; Ghosh, M.; Schnaider, L.; Adadi, N.; Ji, W.; Bychenko, D.; Dvir, T.; Adler-Abramovich, L.; Gazit, E. Composite of Peptide-Supramolecular Polymer and Covalent Polymer Comprises a New Multifunctional, Bio-Inspired Soft Material. *Macromol. Rapid Commun.* **2019**, *40*, 1900175. [CrossRef] [PubMed]
23. Wang, C.; Sun, Y.; Wang, J.; Xu, H.; Lu, J. Copper (II)-Mediated Self-Assembly of Hairpin Peptides and Templated Synthesis of CuS Nanowires. *Chem. Asian J.* **2015**, *10*, 1953–1958. [CrossRef] [PubMed]
24. Sun, P.; Wang, Z.; Sun, D.; Bai, H.; Zhu, Z.; Bi, Y.; Zhao, T.; Xin, X. pH-Guided Self-Assembly of Silver Nanoclusters with Aggregation-Induced Emission for Rewritable Fluorescent Platform and White Light Emitting Diode Application. *J. Colloid Interface Sci.* **2020**, *564*, 235–242. [CrossRef] [PubMed]
25. Bi, Y.; Wang, Z.; Liu, T.; Sun, D.; Godbert, N.; Li, H.; Hao, J.; Xin, X. Supramolecular Chirality from Hierarchical Self-assembly of Atomically Precise Silver Nanoclusters Induced by Secondary Metal Coordination. *ACS Nano* **2021**, *15*, 15910–15919. [CrossRef] [PubMed]
26. Heaven, M.W.; Dass, A.; White, P.S.; Holt, K.M.; Murray, R.W. Crystal Structure of the Gold Nanoparticle $[N(C_8H_{17})_4][Au_{25}(SCH_2CH_2Ph)_{18}]$. *J. Am. Chem. Soc.* **2008**, *130*, 3754–3755. [CrossRef] [PubMed]
27. Bolton, O.; Lee, K.; Kim, H.; Lin, K.; Kim, J. Activating Efficient Phosphorescence from Purely Organic Materials by Crystal Design. *Nat. Chem.* **2011**, *3*, 205–210. [CrossRef]
28. Ni, W.; Qiu, Y.; Li, M.; Zheng, J.; Sun, R.W.Y.; Zhan, S.; Ng, S.W.; Li, D. Metallophilicity-Driven Dynamic Aggregation of a Phosphorescent Gold(I)-Silver(I) Cluster Prepared by Solution-Based and Mechanochemical Approaches. *J. Am. Chem. Soc.* **2014**, *136*, 9532–9535. [CrossRef]
29. Pyo, K.; Thanthirige, V.D.; Kwak, K.; Pandurangan, P.; Ramakrishna, P.; Lee, D. Ultrabright Luminescence from Gold Nanoclusters: Rigidifying the Au(I)-Thiolate Shell. *J. Am. Chem. Soc.* **2015**, *137*, 8244–8250. [CrossRef]
30. Zhao, Y.; Wang, J.; Deng, L.; Zhou, P.; Wang, S.; Wang, Y.; Xu, H.; Lu, J.R. Tuning the Self-Assembly of Short Peptides via Sequence Variations. *Langmuir* **2013**, *29*, 13457–13464. [CrossRef]
31. Xu, Y.; Zhang, Y.; Guo, Z.; Ren, J.; Wang, Y.; Peng, H. Flexible, Stretchable, and Rechargeable Fiber-Shaped Zinc-Air Battery Based on Cross-Stacked Carbon Nanotube Sheets. *Angew. Chem. Int. Ed.* **2015**, *54*, 15390–15394. [CrossRef] [PubMed]
32. Barreiro, E.; Casas, J.S.; Couce, M.D.; Laguna, A.; Lopez-de-Luzuriaga, J.M.; Monge, M.; Sanchez, A.; Sordo, J.; Lopez, E.M.V. A Novel Hexanuclear Silver(I) Cluster Containing a Regular Ag-6 Ring with Short Ag-Ag Distances and an Argentophilic Interaction. *Dalton Trans.* **2013**, *42*, 5916–5923. [CrossRef]
33. Wu, Z.; Du, Y.; Liu, J.; Yao, Q.; Chen, T.; Cao, Y.; Zhang, H.; Xie, J. Aurophilic Interactions in the Self-Assembly of Gold Nanoclusters into Nanoribbons with Enhanced Luminescence. *Angew. Chem. Int. Ed.* **2019**, *58*, 8139–8144. [CrossRef] [PubMed]
34. Cursino, A.C.T.; Gardolinski, J.E.F.D.; Wypch, F. Intercalation of Anionic Organic Ultraviolet Ray Absorbers into Layered Zinc Hydroxide Nitrate. *J. Colloid Interface Sci.* **2010**, *347*, 49–55. [CrossRef]
35. Moitessier, N.; Chapleur, Y. Modulation of the Relative Reactivities of Carbohydrate Secondary Hydroxyl Groups Modification of the Hydrogen Bond Network. *Tetrahedron Lett.* **2003**, *44*, 1731–1735. [CrossRef]
36. Shen, J.; Wang, Z.; Sun, D.; Liu, G.; Yuan, S.; Kurmoo, M.; Xin, X. Self-Assembly of Water-Soluble Silver Nanoclusters: Superstructure Formation and Morphological Evolution. *Nanoscale* **2017**, *9*, 19191–19200. [CrossRef]
37. Shen, X.; Wang, Y.; Zhao, E.; Yuan, W.; Liu, Y.; Lu, P.; Qin, A.; Ma, Y.; Sun, J.; Tang, B. Effects of Substitution with Donor-Acceptor Groups on the Properties of Tetraphenylethene Trimer: Aggregation-Induced Emission, Solvatochromism, and Mechanochromism. *J. Phys. Chem. C* **2013**, *117*, 7334–7347. [CrossRef]
38. Li, J.; Zhang, H.; Zhang, Y.; Zhou, W.; Liu, Y. Room-Temperature Phosphorescence and Reversible White Light Switch Based on a Cyclodextrin Polypseudorotaxane Xerogel. *Adv. Opt. Mater.* **2019**, *7*, 1900589. [CrossRef]
39. Zhu, Y.; Guan, Y.; Niu, Y.; Wang, P.; Chen, R.; Wang, Y.; Wang, P.; Xie, H. Ultralong Polymeric Room Temperature Phosphorescence Materials Fabricated by Multiple Hydrogen Bondings Resistant to Temperature and Humidity. *Adv. Opt. Mater.* **2021**, *9*, 2100782. [CrossRef]

40. Wu, H.; Gu, L.; Baryshnikov, G.V.; Wang, H.; Minaev, B.F.; Agren, H.; Zhao, Y. Molecular Phosphorescence in Polymer Matrix with Reversible Sensitivity. *ACS Appl. Mater. Interfaces* **2020**, *12*, 20765–20774. [CrossRef]
41. Gorren, A.C.F.; Schmidt, K.; Mayer, B. Binding of L-arginine and Imidazole Suggests Heterogeneity of Rat Brain Neuronal Nitric Oxide Synthase. *Biochemistry* **2002**, *41*, 7819–7829. [CrossRef] [PubMed]
42. Senkevitch, E.; Cabrera-Luque, J.; Morizono, H.; Caldovic, L.; Tuchman, M. A Novel Biochemically Salvageable Animal Model of Hyperammonemia Devoid of *N*-acetylglutamate Synthase. *Mol. Genet. Genom.* **2012**, *106*, 160–168. [CrossRef] [PubMed]
43. Kapila, S.; Saba, M.; Lin, C.; Bawle, E.V. Arginine Deficiency-Induced Hyperammonemia in a Home Total Parenteral Nutrition-Dependent Patient: A Case Report. *JPEN J. Parenter. Enter. Nutr.* **2001**, *25*, 286–288. [CrossRef] [PubMed]

nanomaterials

MDPI

Article

Advancement of Fluorescent and Structural Properties of Bovine Serum Albumin-Gold Bioconjugates in Normal and Heavy Water with pH Conditioning and Ageing

Bence Fehér [1,2], Judith Mihály [1], Attila Demeter [1,*], László Almásy [2], András Wacha [1], Zoltán Varga [1], Imre Varga [3], Jan Skov Pedersen [4] and Attila Bóta [1,*]

1 Research Centre for Natural Sciences, Institute of Materials and Environmental Chemistry, Magyar Tudósok Körútja 2, 1117 Budapest, Hungary; feher.bence@ek-cer.hu (B.F.); mihaly.judith@ttk.hu (J.M.); wacha.andras@ttk.hu (A.W.); varga.zoltan@ttk.hu (Z.V.)

2 Neutron Spectroscopy Department, Centre for Energy Research, Konkoly-Thege M. út 29-33, 1121 Budapest, Hungary; almasy.laszlo@ek-cer.hu

3 Institute of Chemistry, Eötvös Loránd University (ELTE), Pázmány Péter sétány 1/A, 1117 Budapest, Hungary; varga.imre@ttk.elte.hu

4 Department of Chemistry and Interdisciplinary Nanoscience Center (iNANO), Aarhus University, Gustav Wieds Vej 14, 8000 Aarhus C, Denmark; jsp@chem.au.dk

* Correspondence: demeter.attila@ttk.hu (A.D.); bota.attila@ttk.hu (A.B.)

Abstract: The red-emitting fluorescent properties of bovine serum albumin (BSA)–gold conjugates are commonly attributed to gold nanoclusters formed by metallic and ionized gold atoms, stabilized by the protein. Others argue that red fluorescence originates from gold cation–protein complexes instead, not gold nanoclusters. Our fluorescence and infrared spectroscopy, neutron, and X-ray small-angle scattering measurements show that the fluorescence and structural behavior of BSA–Au conjugates are different in normal and heavy water, strengthening the argument for the existence of loose ionic gold–protein complexes. The quantum yield for red-emitting luminescence is higher in heavy water (3.5%) than normal water (2.4%), emphasizing the impact of hydration effects. Changes in red luminescence are associated with the perturbations of BSA conformations and alterations to interatomic gold–sulfur and gold–oxygen interactions. The relative alignment of domains I and II, II and III, III and IV of BSA, determined from small-angle scattering measurements, indicate a loose ("expanded-like") structure at pH 12 (pD ~12); by contrast, at pH 7 (pD ~7), a more regular formation appears with an increased distance between the I and II domains, suggesting the localization of gold atoms in these regions.

Keywords: red-fluorescence; protein–gold conjugates; change in protein conformation; fluorescence; small angle X-ray scattering; small angle neutron scattering

1. Introduction

Biocompatible red fluorescent BSA–gold conjugates offer favorable biological applications because their red fluorescence significantly differs from tissue materials' blue or green auto-fluorescence [1]. An elegant and simple one-pot aqueous synthesis of fluorescent bovine serum albumin–gold (BSA–Au) complexes was first described by Xie et al., a pioneering work in the field [2]. In the presence of hydrogen tetrachloroaurate ($HAuCl_4$), the interaction between the BSA and gold salt is driven in the alkaline medium (pH ~12), resulting in a significant red luminescence at around 670 nm, which slightly increases when the system neutralizes. Despite intensive studies, there is no consensus about the origin of red emittance. The question is still open—what are the red-emitting parts of the BSA–Au conjugates, and what do they look like? After the inventors, it has been widely accepted that the fluorescent properties result from gold nanoclusters composed of 25 atoms and stabilized by BSA, which were already known to show red fluorescence [2].

It was reported that nanoclusters with 25 atoms have a core-shell structure consisting of an icosahedral core of 12–13 metallic gold atoms surrounded by six $Au_2(SR)_3$ staples covalently bonded to a BSA molecule via the sulfuric groups [2–7]. This structural explanation was deduced from matrix-assisted laser desorption-ionization time-of-flight (MALDI-TOF) and X-ray photoelectron spectroscopy (XPS) methods where, unfortunately, water, as a characteristic structural element, is expelled from the system. Other methods, such as Nuclear Magnetic Resonance (NMR), infrared spectroscopy (IR), and small-angle X-ray scattering (SAXS), show that BSA is a dynamic, "living" system assuming different conformations reversibly in an extensive pH range. The aqueous solution of BSA has five different conformations as the function of pH: expanded, fast, normal, basic, and aged forms [8]. In the presence of $HAuCl_4$, pH plays an important role. At pH = 12, the BSA-molecules become negatively charged and undergo conformational changes while attracting Au(III) ions, predominantly in regions rich in reducing tyrosine and histidine residues. The key to understanding the red emission's origin is clarifying the interactions between the neighboring Au atoms or ions [9]. Dixon and Egussa observed that these gold cation–protein complexes caused their red emissions after a further reduction process using sodium borohydride, whereby metal nanoparticles were formed. Therefore, they supposed that the BSA–gold compounds, described by Xie et al. as Au_{25} nanoclusters, were BSA–cationic gold complexes [10]. Moreover, they observed that conformational changes of BSA influence the fluorescence behavior of complexes. They also suggested that the origin of the red fluorescence involves an energy transfer among chromophores formed by the gold and protein residues. The same authors concluded that cysteine, 34 of which form disulfide bonds in BSA, is the binding site of Au(III) but not the location of the red-emitting fluorophore. Recently, the same research group identified the Au(III) binding domain of BSA and localized the origin of red fluorescence within the N-terminal domain using limited proteolysis and molecular cloning techniques based on luminescence measurements [11]. However, the changes and evolution in fluorescent properties are primarily connected to global changes in BSA–gold complexes dependent on the pH value and accompanied by significant structural changes on the atomic level as observed by small-angle X-ray scattering and infrared spectroscopy [12]. The conformation of BSA is not reversible after a neutral–alkali–neutral cycle, and its consequences in fluorescence can be observed. The irreversibility is more pronounced in the presence of Au(III) ions, indicating the importance of hydration effects. Besides the advantages of heavy water use in the neutron scattering techniques, it provides a solvent milieu different from normal water, enabling the observation of hydration-induced structural and conformational changes. It is well-known that the deuterium–hydrogen isotope effect causes significant changes in the folding–unfolding processes of proteins [13,14]. In this work, we show that heavy water, compared to normal water, induces more decided effects in both global and fine structures and that these changes bring a significantly increased red fluorescence than that observed in normal water.

2. Materials and Methods

2.1. Materials

Bovine serum albumin (BSA, >99%) and $HAuCl_4 \cdot 3H_2O$ (99.99%) were purchased from Sigma-Aldrich (Sigma-Aldrich, St. Louis, MO, USA) and used as received. The pH of the solutions was adjusted with HCl (Sigma-Aldrich, St. Louis, MO, USA) and NaOH (Sigma-Aldrich, St. Louis, MO, USA). All solutions were prepared in ultra-pure Milli-Q water (total organic content $\leq$ 4 ppb; resistivity $\geq$ 18 MΩcm) and heavy water (deuterium oxid for NMR, 99.8% D) purchased from Acros Organics (Morris Plains, NJ, USA).

2.2. Sample Preparation

A total of 1.67 w% BSA–Au(III) stock solutions in normal and heavy water were prepared by mixing 2.5 w% BSA solutions with 10 mM $HAuCl_4$ solutions in 2:1 volume ratio (corresponding with 13:1 ion–protein molar ratio) at moderate stirring rate (600 rpm).

After mixing, the pH of the stock solutions was set to 12 (pD to ≈11.6) by dilute NaOH solutions (1 M, both in normal and heavy water) under stirring. After storage at room temperature for two days, the BSA−Au(III) systems were heat-treated at 37 °C for 2 h, then neutralized and stored at room temperature. The beginning and transitional states at pH = 12 (pD ≈ 11.6) and neutral pH (pD) were measured with fluorescence, infrared spectroscopy, and DLS methods. The same preparation protocol was used for the X-ray and neutron scattering methods.

2.3. Fluorescence Spectroscopy

The luminescence of the BSA–Au conjugates was measured using a Jasco FP8500 spectrofluorometer (Jasco International Co., Ltd., Tokyo, Japan) at 25 °C in MQ water and 360 ± 5 nm excitation in the 380–750 nm range. The spectral correction function was assessed by the Maroncelly dye setup [15]. The luminescence quantum yields were determined relative to the 0.546 value of quinine sulfate in 1N sulfuric acid [16]. It must be noted that considering the low sensitivity of the fluorometer in the red region, the red band's maxima, and the corresponding fluorescence yields, may be slightly underestimated. The air-saturated samples were measured in a 3 mm × 3 mm × 40 mm quartz cuvette with an optical density at the excitation wavelength around 1.4.

2.4. Infrared Spectroscopy

Attenuated total reflection-Fourier transform infrared (ATR-FTIR) spectroscopy measurements were conducted using a Varian 2000 FTIR Scimitar spectrometer (Varian Inc., Palo Alto, CA, USA) fitted with a liquid nitrogen-cooled mercury cadmium telluride (MCT) detector and a 'Golden Gate' single reflection diamond ATR accessory (Specac Ltd., Orpington, UK). A sample amount of 5 μL was dropped onto the diamond ATR surface, and the dry film spectrum was collected (at 2 cm^{-1} resolution and 64 scans) after the slow evaporation of the solvent under ambient conditions. Each data acquisition was followed by ATR correction. Spectral deconvolution was performed using the GRAMS/AI (7.02) spectroscopy software (Thermo Galactic, Walthman, MA, USA). Band positions for curve fitting were established using the second derivative and were fixed during the fitting procedure. Band shapes were approximated by Lorentzian functions. The intensities and the bandwidth of each component were allowed to vary until the minimal χ^2 parameter was reached. After the fitting procedure, the relative contribution of a particular band component was calculated from the integrated areas of the individual components [17].

2.5. Dynamic Light Scattering

Dynamic light scattering (DLS) of the samples was measured at 20 °C using a W130i dynamic light scattering device (Avid Nano Ltd., High Wycombe, UK) with a diode laser (660 nm) and a photodiode detector. Eppendorf disposable cuvettes (50–2000 μL, UVette routine pack, Vienna, Austria) with a 1 cm path-length were used [18]. Samples containing approx. 10 μM BSA were measured in a final volume of 80 μL in MQ water. We measured the time-dependent autocorrelation function for 10 s, repeated it ten times, and reported the average distributions. A data analysis yielding the mean hydrodynamic diameter (D_h) and polydispersity (%) was performed with iSize 3.0 software supplied with the device.

2.6. Small-Angle X-ray Scattering

Small-angle X-ray scattering measurements were performed using CREDO, an in-house transmission geometry setup [19,20]. Thin-walled quartz capillaries with a 1.5 mm average outer diameter were filled with samples. After proper sealing, these were placed into a temperature-controlled aluminum block inserted into the vacuum space of the sample chamber. Measurements were recorded using monochromatized and collimated Cu Kα radiation (1.542 Å wavelength); the scattering pattern was recorded in the range of 0.02–0.5 $Å^{-1}$ in terms of the scattering variable/momentum transfer ($q = (4\pi/\lambda)\sin\theta$, where 2θ is the scattering angle, and λ is the X-ray wavelength). The total measurement time was 7.5 h for each sample. In order to assess sample and instrument stability during the experiment, the exposures were recorded in 5-min units, with frequent sample change and reference measurements. These individual exposures were corrected for beam flux, geometric effects, sample self-absorption, and instrumental background, as well as calibrated into physical units of momentum transfer and volume-normalized differential scattering cross-sections (absolute intensity, $cm^{-1} \times sr^{-1}$). The corrected and calibrated 5-min scattering patterns were azimuthally averaged to yield a single one-dimensional scattering curve for each sample.

2.7. Small-Angle Neutron Scattering

Small-angle neutron scattering measurements were recorded with the "*Yellow Submarine*" diffractometer operating at the Budapest Neutron Centre [21,22]. Two sample-detector distances of 1.2 and 5.3 m and a quasi-monochromatic neutron wavelength of 0.42 nm allowed us to cover a q-range of 0.01–0.5 $Å^{-1}$. Liquid samples were filled in quartz cells with a 2 mm path length, and the measurements were recorded at 25 °C. The raw data were corrected for sample transmission, cell and room background scattering, and the absolute intensity scale was calibrated by the level of incoherent scattering from an H_2O sample.

3. Results and Discussion

3.1. Red Emission Characterized with Fluorescence Spectroscopy

The luminescence spectra of BSA–Au conjugates were studied in normal and heavy water solutions. Two hours after mixing the two basic (BSA and $HAuCl_4$) solutions (at pH = 12 and pD $\approx$ 11.6), luminescence appears in both normal and heavy water systems. The emission, however, is very low in terms of quantum yield, and the moderate deviation of the emission spectra indicates different characteristics for the two systems, as shown in Figure 1A. Indeed, a simple visual observation of the systems, with a blue laser pointer, already indicates the rapid evolution of red emission in H_2O, whereas the amount of time required for D_2O is more prolonged, at least one day. Blue luminescence (centered at 440 nm) appears in both systems but is relatively larger in normal water. Two days after setting the pH (pD) to neutral (performed two days after the preparation at alkaline conditions), the systems show drastic changes in their luminescence spectra (Figure 1B). The red emission becomes intensive and turns into the prevalent range. The previously observed blue fluorescence is also present but is not the characteristic feature anymore. Interestingly, the red emittance of the D_2O system is significantly higher than H_2O. These differences reflect in the fluorescence quantum yield values, which are 0.035 in the presence of D_2O and 0.024 in the H_2O system.

All spectra in Figure 1 can be described by three distinct bands, with the maxima slightly blue-shifted when the deuterated solvent was used. One may conclude that at pH (pD) $\approx$ 7, after longer conditioning the samples at pH (pD) $\approx$ 12 (shown in Figure 1B), the fluorescence yields increase by a factor of ten, while the red emission bands become much more intensive. The maximum of the lowest energy band is shifted to red with conditioning, especially for D_2O (requiring longer relaxation time). In the case of the conditioned samples, the deuteration of the solvent results in a 50% increase in fluorescence yield, partly resulting from the broadening of the red band. The concomitant blue and red luminescence intensities were followed through several days. The ageing time-course of the red luminescence showed drastically different characters in the two aqueous systems. The red luminescence showed high intensity in normal water after 2 d; then, the values reduced significantly. However, in D_2O, the increase in red luminescence took several days, indicating a longer formation time for the more effective red emission with the concomitant configuration of BSA–Au conjugates.

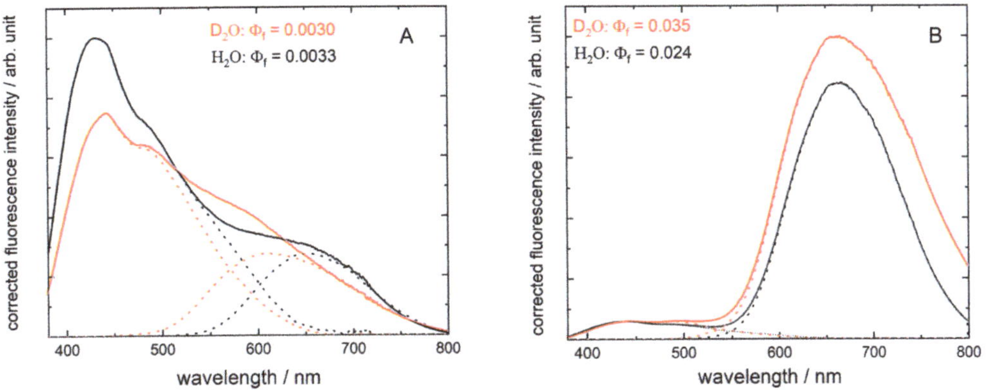

Figure 1. Corrected fluorescence spectra of BSA–Au conjugates in normal (black lines) and heavy water (red lines). In (**A**), the arbitrary intensities are multiplied by a factor of ten, compared with (**B**), the luminescence was detected just 2 h after reaching pH (pD) = 12, whereas in (**B**), after conditioning the samples at pH (pD) = 12 for 2 d, and subsequently neutralizing to pH = 7 (pD ≈ 7) (2 d). (The dotted lines indicate an approximate resolution of the reddest band and the other two bands together.).

3.2. Fine Structural Perturbations Observed by Infrared Spectroscopy

Possible conformational changes of BSA, following preparation steps and luminescence development, were inspected by ATR-FTIR spectroscopy. We focused on the amide I and amide II band regions (from 1750 to 1350 cm^{-1}), belonging to the C=O, and N–H and N–C vibrations of peptide bonds from protein backbones, respectively, as shown in Figure 2A.

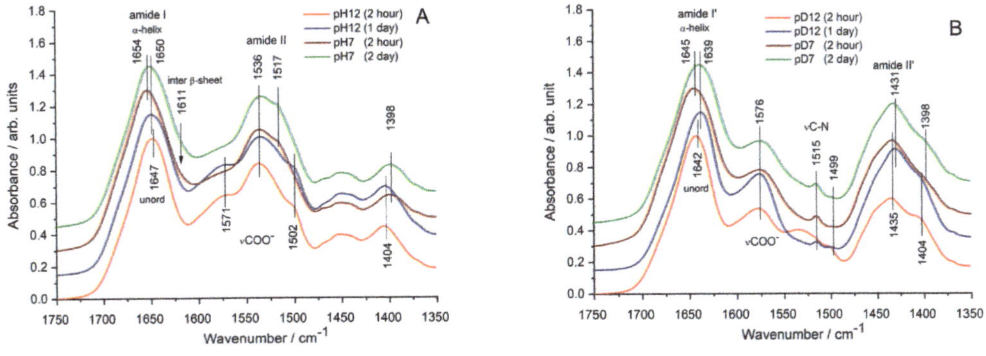

Figure 2. Amide I and amide II regions of IR spectra obtained from BSA–AU bioconjugate samples in H_2O (**A**) and in D_2O (**B**). Spectra are normalized to the highest amide I peak and shifted vertically for improved visualization.

Amide I, composed mainly from the C=O vibration of the peptide bonds, is sensitive to the H-bonding network of the protein backbone. Consequently, by band deconvolution [23,24], the secondary structure of proteins can be deduced (Table 1). At pH = 12, the BSA structure in the BSA–Au bioconjugate is dominated by random coils and β-sheets, resulting in a broad amide I band with a peak at 1647 cm^{-1}. These results align with our previous observation at pH = 12, when the BSA backbone became extended with loose unordered parts [12]. The multiple-band structure of amide II also confirms the open, elongated protein geometry with exposed COO$^-$ groups of amino acid side chains (affirmed by the shoulders at 1571, 1502 cm^{-1}, and the band around 1404 cm^{-1}). Compared to the spectrum of the native BSA [12], however, the relative increase in intensity suggests that Au ions may interact with the exposed, negatively charged carboxy groups of the elongated protein.

Table 1. Secondary structure of BSA–Au conjugate system in H_2O estimated by the deconvolution of amide I IR band (expressed in %, the peak position of the component bands is provided).

	β-Turns 1687 cm^{-1}	α-Helix 1656 m^{-1}	Loose α-Helix 1651 cm^{-1}	Random Coil 1648 cm^{-1}	β-Sheet 1638 cm^{-1}	Inter. β-Sheet 1611 cm^{-1}
pH12	17	-		74	9	-
pH12 (2 d)	35		23	-	42	-
pH7 (2 h)	32	33		-	33	2
pH7 (2 d)	27	-	41	-	29	3

After 1 d, a slight "reorganization" of protein structure can be observed, reflected by the amide I peak shift toward a higher wavenumber (from 1647 to 1650 cm^{-1}). However, more significant spectral changes are witnessed upon adjusting the pH to neutral (pH = 7). The amide I band peak is shifted toward 1654 cm^{-1}, presuming a dominantly helical structure. The shoulder bands in the amide II region, assigned to COO^- groups of the amino acid side-chains also decrease. We assumed that protein refolding in BSA–Au conjugates, forced by H-bonds formation upon pH adjustment, might lead to the development of Au–Au interactions. After 2 d, a "relaxation" occurs in the amide I peak, but no changes occur in the amide II and side-chain band features. A detailed analysis after band deconvolution also revealed the presence of intermolecular β-sheets suggesting that Au association occurs in the final state at pH = 7 (Table 1).

Using D_2O as a solvent, due to the hydrogen/deuterium exchange, the IR spectra of BSA–Au systems are different (Figure 2B). The amide I band of unordered protein conformations appears at 1642 cm^{-1}. The amide II band (N–H deformation vibration of peptide bonds) at 1538 cm^{-1} is suppressed, and a new amide II′ band (N–D deformation vibration of peptide bonds) at 1431 cm^{-1} is raised. It is worth noting that the bands of exposed charged carboxylate groups (at 1576 and 1499 cm^{-1}) are also observable. After 1 day, another slight 'relaxation' can be noticed; however, now the amide I′ peak is shifted in the opposite direction, toward a lower wavenumber. It appears that in D_2O, disordered or sheet-like protein structures are formed in the BSA–Au conjugate. By adjusting the pH (pD) to 7, the alteration in spectral features resembles the BSA–Au/H_2O system with a stronger contribution of helical conformations. After 2 d, however, the sheet-like or disordered structure is favored (Table 2).

Table 2. Secondary structure of BSA–Au conjugate system in D_2O estimated by deconvolution of amide I IR band (expressed in %, with the peak position of the component bands).

	β-Turns 1687 cm^{-1}	α-Helix 1656 cm^{-1}	Loose α-Helix 1651 cm^{-1}	Random Coil 1648 cm^{-1}	β-Sheet 1638 cm^{-1}	Inter. β-Sheet 1611 cm^{-1}
pD12	21	-	-	70	-	9
pD12 (2 d)	35	-	-	-	62	3
pD7 (2 h)	29	1		-	70	-
pD7 (2 d)	34		15	-	51	-

In conclusion, it seems plausible that in alkaline environments, the initial BSA–Au interaction is also affected by the choice of solvent and influenced by the BSA's geometry. In a neutral state (pH = 7 and at pD ~7), a helical structure tends to form to different extents, resulting in slightly different protein secondary structures. This finding aligns with the significant fluorescence changes of BSA–Au conjugates.

3.3. Conformational Changes of BSA–Au Conjugates Observed by Small-Angle X-ray and Neutron Scattering

To receive insight into the global structure of the conjugate prepared in D_2O, with special emphasis on the BSA, we performed small-angle X-ray and neutron scattering experiments. The X-ray scattering length density (SLD) is proportional to the atomic number, whereas the neutron SLD depends on the neutron scattering cross-section of nuclei. Thus, SANS and SAXS provide slightly different information on the two systems; however, the scattering is likely dominated by protein scattering in both cases. Due to the lesser amount of gold ions compared with proteins (13:1 Au ion to BSA molar ratio), the scattering contribution of Au was expected to be negligible, even if concentrated in small nanoclusters.

In Figure 3, the one-dimensional SAXS and SANS curves are presented. The radii of gyration were determined by both the Guinier approximation and indirect Fourier transformation (IFT) and are presented in Table 3 [25,26]. The forward scattering was also determined by IFT. The radius of gyration for BSA in D_2O is in good agreement with the value of BSA in normal water [27]. Adding gold salt did not change this result significantly (pD ~12, SANS result). However, adjusting the pD to 7 results in an increased radius of gyration around 39 Å, which coincides with the literature value for the BSA dimer [27]. SAXS yields slightly larger values than SANS for the same pD states. The reason for this lies in differences between the X-ray and neutron techniques and the uncertainty of R_G determination. The forward scattering of samples at pD ~7 is approximately double that of pD ~12 for both SANS and SAXS, as shown in Figure 3 and Table 3. Since forward scattering is proportional to the molar mass of the scattering objects, we concluded that dimerization occurs during the neutralization because the concentration of the samples is the same. The higher SAXS intensity (compared to SANS) results from the different scattering mechanisms. We noted that dimerization did not occur in normal water, which is a significant difference in the formation of BSA–Au conjugates in the two water systems [12]. Figure 4 presents the Kratky plots of the SAXS curves (taken on D_2O systems). This representation emphasizes the high-q region of the scattering curves, which is only reliable for SAXS because the incoherent scattering of SANS experiments renders the background subtraction slightly imprecise. The sample curve at pD ~12 increases with high q, which is characteristic to (at least partially) unfolded proteins. However, at pD ~7, the Kratky plot exhibits a maximum at low-q and plateau at high-q, indicating that the protein has folded.

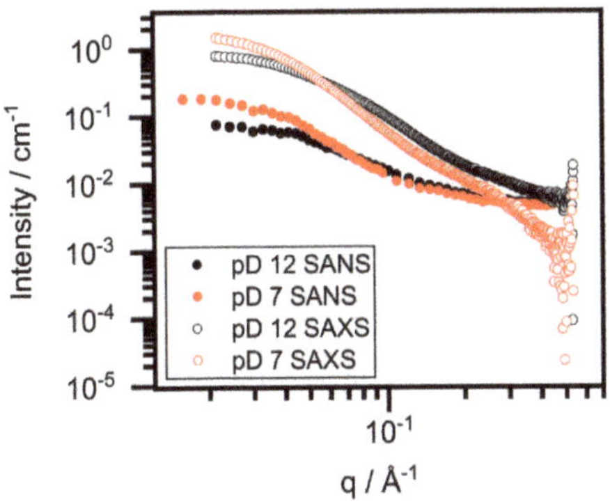

Figure 3. Small-angle X-ray and small-angle neutron scattering curves of BSA–Au conjugates at pD ~12 and pD ~7.

Table 3. Parameters obtained by the data evaluation of SANS and SAXS curves.

	BSA pD7 SANS		BSA–Au pD12 SANS		BSA–Au pD7 SANS		BSA–Au pD12 SAXS		BSA–Au pD7 SAXS	
I(0) (cm^{-1})	$0.037 \pm 1 \times 10^{-3}$		$0.076 \pm 1 \times 10^{-3}$		$0.200 \pm 4 \times 10^{-3}$		$0.948 \pm 1 \times 10^{-3}$		1.970 ± 0.026	
RG (Å) Guinier fit	28.4 ± 0.9		27.1 ± 0.1		38.3 ± 0.4		31.60 ± 0.06		43.0 ± 0.390	
RG (Å) IFT	29.75 ± 1.12		28.47 ± 0.40		39.37 ± 0.46		33.71 ± 0.04		44.98 ± 0.30	
Fitting method	DAMMIF	SASREF	DAMMIF	SASREF	DAMMIF	SASREF	DAMMIF	SASREF	DAMMIF	SASREF
$\chi 2$	1.11	1.09	1.09	1.95	1.63	1.04	1.95	6.56	0.73	1.46

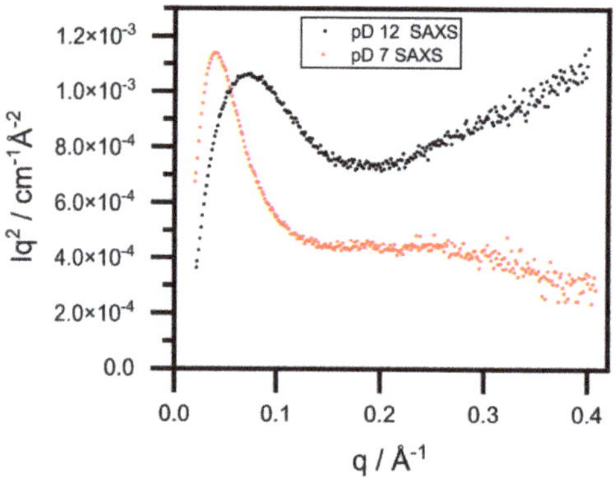

Figure 4. Kratky representation of small-angle X-ray scattering curves of BSA–Au conjugates at pD ~12 and pD ~7.

To reveal protein structural changes in the complex system caused by pH variation, we performed ab initio modeling with DAMMIF and rigid-body refinement with SASREF on curves obtained by SAXS and SANS [28,29]. The native BSA structure was grouped into four linked domains (marked by different colors in Figures 5 and 6: domain I: residue 1–147 (blue), domain II: residue 148–300 (magenta), domain III: residue 301–495 (green), and domain IV: residue 496–583 (orange). Ten runs were performed with DAMMIF and SASREF, and the obtained structures were processed by DAMAVER [30]. For samples with pH (or pD) adjusted to 7 and the presumed dimer structures, a P2 symmetry was imposed. Additionally, the dimer structure was also fitted without requiring symmetry, but the fit did not improve significantly. Therefore, we chose to keep the requirement on the P2 symmetry since it decreases the degrees of freedom and yields more robust results.

The scattering curves of SASREF fits are shown along with measured ones in the graphs on the left-hand side of Figure 6, while the corresponding models obtained by both DAMMIF and SASREF on the right-hand side are aligned and overlaid. The resulting structures from the two different simulation methods are in good agreement. As a reference, the BSA in D_2O without $HAuCl_4$ was also fitted. The structure of BSA in D_2O resembles normal water, as observed in our previous work [12]. From the SASREF fits of SANS curves, we concluded that the four-domain structure satisfactorily describes the scattering data (χ^2 values are presented in Table 3), suggesting that the conformation of the protein is not seriously affected by the gold salt. However, a relative change in the domains' arrangement can be observed by adding $HAuCl_4$ and increasing the pD to ~12. Adjusting the pD to ~7 allows for further change in the domains' arrangement besides dimerization. To receive insight into the motion of domains, we defined the 'main axis' of each domain by the first and last amino acid of the domain. Then, the angles of two adjacent domains were calculated according to their main axis for each sample in all 10 SASREF simulation runs, averaged, and normalized to the appropriate angles of BSA in D_2O without $HauCl_4$. The changes in the domains' arrangement were similar in normal and heavy water. The results presented in Figure 7 show that the relative position of domain I and II at pD 12 changes significantly compared with pure BSA at pD 7. On the other hand, adjusting the BSA–Au system's pD value to ~7, and increasing the angles of two adjacent domains, similar to an opening, is further enhanced. No clear trend was observed in the relative positions of domains II, III, and IV. These results suggest that the domain rich in Au(III) (or Au) is localized between domains I and II. This finding is in good agreement with the recently published work of Dixon et al., who claimed that the fluorophore is located on the fragment of protein from residue 115–312 (almost perfectly covered by domain II in our case) based on proteolysis of the BSA–gold conjugate.

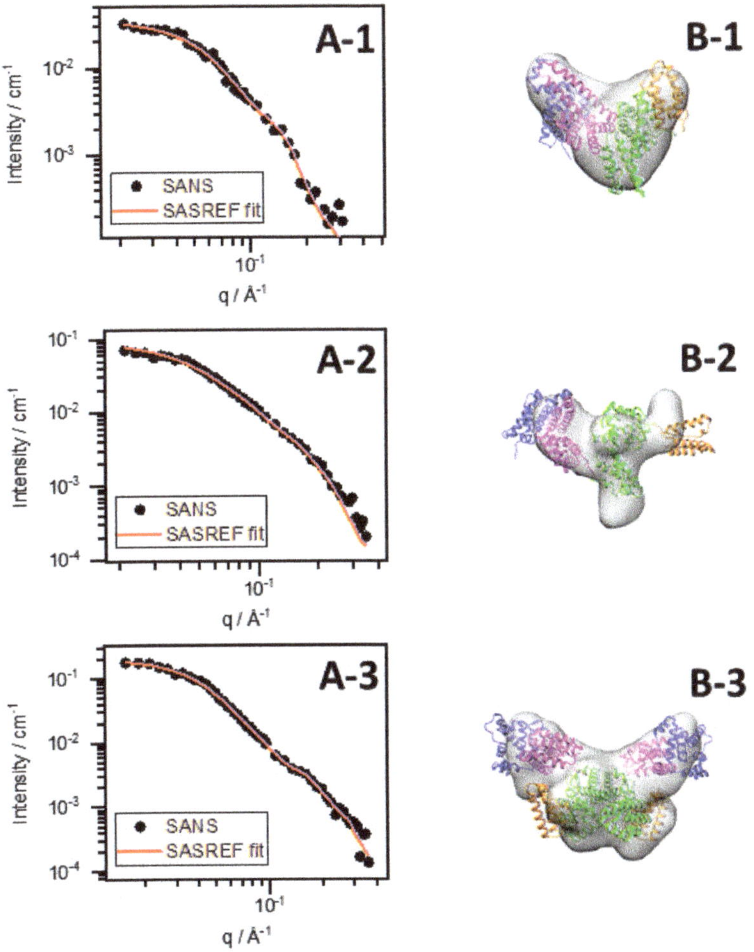

Figure 5. SANS curve of native BSA at pD ~7 and SASREF fit (**A-1**), SANS curve of BSA–Au conjugates at pD ~12 and SASREF fit (**A-2**), SANS curve of BSA–Au at pD ~7 and SASREF fit (**A-3**), obtained SASREF (cartoon) and DAMMIF (surface) structures (**B-1**–**B-3**).

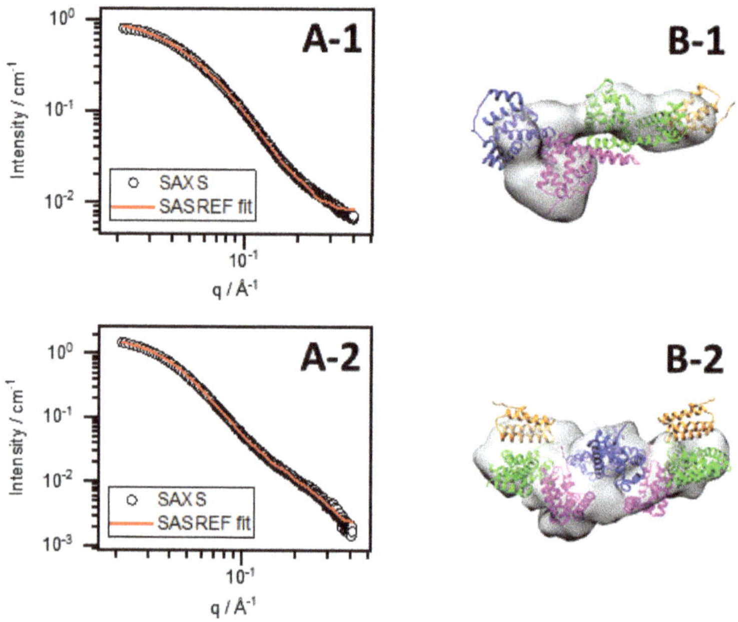

Figure 6. SAXS curve of BSA–Au conjugates at pD ~12 and SASREF fit (**A-1**), SAXS curve of BSA–Au conjugates at pD ~7 and SASREF fit (**A-2**), obtained SASREF (cartoon) and DAMMIF (surface) structures (**B-1**,**B-2**).

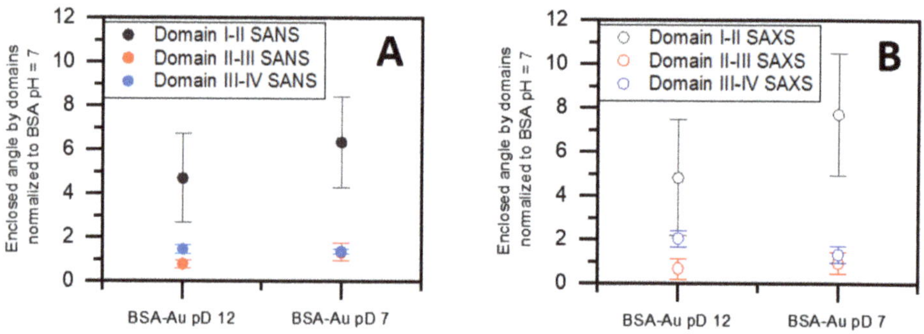

Figure 7. Enclosed angle by domain I–II, II–III, and III–IV obtained by SASREF from SANS (**A**) and SAXS (**B**) curves.

3.4. Dynamic Light Scattering

Apart from some methodological uncertainties and unknown diluting effects inevitable by measuring, dynamic light scattering also shows differences in the solvation effect between normal and heavy water. Although the BSA parent system is not the main subject of our present work, it is reasonable to recall its apparent size difference in the two solvents. In normal water, native BSA shows a small mean hydrodynamic diameter (approx. 37 Å) at pH = 7, extending to a significantly larger value (approx. 99 Å) at pH = 12 due to unfolding. These diameters show similar values in heavy water (38 and 81 Å, respectively). In the presence of Au(III) ions in water, the characteristic value decreases from 131 Å to 115 Å if the pH value changes from 12 to 7. Hydrodynamic diameters show drastically different tendencies in heavy water as the value increases from 126 Å to 233 Å upon neutralization (Table 4). We observed that the latter value was practically halved

(119 Å with 70.7% polydispersity) after a mild ultrasonic treatment, indicating the existence of BSA–Au conjugate dimers which can fall apart in the D_2O matrix.

Table 4. Hydrodynamic radius of the BSA–Au and BSA systems in H_2O and D_2O at given pH (pD) values. The polydispersity index is also given in %.

	BSA pH7	BSA pH12	BSA–Au pH12	BSA–Au pH7	BSA pD7	BSA pD12	BSA–Au pD12	BSA–Au pD7
Diameter, D_h (Å)	37	99	131	115	38	81	126	233/119 *
Polydisp. (%)	57.7	16.7	23.8	27.5	41.0	19.9	21.5	15.9/70.7 *

* after ultrasonic treatment.

Besides the good agreement with model calculations (presented in Figures 5 and 6), comparing radii of gyration to the mean hydrodynamic diameters also provides the same conclusion. One must consider that SAXS and SANS measure the "compact cores" of conjugates uninfluenced by the hydration layer. Radii of gyration increased by about 40% after neutralization in D_2O (Table 4). A decrease was observed in a previous study from 44.7 Å to 37.5 Å in H_2O, which aligns with our present DLS data and strengthens the fact that dimerization does not occur in the normal water system. The fluorescence measurements met this finding since the red emission spectrum in the D_2O system is wider than that in H_2O, emphasizing structural differences in the two systems. Supposedly, the extended dimer conjugates contribute to energy transfer mechanisms and, finally, increase the quantum yield in heavy water.

4. Conclusions

Our experimental studies revealed that red fluorescence's evolution is not exclusively related to protein–gold interactions and is strongly influenced by hydration effects. The perturbation differences in hydration processes induced by the hydrogen–deuterium isotope effect were revealed by fluorescence and infrared spectroscopies. The fine structural changes observed were manifested in the luminescence behavior. Since the conformation of the protein is highly affected by the (actual) H(D)-bond network, the medium appears to play an important role in the conformation-related optical properties. Indeed, the fluorescence quantum yield of the BSA–Au conjugates in heavy water increased significantly (50%) compared with normal water-containing matrix. Collaterally, the alterations of fine atomic structures observed with IR spectroscopy are in rapport with the conformational changes appearing at significantly larger dimensions. The use of a single rigid-body model is admittedly a crude one; however, the changes in the scattering curves induced by choice of medium (H_2O or D_2O), as well as the pH (or pD), were found to be much larger than the effects of conformational dynamics of the protein, justifying the use of a single, representative conformational model for the dynamic ensemble in all cases [31].

The choice of medium slightly affects the overall protein conformation. The trend of domain motion upon adjusting the pH in alkaline and neutral intervals is the same in both water systems, with only the first two domains affected by the interaction with Au. Dimer formation in BSA–gold conjugates at pH = 7 and even pD ~7 was observed with scattering techniques. The connected fine structural, configurational, and optical features may indicate that the developed BSA–gold ion bioconjugates (sensitive to several kinds of external perturbations) are similar to associates of polyelectrolyte and gold ions, rather than "robust" compact nanoclusters attached to the BSA protein. The versatility of the occurring energy transfer in these associates may explain the observed complex luminescence features.

Author Contributions: Conceptualization: A.B., B.F.; methodology: B.F., A.B., J.M., A.D., L.A., A.W., Z.V., J.S.P.; software: B.F., A.W., J.S.P.; validation: A.W., B.F., A.D., Z.V.; formal analysis: B.F., J.M, A.W., J.S.P.; investigations: A.B., J.M., B.F., A.D.; data curation: A.B., A.W.; writing—original draft preparation: A.B., B.F.; writing—review and editing: A.W., A.D., J.M.; visualization: B.F., J.M., A.B.;

supervision: A.B.; project administration: A.B., I.V., funding acquisition: A.B., J.M., Z.V., B.F., L.A. All authors have read and agreed to the published version of the manuscript.

Funding: This research was funded by the National Research, Development and Innovation Office Hungary under grants K131657 (A. Bóta) and K131594 (J. Mihály) and also by Project no. 2018-1.2.1-NKP-2018-00005 under the 2018-1.2.1-NKP funding scheme (A. Bóta, Z. Varga).

Data Availability Statement: Processed and derived data are available from the first author B.F. on request.

Acknowledgments: We would like to acknowledge the protein research program of the Hungarian Academy of Sciences (MEDinPROT, A. Bóta), and [KIFÜ] for awarding us access to resources based in Hungary at Debrecen for PPU time.

Conflicts of Interest: The authors declare no conflict of interest.

References

1. Peng, H.-S.; Chiu, D.T. Soft fluorescent nanomaterials for biological and biomedical imaging. *Chem. Soc. Rev.* **2015**, *44*, 4699–4722. [CrossRef] [PubMed]
2. Xie, J.; Zheng, Y.; Ying, J.Y. Protein-directed synthesis of highly fluorescent gold nanoclusters. *J. Am. Chem. Soc.* **2009**, *131*, 888–889. [CrossRef] [PubMed]
3. Wu, Z.; Jin, R. On the ligand's role in the fluorescence of gold nanoclusters. *Nano Lett.* **2010**, *10*, 2568–2573. [CrossRef] [PubMed]
4. Wen, X.; Yu, P.; Toh, Y.-R.; Tang, J. Structure-correlated dual fluorescent bands in BSA-protected Au_{25} nanoclusters. *J. Phys. Chem. C* **2012**, *116*, 11830–11836. [CrossRef]
5. Mohanty, J.S.; Chaudhari, K.; Sudhakar, C.; Pradeep, T. Metal-ion-induced luminescence enhancement in protein protected gold clusters. *J. Phys. Chem. C* **2019**, *123*, 28969–28976. [CrossRef]
6. Le Guével, X.; Hötzer, B.; Jung, G.; Hollemeyer, K.; Trouillet, V.; Schneider, M. Formation of fluorescent metal (Au, Ag) nanoclusters capped in bovine serum albumin followed by fluorescence and spectroscopy. *J. Phys. Chem. C* **2011**, *115*, 10955–10963. [CrossRef]
7. Chaudhari, K.; Xavier, P.L.; Pradeep, T. Understanding the evolution of luminescent gold quantum clusters in protein templates. *ACS Nano* **2011**, *5*, 8816–8827. [CrossRef]
8. Sadler, P.J.; Tucker, A. PH-induced structural transitions of bovine serum albumin. Histidine PKa values and unfolding of the N-terminus during the N to F transition. *Eur. J. Biochem.* **1993**, *212*, 811–817. [CrossRef]
9. Lin, H.; Imakita, K.; Fujii, M.; Sun, C.; Chen, B.; Kanno, T.; Sugimoto, H. New insights into the red luminescent bovine serum albumin conjugated gold nanospecies. *J. Alloys Compd.* **2017**, *691*, 860–865. [CrossRef]
10. Dixon, J.M.; Egusa, S. Conformational change-induced fluorescence of bovine serum albumin–gold complexes. *J. Am. Chem. Soc.* **2018**, *140*, 2265–2271. [CrossRef]
11. Dixon, J.M.; Tomida, J.; Egusa, S. Identifying the red-luminophore-forming domain in serum albumin–gold complexes. *J. Phys. Chem. Lett.* **2020**, *11*, 3345–3349. [CrossRef]
12. Fehér, B.; Lyngsø, J.; Bartók, B.; Mihály, J.; Varga, Z.; Mészáros, R.; Pedersen, J.S.; Bóta, A.; Varga, I. Effect of PH on the conformation of bovine serume albumin–gold bioconjugates. *J. Mol. Liq.* **2020**, *309*, 113065. [CrossRef]
13. Grdadolnik, J.; Maréchal, Y. Hydrogen–deuterium exchange in bovine serum albumin protein monitored by fourier transform infrared spectroscopy, Part I: Structural studies. *Appl. Spectrosc.* **2005**, *59*, 1347–1356. [CrossRef]
14. Reslan, M.; Kayser, V. The effect of deuterium oxide on the conformational stability and aggregation of bovine serum albumin. *Pharm. Dev. Technol.* **2018**, *23*, 1030–1036. [CrossRef]
15. Gardecki, J.A.; Maroncelli, M. Set of secondary emission standards for calibration of the spectral responsivity in emission spectroscopy. *Appl. Spectrosc.* **1998**, *52*, 1179–1189. [CrossRef]
16. Crosby, G.A.; Demas, J.N. Measurement of photoluminescence quantum yields. Review. *J. Phys. Chem.* **1971**, *75*, 991–1024. [CrossRef]
17. Varga, Z.; Mihály, J.; Berényi, S.; Bóta, A. Structural characterization of the poly(ethylene glycol) layer of sterically stabilized liposomes by means of FTIR spectroscopy. *Eur. Polym. J.* **2013**, *49*, 2415–2421. [CrossRef]
18. CENIMAT. Dynamic Light Scattering (DLS)—AvidNano W130i. Available online: https://www.cenimat.fct.unl.pt/services/laboratory-electronic-and-optoelectronic-materials-and-devices/dynamic-light-scattering-dls-avidnano-w130i (accessed on 9 December 2021).
19. Wacha, A.; Varga, Z.; Bóta, A. CREDO: A new general-purpose laboratory instrument for small-angle X-ray scattering. *J. Appl. Cryst.* **2014**, *47*, 1749–1754. [CrossRef]
20. Wacha, A. Optimized pinhole geometry for small-angle scattering. *J. Appl. Cryst.* **2015**, *48*, 1843–1848. [CrossRef]
21. SANS. Budapest Neutron Centre. Available online: https://www.bnc.hu/?q=ys-sans (accessed on 19 December 2021).
22. Almásy, L. New measurement control software on the yellow submarine SANS instrument at the Budapest Neutron Centre. *J. Surf. Investig. X-ray Synchrotron Neutron Tech.* **2021**, *15*, 527–531. [CrossRef]
23. Barth, A. Infrared spectroscopy of proteins. *Biochim. Biophys. Acta-Bioenerg.* **2007**, *1767*, 1073–1101. [CrossRef] [PubMed]

24. Lu, R.; Li, W.-W.; Katzir, A.; Raichlin, Y.; Yu, H.-Q.; Mizaikoff, B. Probing the secondary structure of bovine serum albumin during heat-induced denaturation using mid-infrared fiberoptic sensors. *Analyst* **2015**, *140*, 765–770. [CrossRef] [PubMed]
25. Glatter, O. A new method for the evaluation of small-angle scattering data. *J. Appl. Cryst.* **1977**, *10*, 415–421. [CrossRef]
26. Svergun, D.I. Determination of the regularization parameter in indirect-transform methods using perceptual criteria. *J. Appl. Cryst.* **1992**, *25*, 495–503. [CrossRef]
27. Jeffries, C.M.; Graewert, M.A.; Blanchet, C.E.; Langley, D.B.; Whitten, A.E.; Svergun, D.I. Preparing monodisperse macromolecular samples for successful biological small-angle X-ray and neutron-scattering experiments. *Nat. Protoc.* **2016**, *11*, 2122–2153. [CrossRef] [PubMed]
28. Franke, D.; Svergun, D.I. DAMMIF, a program for rapid *ab-initio* shape determination in small-angle scattering. *J. Appl. Crystallogr.* **2009**, *42*, 342–346. [CrossRef]
29. Petoukhov, M.V.; Svergun, D.I. Global rigid body modeling of macromolecular complexes against small-angle scattering data. *Biophys. J.* **2005**, *89*, 1237–1250. [CrossRef]
30. Volkov, V.V.; Svergun, D.I. Uniqueness of *ab initio* shape determination in small-angle scattering. *J. Appl. Cryst.* **2003**, *36*, 860–864. [CrossRef]
31. Tria, G.; Mertens, H.D.T.; Kachala, M.; Svergun, D.I. Advanced ensemble modelling of flexible macromolecules using X-ray solution scattering. *IUCrJ* **2015**, *2*, 207–217. [CrossRef]

nanomaterials

Article

Ligand Effects on Intramolecular Configuration, Intermolecular Packing, and Optical Properties of Metal Nanoclusters

Sainan Wu, Xiao Wei, Hao Li, Honglei Shen, Jiaojiao Han, Xi Kang * and Manzhou Zhu *

Department of Chemistry and Centre for Atomic Engineering of Advanced Materials, Key Laboratory of Structure and Functional Regulation of Hybrid Materials of Ministry of Education, Institutes of Physical Science and Information Technology and Anhui Province Key Laboratory of Chemistry for Inorganic/Organic Hybrid Functionalized Materials, Anhui University, Hefei 230601, China; WSN_chem@163.com (S.W.); weixiao_chem@163.com (X.W.); speechless95@outlook.com (H.L.); shenhonglei_chem@163.com (H.S.); hjj_chem@163.com (J.H.)

* Correspondence: kangxi_chem@ahu.edu.cn (X.K.); zmz@ahu.edu.cn (M.Z.)

Abstract: Surface modification has served as an efficient approach to dictate nanocluster structures and properties. In this work, based on an Ag_{22} nanocluster template, the effects of surface modification on intracluster constructions and intercluster packing modes, as well as the properties of nanoclusters or cluster-based crystallographic assemblies have been investigated. On the molecular level, the Ag_{22} nanocluster with larger surface steric hindrance was inclined to absorb more small-steric chlorine but less bulky thiol ligands on its surface. On the supramolecular level, the regulation of intramolecular and intermolecular interactions in nanocluster crystallographic assemblies rendered them CIEE (crystallization-induced emission enhancement)-active or -inactive nanomaterials. This study has some innovation in the molecular and intramolecular tailoring of metal nanoclusters, which is significant for the preparation of new cluster-based nanomaterials with customized structures and enhanced performances.

Keywords: atomically precise nanoclusters; surface modification; intramolecular configuration; intermolecular packing; optical properties

1. Introduction

Metal nanoclusters, bridging between small-sized molecular complexes and large-sized plasmonic nanoparticles, have attracted considerable attention owing to their atomically precise structures and excellent electrical, optical, and catalytical properties directed by the discrete electronic energy level as well as the structure-dependent quantum confinement effect [1–16]. The monodispersed sizes, precise compositions, and accurate configurations of metal nanoclusters make it possible to investigate the relationship between their structures and properties. In addition, the attainable structure–property correlations further enable the rational construction of new nanoclusters with customized performances [17–22]. In this context, the regulatable intramolecular structures and intermolecular packing modes render metal nanoclusters or cluster-based nanocomposites prominent nanomaterials for atomic engineering and further practical applications [23–29].

The past few decades have witnessed great research efforts of the control over intracluster structures/compositions and intercluster aggregates [17–20,23–27]. Specifically, the intramolecular control of nanoclusters touches upon the manipulation of their metal-ligand compositions and bonding environment at the single molecular level, while the intermolecular control of nanoclusters refers to the manipulation over their aggregating patterns among several cluster molecules in amorphous or crystallographic forms [30]. Several control methods, including (i) intracluster approaches (e.g., ligand exchange [31–34], heteroatom alloying [35–39], and molecular charge regulation [40–42]) and (ii) intercluster approaches (e.g., cluster-based metal-organic framework [43–46], aggregation-induced emission [47–49], and intercluster metallophilic reaction [50,51]), have been exploited to control clusters or their assemblies and to dictate their properties. Of note, the intracluster and intercluster controls are not a binary separation, but an interrelated and inseparable whole to regulate the nanocluster system simultaneously. In this context, the intracluster regulation of nanoclusters may alter their aggregating patterns at the supramolecular level, and vice versa [52]. The profound cognition of the correlation between molecular and supramolecular chemistry of nanoclusters offers great opportunities for the fabrication of novel nanoclusters or cluster-based hybrids with customized properties.

Herein, a new Ag_{22} nanocluster, formulated as Ag_{22}(S-Adm)$_{10}$(DPPM)$_4$$Cl_6$ (abbreviated as **Ag_{22}-L1**, where S-Adm = 1-adamantanethiol and DPPM = bis(diphenylphosphino)methane), was synthesized and structure-determined by X-ray single-crystal diffraction. The combination of this Ag_{22} nanocluster and a previously reported Ag_{22}($SPhMe_2$)$_{12}$(DPPE)$_4$$Cl_4$ (abbreviated as **Ag_{22}-L2**, where $SPhMe_2$ = 2,5-dimethyl thiophenol and DPPE = 1,2-bis (diphenylphosphino)ethane) constructed a platform to investigate the effects of surface modification on intramolecular constructions and intermolecular packing modes, as well as the properties of nanoclusters or cluster-based crystallographic assemblies. On the molecular level, because of the larger surface steric hindrance of **Ag_{22}-L1** relative to **Ag_{22}-L2**, the **Ag_{22}-L1** surface contained more small-steric chlorine but fewer bulky thiol ligands. On the supramolecular level, **Ag_{22}-L2** displayed intramolecular and intermolecular interactions in its crystallographic assembly, while these interactions were absent in the **Ag_{22}-L1** crystal. **Ag_{22}-L2** was CIEE (crystallization-induced emission enhancement) active while **Ag_{22}-L1** was CIEE inactive. The optical absorptions and emissions of these two Ag_{22} nanoclusters were also compared.

2. Materials and Methods

Chemicals. All reagents were purchased from Adamas Reagent (Shanghai, China) and used without further purification: silver nitrate ($AgNO_3$, 99%, metal basis), 1-adamantanethiol (HS-Adm, 97%), 2,5-dimethyl thiophenol (HS-$PhMe_2$, 97%), bis(diphenylphosphino)methane (Ph_2P-CH_2-PPh_2, DPPM, 98%), 1,2-bis(diphenylphosphino)ethane (Ph_2P-C_2H_5-PPh_2, DPPE, 98%), sodium cyanoborohydride ($NaBCNH_3$, 99.9%), methylene chloride (CH_2Cl_2, HPLC grade), methanol (CH_3OH, HPLC grade), ethyl ether ((C_2H_5)$_2$O, HPLC grade), and *n*-hexane (Hex, HPLC grade).

Synthesis of Ag_{22}(S-Adm)$_{10}$(DPPM)$_4$$Cl_6$ (Ag_{22}-L1). Specifically, 60 mg of $AgNO_3$ (0.36 mmol) and 40 μL of H_2PtCl_6 (0.2 g/mL; 0.015 mmol) were dissolved in 20 mL of CH_3OH and 1 mL of CH_3CN. Then, 40 mg of DPPM (0.1 mmol) and 30 mg of HS-Adm (0.18 mmol) were added. After stirring for 30 min, 100 mg of $NaBCNH_3$ (1.59 mmol; dissolved in 2 mL of MeOH) was added. The reaction was allowed to proceed for 5 h. After that, the mixture in the organic phase was rotavaporated under vacuum and washed several times by MeOH and Hex. Then, 10 mL of CH_2Cl_2 was used to extract the obtained **Ag_{22}-L1** nanocluster. The yield is 30% based on the Ag element (calculated from $AgNO_3$). Of note, although Pt did not exist in the final **Ag_{22}-L1**, the absence of Pt sources resulted in the failure of the nanocluster synthesis (Figure S1). Such a phenomenon has also been observed in previous works [53].

Synthesis of Ag_{22}(S-$PhMe_2$)$_{12}$(DPPE)$_4$$Cl_4$ (Ag_{22}-L2). The preparation of Ag_{22} (S-$PhMe_2$)$_{12}$(DPPE)$_4$$Cl_4$ was based on the reported method of the Pradeep group [54].

Crystallization of Ag_{22}-L1. In order to accelerate the crystallization process and improve the crystal quality, the counterions (i.e., Cl^-) in the **Ag_{22}-L1** nanocluster were replaced by SbF_6^- [55]. The reaction equation was [Ag_{22}(S-$PhMe_2$)$_{12}$(DPPE)$_4$$Cl_4$]$Cl_2$ + 2 SbF_6^- $\rightarrow$ [Ag_{22}(S-$PhMe_2$)$_{12}$(DPPE)$_4$$Cl_4$]($SbF_6$)$_2$ + 2Cl^-. Nanoclusters were crystallized in a CH_2Cl_2/ether system with a vapor diffusion method (Table S1).

3. Results

The **Ag_{22}-L1** nanocluster was synthesized by directly reducing the Ag-SR-DPPM complexes by $NaBCNH_3$ (Scheme S1; see more details in *Materials and Methods*). The electrospray ionization mass spectrometry (ESI-MS) measurement was performed to verify the molecular composition and to determine the valence state of the **Ag_{22}-L1** nanocluster. As shown in Figure S2, the mass result of the nanocluster exhibited an intense peak at 2897.54 Da. The excellent match of the experimental and simulated isotope patterns illustrated that the measured formula was $[Ag_{22}(S\text{-}Adm)_{10}(DPPM)_4Cl_6]^{2+}$. The "+2" valence state of the nanocluster matched well with the existence of $(SbF_6)^-$ counterions in the crystal lattice, i.e., the molar ratio between the cluster and the counterion was 1:2, as depicted in Figure S3. According to the valence states of **Ag_{22}-L1**, its nominal electron counts was determined as 4e [56], i.e., 22(Ag) − 10(SR) − 6(Cl) − 2(charge) = 4e, the same as that of **Ag_2-L2** [54]. Moreover, the chlorine ligands in **Ag_{22}-L1** were proposed to originate from the H_2PtCl_6 or from the CH_2Cl_2 solvent, which has also been discovered in previously determined nanoclusters [57–60].

Structurally, the **Ag_{22}-L1** nanocluster contained an Ag_{10} kernel which comprised two distorted trigonal bipyramidal Ag_5 units via an edge–edge vertical assembling mode (Figure 1A,B). Then, two $Ag_2(S\text{-}Adm)_3(DPPM)_1Cl_1$ surface units capped the Ag_{10} kernel from the same side via Ag-S or Ag-Cl interactions, giving rise to an $Ag_{14}(S\text{-}Adm)_6(DPPM)_2Cl_2$ structure (Figure 1C,D). The other unprotected side of the Ag_{10} kernel was further stabilized by two $Ag_2(S\text{-}Adm)_2(DPPM)_1Cl_2$ surface units, making up a $Ag_{18}(S\text{-}Adm)_{10}(DPPM)_4Cl_6$ structure (Figure 1E,F). Finally, four Ag atoms acting as bridges linked these surface units via S-Ag-S interactions, yielding the final $Ag_{22}(S\text{-}Adm)_{10}(DPPM)_4Cl_6$ framework (Figure 1G,H). Because of the asymmetry of surface units in **Ag_{22}-L1**, especially the asymmetrical arrangement of peripheral thiol and chlorine ligands, no symmetrical element was observed in the **Ag_{22}-L1** nanocluster framework (Figure 1I and Figure S4).

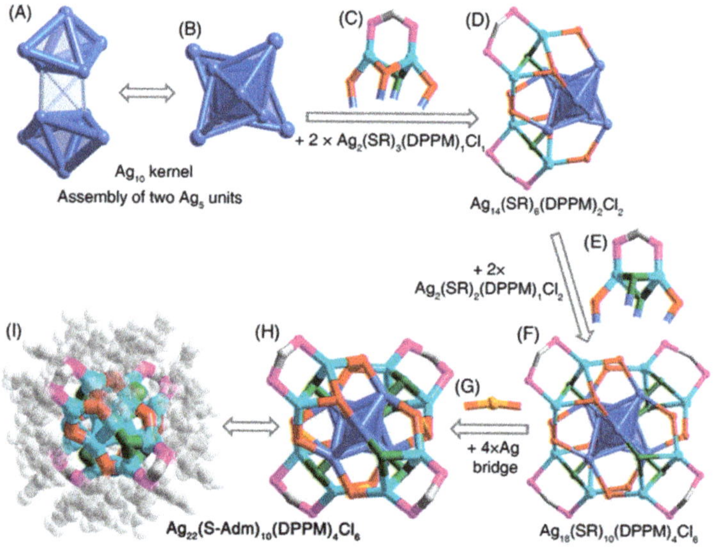

Figure 1. Structural anatomy of the Ag_{22}-L1 nanocluster. **(A,B)** The Ag_{10} kernel, constituted by assembling two Ag_5 units. **(C)** Two $Ag_2(S\text{-}Adm)_3(DPPM)_1Cl_1$ surface units. **(D)** The $Ag_{14}(S\text{-}Adm)_6(DPPM)_2Cl_2$ structure. **(E)** Two $Ag_2(S\text{-}Adm)_2(DPPM)_1Cl_2$ surface units. **(F)** The $Ag_{18}(S\text{-}Adm)_{10}(DPPM)_4Cl_6$ structure. **(G)** Four Ag bridges. **(H,I)** Overall structure of the $Ag_{22}(S\text{-}Adm)_{10}(DPPM)_4Cl_6$ nanocluster. Color codes: blue/light blue/orange sphere, Ag; red sphere, S; magenta sphere, P; green sphere, Cl; grey sphere, C; white sphere, H.

The overall constructions of **Ag_{22}-L1** and **Ag_{22}-L2** nanoclusters were almost the same. However, because of the different steric hindrances of ligands in these two nanoclusters (i.e., S-Adm and DPPM in **Ag_{22}-L1**; S-$PhMe_2$ and DPPE in **Ag_{22}-L2**), these two nanoclusters displayed some structural differences:

(i) For the kernel structure: the average Ag-Ag bond length in bipyramidal Ag_5 of **Ag_{22}-L1** was 2.824 Å, much shorter than that in **Ag_{22}-L2** (i.e., 2.933 Å). In addition, the average Ag-Ag bond lengths between these two Ag_5 bipyramids were 2.870 and 2.937 Å in **Ag_{22}-L1** and **Ag_{22}-L2**, respectively. In this context, due to the larger surface steric hindrance of **Ag_{22}-L1** relative to **Ag_{22}-L2**, the Ag_{10} kernel of the former nanocluster was compressed.

(ii) For the surface environment: the biggest structural difference between the two Ag_{22} nanoclusters lay in their surface ligand environments in terms of the proportion of the chlorine in peripheral ligands. Specifically, the **Ag_{22}-L1** nanocluster contained 10 thiol and 6 chlorine ligands, while **Ag_{22}-L2** included 12 thiol and 4 chlorine ligands (Figure 2). As shown in Figure 2A,B, a thiol ligand at the specific location on the **Ag_{22}-L2** surface was substituted by a chlorine ligand in **Ag_{22}-L1**. Another thiol ligand at the symmetrical position was also replaced by chlorine. Such a substitution from bulky thiol to small-steric chlorine was reasonable by considering that the more compact surface environment on **Ag_{22}-L1**, resulting from the bulkier DPPM and S-Adm ligands relative to DPPE and S-$PhMe_2$, was unable to host as many bulky thiol ligands as **Ag_{22}-L2** (Figure 2C,D). Moreover, several intramolecular noncovalent C-H$\cdots\pi$ and $\pi\cdots\pi$ interactions were observed in the **Ag_{22}-L2** structure, which was advantageous to the compact packing of its surface ligands [54]. By comparison, none of such noncovalent interactions was observed in **Ag_{22}-L1**, which might be another reason that more small-steric chlorine but fewer bulky thiol ligands were arranged on the **Ag_{22}-L1** nanocluster surface.

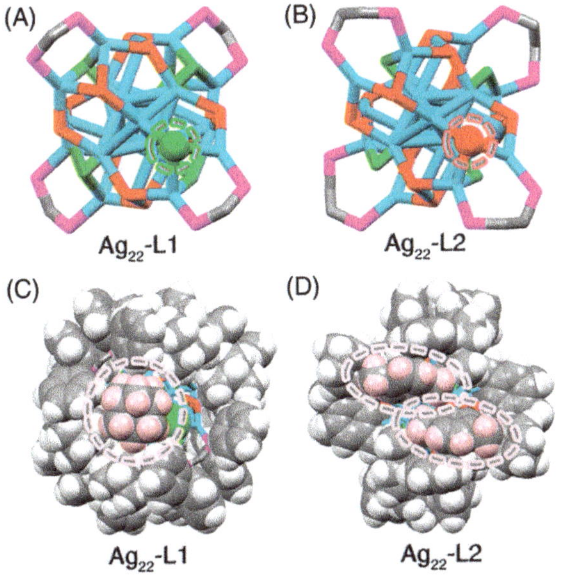

Figure 2. Structure comparison between Ag_{22}-L1 and Ag_{22}-L2 nanoclusters. (**A**) Cluster framework of the **Ag_{22}-L1** nanocluster with Cl ligands at specified locations. (**B**) Cluster framework of the **Ag_{22}-L2** nanocluster with SR ligands at specified locations. (**C**) Spacefill packing of the **Ag_{22}-L1** nanocluster with a S-Adm ligand at the specified surface vacancy. (**D**) Spacefill packing of the **Ag_{22}-L2** nanocluster with two S-$PhMe_2$ ligands at the specified surface vacancy. Color codes: light blue sphere, Ag; red sphere, S; magenta sphere, P; green sphere, Cl; grey sphere, C; pink/white sphere, H.

The **Ag_{22}-L1** cluster entities were crystallized in a triclinic crystal system with a *P*-1 space group, whereas the **Ag_{22}-L2** cluster entities were crystallized in a tetragonal crystal system with an $I4_1/a$ space group. Both nanoclusters followed a lamellar eutectic packing pattern between *R*-nanocluster and *S*-nanocluster enantiomers in the crystal lattice; however, due to their distinct crystal systems, the interlayer distances were different: 26.561 Å of **Ag_{22}-L1**, and 28.957 Å of **Ag_{22}-L2** (Figure 3 and Figure S5). Of note, there are equal *R*-nanocluster and *S*-nanocluster enantiomers in the crystal lattice, and the crystalline material of the nanocluster was racemic. Furthermore, owing to the existence of several benzene-rings in the **Ag_{22}-L2** nanoclusters, strong intracluster and intercluster interactions occurred, including C-H$\cdots\pi$ interaction and π-π stacking [54]. In vivid contrast, these interactions were absent within the **Ag_{22}-L1** nanocluster or among **Ag_{22}-L1** cluster entities (Figure S6).

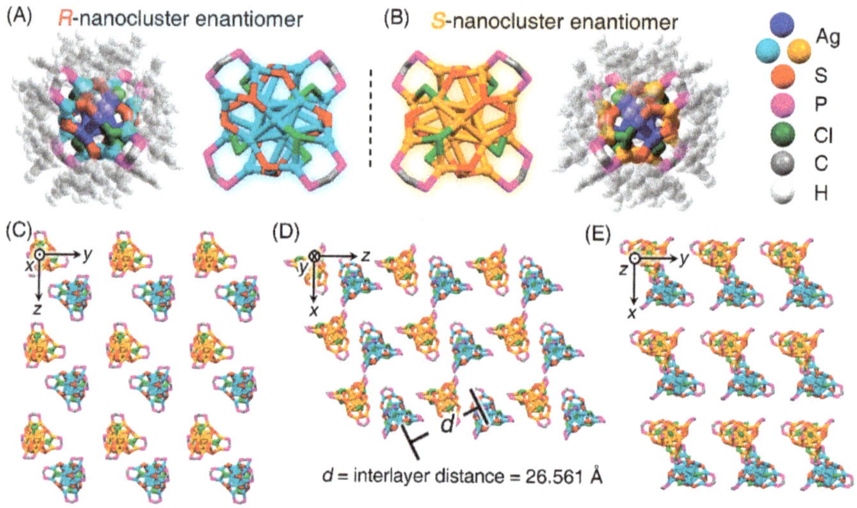

Figure 3. Crystalline packing of the **Ag_{22}-L1** nanocluster molecules. (**A**) Structure of the *R*-nanocluster enantiomer. (**B**) Structure of the *S*-nanocluster enantiomer. (**C–E**) Packing of the **Ag_{22}-L1** molecules in the crystal lattice: view from the *x* axis (**C**), *y* axis (**D**), and *z* axis (**E**). The inter-layer distance along with the *z* axis is 25.561 Å. Color codes: blue/light blue sphere, Ag in *R*-nanocluster enantiomer; blue/orange sphere, Ag in *S*-nanocluster enantiomer; red sphere, S; magenta sphere, P; green sphere, Cl; grey sphere, C; white sphere, H.

The **Ag_{22}-L1** nanocluster (dissolved in CH_2Cl_2) exhibited three intense absorptions centered at 368, 494, and 635 nm (Figure 4A). By comparison, the UV-vis spectrum of **Ag_{22}-L2** displayed several peaks at 445, 512, and 670 nm (Figure 4A). The blue shifts in the optical absorptions of **Ag_{22}-L1** relative to **Ag_{22}-L2** resulted from the different electronic structures of the two Ag_{22} nanoclusters. The CH_2Cl_2 solution of **Ag_{22}-L1** emitted at 650 nm, while the emission of **Ag_{22}-L2** was located around 670 nm (Figure 4B). The 20 nm blue-shift and 1.2-fold enhancement of the emission of **Ag_{22}-L1** relative to that of **Ag_{22}-L2** resulted from their different electronic structures. Indeed, these two nanoclusters displayed different optical absorptions, demonstrating their distinguishable electronic excitations and HOMO-LUMO energy gaps (HOMO: the highest occupied molecular orbital; LUMO: the lowest unoccupied molecular orbital). In addition, the different electronic excitations endowed these two nanoclusters with distinct emissions.

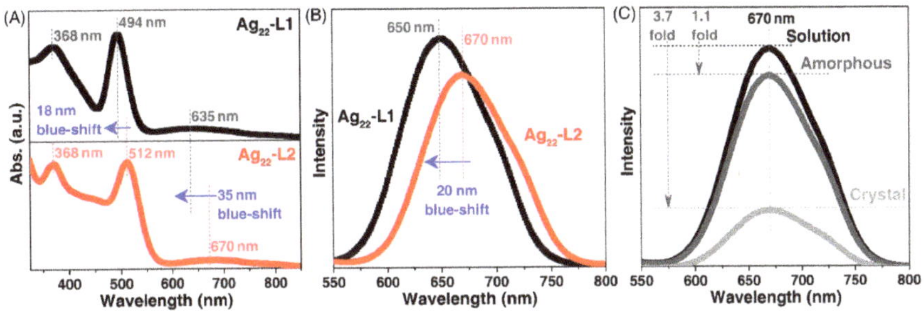

Figure 4. Comparison of optical properties between two Ag_{22} nanoclusters. (**A**) Comparison of optical absorptions between **Ag_{22}-L1** (black line) and **Ag_{22}-L2** (red line). (**B**) Comparison of emissions between **Ag_{22}-L1** (black line) and **Ag_{22}-L2** (red line). (**C**) Emission spectra of **Ag_{22}-L1** in the solution (black line), amorphous (red line), and crystalline (blue line) states.

The **Ag_{22}-L2** nanocluster was CIEE active owing to the presence of extensive intramolecular and intermolecular interactions in its crystal lattice [54]. In this context, the emission intensity of **Ag_{22}-L2** in the crystalline state was remarkably higher than that of the nanocluster in the solution or the amorphous state. By comparison, the **Ag_{22}-L1** was CIEE inactive since no significant enhancement in emission intensity was observed (Figure 4C). Actually, the **Ag_{22}-L1** in the amorphous or crystalline state was almost non-emissive. Such a striking contrast was reasonable considering that the intramolecular and intermolecular interactions were absent in the crystal lattice of **Ag_{22}-L1**, as mentioned above. The investigation of the Ag_{22} nanocluster system promoted the understanding of the crystalline packing mode and the CIEE of cluster-based nanomaterials.

4. Conclusions

In summary, a new Ag_{22} nanocluster, formulated as $Ag_{22}(S\text{-}Adm)_{10}(DPPM)_4Cl_6$, has been synthesized and structurally determined, which constituted an Ag_{22} cluster system together with the previously reported $Ag_{22}(S\text{-}PhMe_2)_{12}(DPPE)_4Cl_4$. Based on this Ag_{22} cluster system, the effects of surface modification on intracluster constructions and intercluster packing modes, as well as the properties of nanoclusters or cluster-based crystallographic assemblies were investigated. The Ag_{22} nanocluster with larger surface steric hindrance was inclined to load more small-steric chlorine but fewer bulky thiol ligands on its surface. Moreover, the Ag_{22} nanocluster, which embodied several intramolecular and intermolecular interactions in cluster crystallographic assemblies, was CIEE active; by comparison, the Ag_{22} nanocluster without such interactions was CIEE inactive. This work provides new insight into the surface modification of metal nanoclusters and its effects on intramolecular configuration, intermolecular packing, and optical properties.

Supplementary Materials: The following are available online at https://www.mdpi.com/2079-4991/11//2655/s1, Scheme S1. Synthetic procedure of the nanocluster; Figure S1. Comparison of optical absorptions of the nanocluster synthesis; Figure S2: ESI-MS result of the $[Ag_{22}(SPhMe_2)_{12}(DPPE)_4Cl_4]^{2+}$ nanocluster; Figure S3. Crystalline unit cell of the $[Ag_{22}(SPhMe_2)_{12}(DPPE)_4Cl_4](SbF_6)_2$ nanocluster; Figure S4. Overall structure of the $[Ag_{22}(SPhMe_2)_{12}(DPPE)_4Cl_4](SbF_6)_2$; Figure S5. Crystal unit of **Ag_{22}-L2**; Figure S6. Two adjacent $Ag_{22}(SPhMe_2)_{12}(DPPE)_4Cl_4$ nanocluster molecules in the crystal lattice; Table S1. Crystal data and structure refinement for the $[Ag_{22}(SPhMe_2)_{12}(DPPE)_4Cl_4](SbF_6)_2$ nanocluster.

Author Contributions: X.K. and M.Z. designed the study; S.W., X.W., H.L., H.S. and J.H. performed the experiments and analyzed the data. All authors discussed the results and commented on the manuscript. All authors have read and agreed to the published version of the manuscript.

Funding: We acknowledge the financial support by NSFC (21631001 and 21871001), the Ministry of Education, and the University Synergy Innovation Program of Anhui Province (GXXT-2020-053).

Data Availability Statement: The X-ray crystallographic coordinates for structures reported in this work have been deposited at the Cambridge Crystallographic Data Center (CCDC), under deposition numbers CCDC-2106804. These data can be obtained free of charge from the Cambridge Crystallographic Data Centre via www.ccdc.cam.ac.uk/data_request/cif, which has been mentioned in the article.

Conflicts of Interest: The authors declare no conflict of interest.

References

1. Jin, R.; Zeng, C.; Zhou, M.; Chen, Y. Atomically Precise Colloidal Metal Nanoclusters and Nanoparticles: Fundamentals and Opportunities. *Chem. Rev.* **2016**, *116*, 10346–10413. [CrossRef] [PubMed]
2. Chakraborty, I.; Pradeep, T. Atomically Precise Clusters of Noble Metals: Emerging Link between Atoms and Nanoparticles. *Chem. Rev.* **2017**, *117*, 8208–8271. [CrossRef] [PubMed]
3. Li, Y.; Zhou, M.; Jin, R. Programmable Metal Nanoclusters with Atomic Precision. *Adv. Mater.* **2021**. [CrossRef]
4. Bhattarai, B.; Zaker, Y.; Atnagulov, A.; Yoon, B.; Landman, U.; Bigioni, T.P. Chemistry and Structure of Silver Molecular Nanoparticles. *Acc. Chem. Res.* **2018**, *51*, 3104–3113. [CrossRef] [PubMed]
5. Yan, J.; Teo, B.K.; Zheng, N. Surface Chemistry of Atomically Precise Coinage-Metal Nanoclusters: From Structural Control to Surface Reactivity and Catalysis. *Acc. Chem. Res.* **2018**, *51*, 3084–3093. [CrossRef] [PubMed]
6. Kurashige, W.; Niihori, Y.; Sharma, S.; Negishi, Y. Precise Synthesis, Functionalization and Application of Thiolate-Protected Gold Clusters. *Coord. Chem. Rev.* **2016**, *320*, 238–250. [CrossRef]
7. Cook, A.W.; Hayton, T.W. Case Studies in Nanocluster Synthesis and Characterization: Challenges and Opportunities. *Acc. Chem. Res.* **2018**, *51*, 2456–2464. [CrossRef] [PubMed]
8. Takano, S.; Hasegawa, S.; Suyama, M.; Tsukuda, T. Hydride Doping of Chemically Modified Gold-Based Superatoms. *Acc. Chem. Res.* **2018**, *51*, 3074–3083. [CrossRef] [PubMed]
9. Tang, Q.; Hu, G.; Fung, V.; Jiang, D.-e. Insights into Interfaces, Stability, Electronic Properties, and Catalytic Activities of Atomically Precise Metal Nanoclusters from First Principles. *Acc. Chem. Res.* **2018**, *51*, 2793–2802. [CrossRef]
10. Kang, X.; Chong, H.; Zhu, M. $Au_{25}(SR)_{18}$: The Captain of the Great Nanocluster Ship. *Nanoscale* **2018**, *10*, 10758–10834. [CrossRef] [PubMed]
11. Kang, X.; Zhu, M. Cocrystallization of Atomically Precise Nanoclusters. *ACS Mater. Lett.* **2020**, *2*, 1303–1314. [CrossRef]
12. Sakthivel, N.A.; Dass, A. Aromatic Thiolate-Protected Series of Gold Nanomolecules and a Contrary Structural Trend in Size Evolution. *Acc. Chem. Res.* **2018**, *51*, 1774–1783. [CrossRef] [PubMed]
13. Hussain, R.; Hussain, A.I.; Chatha, S.A.S.; Mansha, A.; Ayub, K. Density Functional Theory Study of Geometric and Electronic Properties of Full Range of Bimetallic Ag_nY_m (n + m = 10) Clusters. *J. Alloys Compd.* **2017**, *705*, 232–246. [CrossRef]
14. Hussain, R.; Hussain, A.I.; Chatha, S.A.S.; Hussain, R.; Hanif, U.; Ayub, K. Density Functional Theory and Surface Reactivity Study of Bimetallic Ag_nY_m (n + m = 10) Clusters. *Solid State Sci.* **2018**, *80*, 46–64. [CrossRef]
15. Jadoon, T.; Carter-Fenk, K.; Siddique, M.B.A.; Herbert, J.M.; Hussain, R.; Iqbal, S.; Iqbal, J.; Ayub, K. Silver Clusters Tune Up Electronic Properties of Graphene Nanoflakes: A Comprehensive Theoretical Study. *J. Mol. Liq.* **2020**, *297*, 111902. [CrossRef]
16. Jadoon, T.; Ahsin, A.; Ullah, F.; Mahmood, T.; Ayub, K. Adsorption Mechanism of *p*-Aminophenol over Silver-Graphene Composite: A First Principles Study. *J. Mol. Liq.* **2021**, *341*, 117415. [CrossRef]
17. Agrachev, M.; Ruzzi, M.; Venzo, A.; Maran, F. Nuclear and Electron Magnetic Resonance Spectroscopies of Atomically Precise Gold Nanoclusters. *Acc. Chem. Res.* **2019**, *52*, 44–52. [CrossRef]
18. Kwak, K.; Lee, D. Electrochemistry of Atomically Precise Metal Nanoclusters. *Acc. Chem. Res.* **2019**, *52*, 12–22. [CrossRef]
19. Kang, X.; Zhu, M. Tailoring the Photoluminescence of Atomically Precise Nanoclusters. *Chem. Soc. Rev.* **2019**, *48*, 2422–2457. [CrossRef]
20. Gan, Z.; Xia, N.; Wu, Z. Discovery, Mechanism, and Application of Antigalvanic Reaction. *Acc. Chem. Res.* **2018**, *51*, 2774–2783. [CrossRef]
21. Lei, Z.; Wan, X.-K.; Yuan, S.-F.; Guan, Z.-J.; Wang, Q.-M. Alkynyl Approach toward the Protection of Metal Nanoclusters. *Acc. Chem. Res.* **2018**, *51*, 2465–2474. [CrossRef] [PubMed]
22. Konishi, K.; Iwasaki, M.; Shichibu, Y. Phosphine-Ligated Gold Clusters with Core+*exo* Geometries: Unique Properties and Interactions at the Ligand-Cluster Interface. *Acc. Chem. Res.* **2018**, *51*, 3125–3133. [CrossRef] [PubMed]
23. Chakraborty, P.; Nag, A.; Chakraborty, A.; Pradeep, T. Approaching Materials with Atomic Precision Using Supramolecular Cluster Assemblies. *Acc. Chem. Res.* **2019**, *52*, 2–11. [CrossRef] [PubMed]
24. Zhao, S.; Jin, R.; Jin, R. Opportunities and Challenges in CO_2 Reduction by Gold- and Silver-Based Electrocatalysts: From Bulk Metals to Nanoparticles and Atomically Precise Nanoclusters. *ACS Energy Lett.* **2018**, *3*, 452–462. [CrossRef]
25. Nasaruddin, R.R.; Chen, T.; Yan, N.; Xie, J. Roles of Thiolate Ligands in the Synthesis, Properties and Catalytic Application of Gold Nanoclusters. *Coord. Chem. Rev.* **2018**, *368*, 60–79. [CrossRef]

26. Liu, Y.; Chai, X.; Cai, X.; Chen, M.; Jin, R.; Ding, W.; Zhu, Y. Central Doping of a Foreign Atom into the Silver Cluster for Catalytic Conversion of CO_2 toward C-C Bond Formation. *Angew. Chem. Int. Ed.* **2018**, *57*, 9775–9779. [CrossRef]
27. Hu, X.; Zheng, Y.; Zhou, J.; Fang, D.; Jiang, H.; Wang, X. Silver-Assisted Thiolate Ligand Exchange Induced Photoluminescent Boost of Gold Nanoclusters for Selective Imaging of Intracellular Glutathione. *Chem. Mater.* **2018**, *30*, 1947–1955. [CrossRef]
28. Yuan, X.; Luo, Z.; Yu, Y.; Yao, Q.; Xie, J. Luminescent Noble Metal Nanoclusters as an Emerging Optical Probe for Sensor Development. *Chem. Asian. J.* **2013**, *8*, 858–871. [CrossRef] [PubMed]
29. Liu, X.; Astruc, D. Atomically Precise Copper Nanoclusters and Their Applications. *Coord. Chem. Rev.* **2018**, *359*, 112–126. [CrossRef]
30. Kang, X.; Zhu, M. Intra-Cluster Growth Meets Inter-Cluster Assembly: The Molecular and Supramolecular Chemistry of Atomically Precise Nanoclusters. *Coord. Chem. Rev.* **2019**, *394*, 1–38. [CrossRef]
31. Kang, X.; Zhu, M. Transformation of Atomically Precise Nanoclusters by Ligand-Exchange. *Chem. Mater.* **2019**, *31*, 9939–9969. [CrossRef]
32. Gan, Z.; Chen, J.; Liao, L.; Zhang, H.; Wu, Z. Surface Single-Atom Tailoring of a Gold Nanoparticle. *J. Phys. Chem. Lett.* **2018**, *9*, 204–208. [CrossRef] [PubMed]
33. Kang, X.; Huang, L.; Liu, W.; Xiong, L.; Pei, Y.; Sun, Z.; Wang, S.; Wei, S.; Zhu, M. Reversible Nanocluster Structure Transformation between Face-Centered Cubic and Icosahedral Isomers. *Chem. Sci.* **2019**, *10*, 8685–8693. [CrossRef] [PubMed]
34. Sels, A.; Salassa, G.; Cousin, F.; Lee, L.-T.; Bürgi, T. Covalently Bonded Multimers of $Au_{25}(SBut)_{18}$ as a Conjugated System. *Nanoscale* **2018**, *10*, 12754–12762. [CrossRef]
35. Ghosh, A.; Mohammed, O.F.; Bakr, O.M. Atomic-Level Doping of Metal Clusters. *Acc. Chem. Res.* **2018**, *51*, 3094–3103. [CrossRef]
36. Kang, X.; Li, Y.; Zhu, M.; Jin, R. Atomically Precise Alloy Nanoclusters: Syntheses, Structures, and Properties. *Chem. Soc. Rev.* **2020**, *49*, 6443–6514. [CrossRef]
37. Kang, X.; Wei, X.; Jin, S.; Yuan, Q.; Luan, X.; Pei, Y.; Wang, S.; Zhu, M.; Jin, R. Rational Construction of a Library of M_{29} Nanoclusters from Monometallic to Tetrametallic. *Proc. Natl. Acad. Sci. USA* **2019**, *116*, 18834–18840. [CrossRef]
38. Fei, W.; Antonello, S.; Dainese, T.; Dolmella, A.; Lahtinen, M.; Rissanen, K.; Venzo, A.; Maran, F. Metal Doping of $Au_{25}(SR)_{18}^-$ Clusters: Insights and Hindsights. *J. Am. Chem. Soc.* **2019**, *141*, 16033–16045. [CrossRef]
39. Lee, S.; Bootharaju, M.S.; Deng, G.; Malola, S.; Häkkinen, H.; Zheng, N.; Hyeon, T. $[Pt_2Cu_{34}(PET)_{22}Cl_4]^{2-}$: An Atomically Precise, 10-Electron PtCu Bimetal Nanocluster with a Direct Pt-Pt Bond. *J. Am. Chem. Soc.* **2021**, *143*, 12100–12107. [CrossRef]
40. Zhu, M.; Aikens, C.M.; Hendrich, M.P.; Gupta, R.; Qian, H.; Schatz, G.C.; Jin, R. Reversible Switching of Magnetism in Thiolate-Protected Au_{25} Superatoms. *J. Am. Chem. Soc.* **2009**, *131*, 2490–2492. [CrossRef]
41. Kang, X.; Xu, F.; Wei, X.; Wang, S.; Zhu, M. Valence Self-Regulation of Sulfur in Nanoclusters. *Sci. Adv.* **2019**, *5*, eaax7863. [CrossRef]
42. Zeng, C.; Weitz, A.; Withers, G.; Higaki, T.; Zhao, S.; Chen, Y.; Gil, R.R.; Hendrich, M.; Jin, R. Controlling Magnetism of $Au_{133}(TBBT)_{52}$ Nanoclusters at Single Electron Level and Implication for Nonmetal to Metal Transition. *Chem. Sci.* **2019**, *10*, 9684–9691. [CrossRef]
43. Jin, Y.; Zhang, C.; Dong, X.-Y.; Zang, S.-Q.; Mak, T.C.W. Shell Engineering to Achieve Modification and Assembly of Atomically-Precise Silver Clusters. *Chem. Soc. Rev.* **2021**, *50*, 2297–2319. [CrossRef]
44. Huang, R.-W.; Wei, Y.-S.; Dong, X.-Y.; Wu, X.-H.; Du, C.-X.; Zang, S.-Q.; Mak, T.C.W. Hypersensitive Dual-Function Luminescence Switching of a Silver-Chalcogenolate Cluster-Based Metal-Organic Framework. *Nat. Chem.* **2017**, *9*, 689–697. [CrossRef]
45. Lei, Z.; Pei, X.-L.; Jiang, Z.-G.; Wang, Q.-M. Cluster Linker Approach: Preparation of a Luminescent Porous Framework with NbO Topology by Linking Silver Ions with Gold(I) Clusters. *Angew. Chem. Int. Ed.* **2014**, *53*, 12771–12775. [CrossRef]
46. Wei, X.; Kang, X.; Zuo, Z.; Song, F.; Wang, S.; Zhu, M. Hierarchical Structural Complexity in Atomically Precise Nanocluster Frameworks. *Natl. Sci. Rev.* **2021**, *8*, nwaa077. [CrossRef]
47. Goswami, N.; Yao, Q.; Luo, Z.; Li, J.; Chen, T.; Xie, J. Luminescent Metal Nanoclusters with Aggregation-Induced Emission. *J. Phys. Chem. Lett.* **2016**, *7*, 962–975. [CrossRef] [PubMed]
48. Kang, X.; Wang, S.; Zhu, M. Observation of a New Type of Aggregation-induced Emission in Nanoclusters. *Chem. Sci.* **2018**, *9*, 3062–3068. [CrossRef] [PubMed]
49. Wu, Z.; Liu, H.; Li, T.; Liu, J.; Yin, J.; Mohammed, O.F.; Bakr, O.M.; Liu, Y.; Yang, B.; Zhang, H. Contribution of Metal Defects in the Assembly Induced Emission of Cu Nanoclusters. *J. Am. Chem. Soc.* **2017**, *139*, 4318–4321. [CrossRef]
50. Yuan, P.; Zhang, R.; Selenius, E.; Ruan, P.; Yao, Y.; Zhou, Y.; Malola, S.; Häkkinen, H.; Teo, B.K.; Cao, Y.; et al. Solvent-Mediated Assembly of Atom-Precise Gold-Silver Nanoclusters to Semiconducting One-Dimensional Materials. *Nat. Commun.* **2020**, *11*, 2229. [CrossRef] [PubMed]
51. Wu, Z.; Du, Y.; Liu, J.; Yao, Q.; Chen, T.; Cao, Y.; Zhang, H.; Xie, J. Aurophilic Interactions in the Self-Assembly of Gold Nanoclusters into Nanoribbons with Enhanced Luminescence. *Angew. Chem. Int. Ed.* **2019**, *58*, 8139–8144. [CrossRef] [PubMed]
52. Huang, J.-H.; Si, Y.; Dong, X.-Y.; Wang, Z.-Y.; Liu, L.-Y.; Zang, S.-Q.; Mak, T.C.W. Symmetry Breaking of Atomically Precise Fullerene-like Metal Nanoclusters. *J. Am. Chem. Soc.* **2021**, *143*, 12439–12444. [CrossRef] [PubMed]
53. Tian, F.; Chen, R. Pd-Mediated Synthesis of Ag_{33} Chiral Nanocluster with Core–Shell Structure in T Point Group. *J. Am. Chem. Soc.* **2019**, *141*, 7107–7114. [CrossRef] [PubMed]

54. Khatun, E.; Bodiuzzaman, M.; Sugi, K.S.; Chakraborty, P.; Paramasivam, G.; Dar, W.A.; Ahuja, T.; Antharjanam, S.; Pradeep, T. Confining an Ag_{10} Core in an Ag_{12} Shell: A Four-Electron Superatom with Enhanced Photoluminescence upon Crystallization. *ACS Nano* **2019**, *13*, 5753–5759. [CrossRef] [PubMed]
55. Kang, X.; Jin, S.; Xiong, L.; Wei, X.; Zhou, M.; Qin, C.; Pei, Y.; Wang, S.; Zhu, M. Nanocluster Growth via "Graft-Onto": Effects on Geometric Structures and Optical Properties. *Chem. Sci.* **2020**, *11*, 1691–1697. [CrossRef]
56. Walter, M.; Akola, J.; Lopez-Acevedo, O.; Jadzinsky, P.D.; Calero, G.; Ackerson, C.J.; Whetten, R.L.; Grönbeck, H.; Häkkinen, H. A Unified View of Ligand-Protected Gold Clusters as Superatom Complexes. *Proc. Natl. Acad. Sci. USA* **2008**, *105*, 9157–9162. [CrossRef]
57. Hu, F.; Li, J.-J.; Guan, Z.-J.; Yuan, S.-F.; Wang, Q.-M. Formation of an Alkynyl-Protected Ag_{112} Silver Nanocluster as Promoted by Chloride Released in Situ from CH_2Cl_2. *Angew. Chem. Int. Ed.* **2020**, *59*, 5312–5315. [CrossRef] [PubMed]
58. Zeng, J.-L.; Guan, Z.-J.; Du, Y.; Nan, Z.-A.; Lin, Y.-M.; Wang, Q.-M. Chloride-Promoted Formation of a Bimetallic Nanocluster $Au_{80}Ag_{30}$ and the Total Structure Determination. *J. Am. Chem. Soc.* **2016**, *138*, 7848–7851. [CrossRef]
59. Zou, X.; Jin, S.; Wei, X.; Li, X.; Zhou, M.; Wang, S.; Zhu, M. Overall Structures of Two Metal Nanoclusters: Chloride as a Bridge Fills the Space between the Metal Core and the Metal Shell. *Inorg. Chem.* **2020**, *59*, 11905–11909. [CrossRef]
60. Wei, X.; Shen, H.; Xu, C.; Li, H.; Jin, S.; Kang, X.; Zhu, M. Ag_{48} and Ag_{50} Nanoclusters: Toward Active-Site Tailoring of Nanocluster Surface Structures. *Inorg. Chem.* **2021**, *60*, 5931–5936. [CrossRef]

nanomaterials

MDPI

Review

Viewing Aggregation-Induced Emission of Metal Nanoclusters from Design Strategies to Applications

Tingting Li [1], Haifeng Zhu [2,*] and Zhennan Wu [2,*]

1 School of Materials Science and Engineering, Jilin Jianzhu University, Changchun 130018, China
2 State Key Laboratory of Integrated Optoelectronics, College of Electronic Science and Engineering, Jilin University, Changchun 130021, China
* Correspondence: zhuhf0102@163.com (H.Z.); wuzn@jlu.edu.cn (Z.W.)

Abstract: Aggregation-induced emission (AIE)-type metal nanoclusters (NCs) represent an innovative type of luminescent metal NCs whose aggregates exhibit superior performance over that of individuals, attracting wide attention over the past decade. Here, we give a concise overview of the progress made in this area, from design strategies to applications. The representative design strategies, including solvent-induction, cation-induction, crystallization-induction, pH-induction, ligand inheritance, surface constraint, and minerals- and MOF-confinement, are first discussed. We then present the typical practical applications of AIE-type metal NCs in the various sectors of bioimaging, biological diagnosis and therapy (e.g., antibacterial agents, cancer radiotherapy), light-emitting diodes (LEDs), detection assays, and circularly polarized luminescence (CPL). To this end, we present our viewpoints on the promises and challenges of AIE-type metal NCs, which may shed light on the design of highly luminescent metal NCs, stimulating new vitality and serving as a continuous boom for the metal NC community in the future.

Keywords: aggregation-induced emission; metal nanoclusters; luminescent nanoclusters

1. Introduction

Metal nanoclusters (NCs) typically possess sub-2 nm metal cores consisting of a few to a hundred metal atoms that are protected by an organic ligand monolayer [1–3]. They are a rising star in the nanoparticle community due to their significance in both basic and applied research (e.g., as the missing link between atoms and nanocrystals, for the utilization of their intriguing molecular-like properties, including discrete electronic transitions, and for quantized charging) [4–7]. Among various properties, luminescence is the most attractive property of metal NCs. Because of their facile preparation, ultrafine size, low toxicity, high renal clearance, and excellent photostability, luminescent metal NCs have recently emerged as a novel class of chromophores, holding great promise for practical applications in lighting, imaging, sensing, and so on [8–13].

However, highly luminescent metal NCs are curtailed by the comparatively limited knowledge of their fundamental aspects. In particular, the complexity, diversity, and mutability in terms of their total structures preclude an in-depth understanding of their emission origin. Because of their unclear size/composition effects and the puzzling underlying mechanism of their luminescence, there has been limited success in attaining the desired improvement and tailoring the luminescence property of metal NCs, thus greatly limiting their boom in practical applications [14–16]. In this scenario, with continuous efforts in the community, Xie et al. first presented the aggregation-induced emission (AIE) of metal nanoclusters in 2012, which allows for an extraordinary strength in their emission intensity and stability [17]. In detail, the authors designed a new family of ultrabright Au(0)@Au(I)-SR (SR: deprotonated thiol ligands) core–shell NCs by preserving a high content of Au(I)-SR complexes in the protective shell. In this way, the intra-/intermolecular vibration and rotation of surface motifs can be effectively suppressed, giving rise to a minimization of non-radiative decay and hence the improved luminescence performance of metal NCs.

Thereafter, AIE-type metal NCs have been a recognized concept and a well-developed topic in both the fundamental and practical sectors of luminescent metal NCs (Figure 1). In the scheme of AIE-type metal NCs, over the past decade, several significant design strategies have been identified at and beyond the single-cluster level [18–20]. Their practical applications have been well proven in broad fields and cross-disciplines. Therefore, we deploy this review by viewing AIE-type metal NCs from their design strategies to their practical applications (Figure 2). Together with our overviews on the promises and challenges of AIE-type metal NCs, we try to offer a brief and in-depth discussion on AIE-type metal NCs in order to increase the acceptance of AIE-type metal NCs in various related communities in chemistry and materials science.

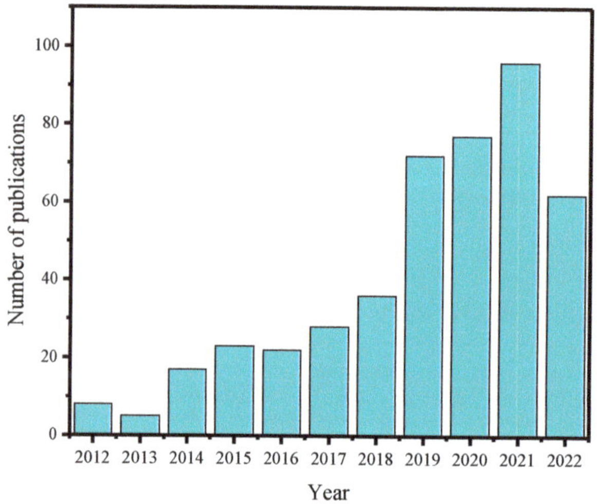

Figure 1. The increasing number of publications on AIE-type metal NCs over the years between 2012 and 2022 (data statistics from the Web of Science in January 2023, using keywords of AIE and metal nanoclusters).

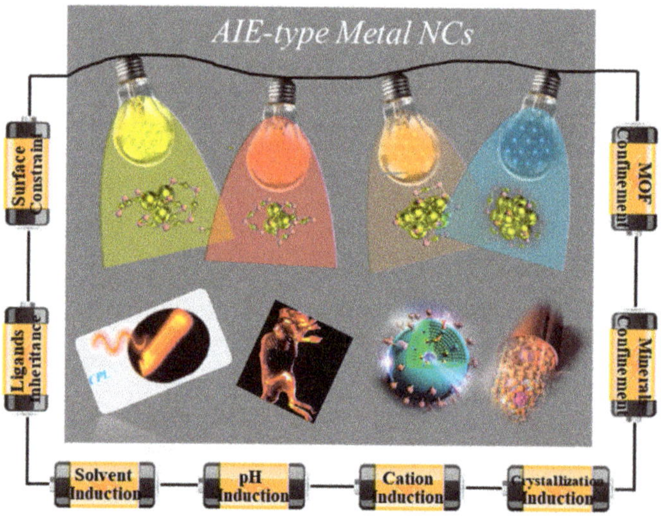

Figure 2. Schematic illustration of AIE-type metal NCs from the design strategies to the applications.

2. Design Strategies of AIE-Type Metal NCs

AIE is an extraordinary luminescence concept, which was first proposed by Tang in 2001 to illuminate an interesting photophysical phenomenon: the polymeric state is brighter than the dispersed state [21]. In particular, restricted intramolecular motion (RIM) has been proved to be the general mechanism of AIE [22]. Normally, many AIE molecules contain molecular rotors and/or vibrators. In a solution state, these rotors and vibrators can rotate and vibrate flexibly to consume the excited state energy, resulting in AIE molecules being nonemissive or weakly emissive. Meanwhile, in an aggregate state, intermolecular interactions restrict the rotation and vibration, causing radiative decay to dominate the energy dissipation of the excited state [23–25].

Since Xie et al. introduced the concept of AIE into metal NCs in 2012 [17–19,26], the family of AIE-type metal NCs has grown over the past decade: from a single-cluster level to the beyond-single-cluster level, from small molecules to macromolecules, from the solution state to the crystalline state, from condensed oligomeric complexes to scaffold confinement-induced materials, etc. Accordingly, in this section, we rationally categorize and discuss AIE-type metal NCs by focusing on their construction strategies.

2.1. Solvent-Induced AIE

Solvent-induced AIE is a representative strategy to trigger the aggregates from isolated metal NCs by generally altering the solvent polarity. Xie et al. successfully designed the first AIE-Au(0)@Au(I)-SR NCs based on this strategy by introducing ethanol (a weakly polar solvent) into water [17]. The formation of aggregates can be attributed to two reasons: (i) the addition of the ethanol destabilizes the Au(I)−thiolate compound, resulting in charge neutralization; (ii) the amplification of the Au(I)···Au(I) interaction to provide momentum for aggregation. The luminescence properties of aggregates strongly depend on the aggregation degree, which can be expressed by the formula $f_e = vol_{ethanol}/vol_{ethanol+water}$. The solution exhibited non-emission before $f_e > 75\%$; when f_e increases from 75% to 95%, the aggregates began to form, emitting weak red radiation; finally, the solution turned clarified and emitted very strong yellow luminescence at $f_e > 95\%$, as shown in Figure 3a–c. Upon aggregate formation, owing to the multiple intra- and inter-complex interactions (Au(I)···Au(I) interactions and van der Waals forces), intramolecular rotations and vibrations are limited, therefore enhancing the luminescence. Moreover, the blueshift mechanism is interpreted as the increase in Au(I)···Au(I) distance, because the inter-complex aurophilic interactions are stronger than intra-complex interactions.

$Au(I)_x(SR)_{x+1}$ (SR = glutathione), recently proposed by Xie et al., is another typical case to demonstrate the importance of solvents for the preparation of AIE-type metal NCs [26]. Serial AIE-type Au NCs were reported in this work, that is, the aggregates of $Au_{10}(SR)_{10}$, $Au_{15}(SR)_{13}$, $Au_{18}(SR)_{14}$, and $[Au_{25}(SR)_{18}]^-$ (Figure 3d). Except for $[Au_{25}(SR)_{18}]^-$ NCs, the quantum yield of $Au_{10}(SR)_{10}$, $Au_{15}(SR)_{13}$, and $Au_{18}(SR)_{14}$ was heightened at 11.4%, 4.9%, and 3.7%, respectively, as f_e increased from 75% to 95%. The strong emission can be attributed to the restricted molecular motion due to the adequate interactions between Au(I)-thiolate motifs. As for emission energy, $Au_{10}(SR)_{10}$, $Au_{15}(SR)_{13}$, and $Au_{18}(SR)_{14}$ blue-shifted by 15, 9, and 6 nm, respectively, and $[Au_{25}(SR)_{18}]^-$ red-shifted by 167 nm. The average Au^I-Au^I distance is affected by the competition with intra- and inter-motif aurophilic interactions, which ultimately influences the emission peak position. A similar phenomenon was observed in Cu NCs by the Patra group, where the solvent-induced aggregation strategy was adopted to tune the emission peak and enhance the emission intensity of $Cu_{34-32}(SG)_{16-13}$ (Figure 3e,f) [27]. When $f_e = 90\%$, the emission intensity enhanced 36-fold, which was associated with a 28 nm blue-shifting (624 → 597 nm).

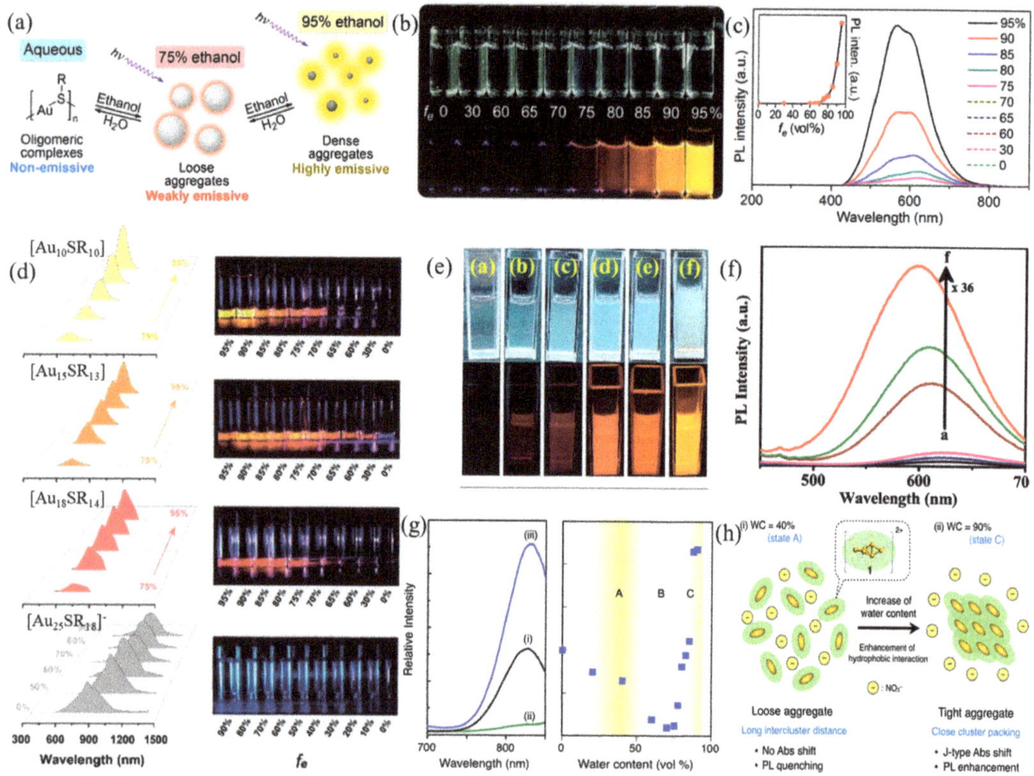

Figure 3. (**a**) Mechanism model of solvent-induced AIE phenomenon of Au(0)@Au(I)-SR NCs. (**b**) Digital photos and (**c**) photoemission spectra of Au(0)@Au(I)-SR with different f_e. Reprinted/adapted with permission from Ref. [17]. 2012, copyright American Chemical Society. (**d**) Luminescence spectra (**left**) and the corresponding digital photos (**right**) of $Au_{10}SR_{10}$, $Au_{15}SR_{13}$, $Au_{18}SR_{14}$, and $[Au_{25}SR_{18}]^-$ with a different fraction of ethanol and water. Reprinted/adapted with permission from Ref. [26]. 2020, copyright Wiley-VCH. (**e**) Digital photos and (**f**) PL spectra of $Cu_{34-32}(SG)_{16-13}$ with different f_e. Reprinted/adapted with permission from Ref. [27]. 2019, copyright American Chemical Society. (**g**) Photoluminescence spectra of (i) MeOH, (ii) MeOH/water (30/70), and (iii) MeOH/water (10/90). (**h**) Mechanism model of possible aggregate forms of MeOH/water at low (i) and high (ii) WC. Reprinted/adapted with permission from Ref. [28]. 2020, copyright American Chemical Society.

Sugiuchi et al. reported the AIE phenomenon on a diphosphine-ligated $[Au_6]^{2+}$ cluster by regulating the water contents (WC) [28]. The emission intensity decreased first and then increased by increasing the WC from 0% to 90%. Specifically, the aggregation at the water-rich conditions increased luminescence by 20 times (Figure 3g). The luminescent performance of the aggregates is highly dependent on the WC, which can be divided into the following three zones: (i) Zone A (WC < 40%), in which the cluster molecules separate off from each other and randomly orient themselves due to the weak attractive hydrophobic interactions (Figure 3h(i)). The concentration quenching would occur at this stage. (ii) Zone B (40% < WC < 70%) is the transition stage of zone A to zone C, in which the cluster molecules will ulteriorly agglomerate, resulting in severe concentration quenching. (iii) In zone C (70% < WC < 90%), the clustered molecules are tightly packed, thus suppressing the molecular motion and vibration to enhance luminescence (Figure 3h(ii)). Very recently, the Li group also discovered the AIE phenomenon in Cu(I)-based cyclic trinuclear complexes (Cu(I)-CTC) [29]. These NCs exhibit no photoluminescence in pure tetrahydrofuran until water is added. In particular, Cu(I)-CTC with a near-unity PLQY was obtained when $vol_{water}/vol_{water+THF} = 90\%$. As a result, the molecular forces including C−H···C/C−H···π interaction are the main reasons for the cluster aggregation and thus the non-radiative decay reduction.

2.2. Cation-Induced AIE

Like solvent-induced AIE, cation-induced AIE of metal NCs has also been well developed, due to the variations, rotations, and other movements of ligands and complexes that can be restrained by surface charge neutralization and crosslinking. In particular, silver-doped Au(0)@Au(I)-thiolate NCs are an important example of cation-induced AIE; their emission is greatly enhanced by four times [30]. The Ag-doped Au NCs' lifetimes are 2.03, 5.16, and 52.2 μs, which is longer than pristine Au NCs, demonstrating that the additive silver can crosslink Au^{I}-GSH motifs to facilitate the evolution of denser aggregates. A similar phenomenon was observed in the system of $[Ag_xAu_{25-x}(PPh_3)_{10}(SC_2H_4Ph)_5Cl_2]^{2+}$ (200-fold quantum yield boosting) [31] and Au-doped $Ag_{29}(BDT)_{12}$ $(TPP)_4$ (26-fold quantum yield boosting) NCs [6].

In addition to coinage metal ions, group IIB metal ions (Cd^{2+} and Zn^{2+}) have also been used to promote the AIE process. Xie et al. utilized Cd^{2+} to induce the aggregation of Au(I)-thiolate complexes based on the electrostatic and coordination interactions between Cd^{2+} and GSH [17]. Kuppan et al. observed that non-luminescent 3-mercaptopropionic acid Au NCs can achieve strong yellow emission (10^6-fold improvement) by the incorporation of Zn^{2+} [32]. As depicted in Figure 4a, the self-assembly of Au NCs was efficiently induced by Zn^{2+} through two main ways: (i) the strong complexation of Zn^{2+} with negatively charged carboxyl (i.e., COO^-) groups by the charge screening effect, enhancing the rigidity of the Au(I) thiolate shell; (ii) Zn^{2+} crosslinking with surrounding Au NCs to form tight aggregates upon the force of van der Waals interactions. These interactions hinder the vibrations and rotations of ligands, significantly decreasing the non-radiative relaxation. Another example of a Zn^{2+} cation-induced AIE is based on $Au_4(MHA)_4$ (MHA = 6-mercaptohexanoic acid) NCs [16]. The aggregated $Au_4(MHA)_4$ exhibits ultrabright greenish-blue emission centered at 485 nm, with a narrow FWHM of 25 nm and a high QY of 90%. Similarly, the assembled mechanism is attributed to the interactions between the COO^- in the MHA ligand and Zn^{2+}, which rigidify the surrounding environment of $Au_4(MHA)_4$ NCs.

Recently, researchers have focused their attention on the rare earth ions. Luminogen ATT-AuNCs (ATT: 6-aza-2-thiothymine) are discussed here as a representation [33]. It was discovered that the luminescence properties of ATT-Au NCs significantly depend on the pH. Decreasing the pH results in the red-shift of emission spectra (from 520 to 535 nm) and significant PL quenching (Figure 4b). Sc^{3+} exhibits Lewis acidity and can stabilize the ICT state, thus being used to induce aggregation of ATT-Au NCs. The addition of Sc^{3+} not only makes the spectrum red-shift, but also promotes the PLQY enhancement (from 0.2 to 5.5%). As a result, the role of Sc^{3+} mainly includes three aspects: enhancing the ICT state; triggering the interaction between Sc^{3+} and the surface ligand to form the RIM process; constructing a donor–bridge–acceptor (DBA) structure by linking ATT-AuNCs and minocycline (Figure 4c,d).

2.3. Crystallization-Induced AIE

Crystallization-induced emission (CIE) may be treated as a derivative of AIE. In general, the active intramolecular motion (vibrations and rotations) in the solution state dissipates the energy of the excited state through non-radiative relaxation channels, leading to luminescence quenching. However, the intramolecular motion is restricted in the solid state, which minimizes energy consumption [34]. Crystallization can achieve a transition from liquid to solid, that is, by physical constraints and intermolecular interactions to lock the molecular conformations and hold them together. Zhu et al. presented the first CIE metal clusters in 2017. The bimetallic NCs were protected by (diphenylphosphino)methane (DPPM) and 2, 5-dimethylbenzenethiol (SR), which provided a guiding strategy to build up the AIE-type metal NCs [35]. The solution of $Au_4Ag_{13}(DPPM)_3(SR)_9$ is almost non-emissive (QY: $\sim 4.1 \times 10^{-5}$), whether in CH_2Cl_2 or MeOH, but its amorphization and crystallization are fluorescent. The QY value of crystallization and amorphization is 653-fold and 278-fold higher than that of the solution state, respectively. In the solution state, the weak interactions and random collisions of the solvent dissipate the exciton energy and reduce luminescence. In the crystalline state, however, the inherent tri-blade fan configuration and multiple C-H$\cdots\pi$ interactions effectively restrict the molecular vibrations and rotations. CIE sliver-based NCs, $[Ag_{22}(dppe)_4(SR)_{12}Cl_4]^{2+}$, were discovered in subsequent years by the Pradeep group. There exist multiple C−H$\cdots\pi$ and $\pi\cdots\pi$ interactions between phenyl groups in these NCs' crystals, enhancing their luminescence in the crystalline state by 12 times compared to the amorphous state [36]. Two examples of CIE copper-based NCs, $Cu_2(C_3H_3N)_6I_2$ (1) and $Cu(C_9H_8N)_3P$ (2), are shown in Figure 4e. After grinding, the intermolecular $\pi\cdots\pi$ interactions between adjacent ligands were breached, leading to a 42.2% and 13.4% decrease in QY of complexes 1 and 2, respectively [37]. Recently, Bakr et al. have observed a crystallization-induced emission enhancement phenomenon in $[Cu_{15}(PPh_3)_6(PET)_{13}]^{2+}$ (PET = 2-phenylethanthiol, PPh_3 = triphenylphosphine) [38]. When Cu NCs are dissolved, they present almost non-luminescence; however, crystallization makes the molecules luminescent (Figure 4f). The QY value of the crystalline state is 3.2%, which is much higher than that of the solution state (0.1%). It is worth pointing out that the "tri-blade fan" structure configuration plays an important effect in controlling the structural rigidity. In addition, the compact C-H$\cdots\pi$ and $\pi\cdots\pi$ interactions of ligands jointly contributed to the AIE performance.

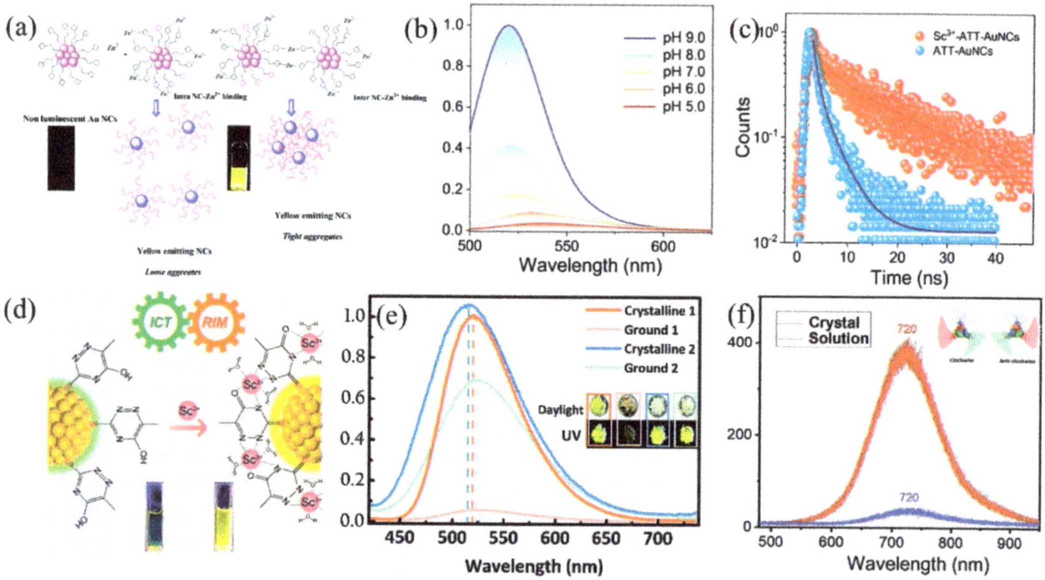

Figure 4. (**a**) Graphic illustration of self-assembly based on coordination effect of Zn^{2+} ions with different binding levels. Reprinted/adapted with permission from Ref. [32]. 2017, copyright Royal Society of Chemistry. (**b**) Photoemission spectra of ATT-AuNCs in different pH buffers. (**c**) Photoluminescence decay of ATT-AuNCs and Sc^{3+}-ATT-AuNCs. (**d**) The binding model of Sc^{3+} to ATT-AuNCs. Reprinted/adapted with permission from Ref. [33]. 2022, American Chemical Society. (**e**) PL spectra of crystalline and ground samples of complexes 1 and 2. Reprinted/adapted with permission from Ref. [37]. 2019, copyright Royal Society of Chemistry. (**f**) PL spectra of $[Cu_{15}(PPh_3)_6(PET)_{13}]^{2+}$ NCs in crystalline and solution state. Reprinted/adapted with permission from Ref. [38]. 2021, copyright Wiley-VCH.

2.4. pH-Induced AIE

The pKa of ionizable groups strongly affects the structure and activity of molecules, which determines the state of the complex in water: solution, gel, or suspension. In other words, the aggregation of metal NCs would be pH-dependent in most cases. Bimetallic Au-Cu GSH NCs (GSH: glutathione reduced), for example, show a typical pH-dependent AIE phenomenon [39]. They are weakly fluorescent in neutral or alkaline EtOH/aqueous solution, whereas regulating the pH < 5 produces intense radiation at 584 nm. This aggregation is assigned to the COO^- and NH^{3+} interactions between electron-donating groups in a lower pH. However, in neutral or alkaline conditions, negative charges exhibit strong electrostatic repulsion, resulting in the dispersion of clusters. As a result of structural similarity, the pH-induced AIE phenomenon of GS/C-Au NCs (GS/C: glutathione and citrate) was discovered by Tao's group, which benefits from the changes in surface charge density [40]. In those NCs, three noticeable features were observed in the process of regulating pH: (i) the luminescence intensity of GS/C-Au NCs was positively correlated with their surface charge density at 4.1 < pH < 5.6; (ii) when pH > 8.66, the luminescence intensity began to decrease due to the formation of carboxylate groups; (iii) when pH > 9.12, the Au(I)···Au(I) interaction was destroyed, because the bonds between the thiol group and Au(I) begin to dissociate, resulting in a further decrease in luminescence. Huang et al. explored the AIE mechanism by adjusting the pH to determine the charge state of ligands on Cu NCs [41]. As the pH was adjusted to alkaline, the structure of Cu NCs changed from the aggregate to the solution state, and the luminescence performance decreased significantly. The high pH makes the ligand of Cu NCs form a deprotonated state (COO−), which suppresses the formation of aggregates. Similarly, Ag NCs have been verified to achieve aggregation-induced emission based on the same mechanism. A typical case was exemplified by Zhou et al. in 2018, who demonstrated that the weak emission in TSA-Ag NCs (TSA: thiosalicylic acid) was dramatically improved in an acidic environment [42]. In summary, the pH-induced AIE offers a simple and efficient route to optimize the performance of metal NCs.

2.5. AIE Inheritance from Ligands

Using the AIE properties of organic ligands to endow clusters with AIE activity is an ordinary and remarkable strategy to construct AIE-type metal NCs. 3,5-dimethyl4-(4-(1,2,2-triphenylvinyl)benzyl)-1H-pyrazole (DTBP), a typical AIE luminogen, was selected as an organic ligand for the synthesis of the polymeric clusters $M_3(DTBP)_3$ (M = Cu, Ag or Au) by Mak and co-workers [43]. These clusters showed similar blue-green emission and lifetimes and a strong AIE response in solution. In the solid state, a 7.55-fold QY boosting was found (from 1.59% to 12.01%). From the perspective of structure, each pyrazolyl group connects two metal ions by μ_2–η_1, η_1 ligation type, which constructs a stable triangular structure. Moreover, molecular layers are stacked in an ordered arrangement, with an interlayer spacing of Cu_3, Ag_3, and Au_3 of 3.37, 3.26, and 3.45 Å, respectively (Figure 5a). A stable structure restricts the intermolecular motions, thus reducing the non-radiative decay and improving AIE intensity. 4-(3,5-dimethyl-1H-pyrazol-4-yl) benzaldehyde (HL) was selected as a ligand to form clusters, because its structure is similar to DTBP [44]. Cu^I-HL exhibits AIE behaviors with a 20-fold fluorescent enhancement when it aggregates, which is ascribed to the multiple intermolecular hydrogen bonds and rigid structure. Very recently, Perruchas et al. prepared two $[Cu_4I_4L_4]$ copper iodide clusters, with $PPh_2(C_6H_4$-$CH_2OH)$ and $PPh_2(C_3H_7)$ as ligands, showing AIE photophysical behaviors in solution and obviously enhanced luminescence intensity in the solid state [45]. Interestingly, the decay lifetime was significantly prolonged, and the biexponential was transformed to monoexponential with the formation of a solid-state phase, which can be ascribed to the suppression of non-radiative relaxation in a rigid state.

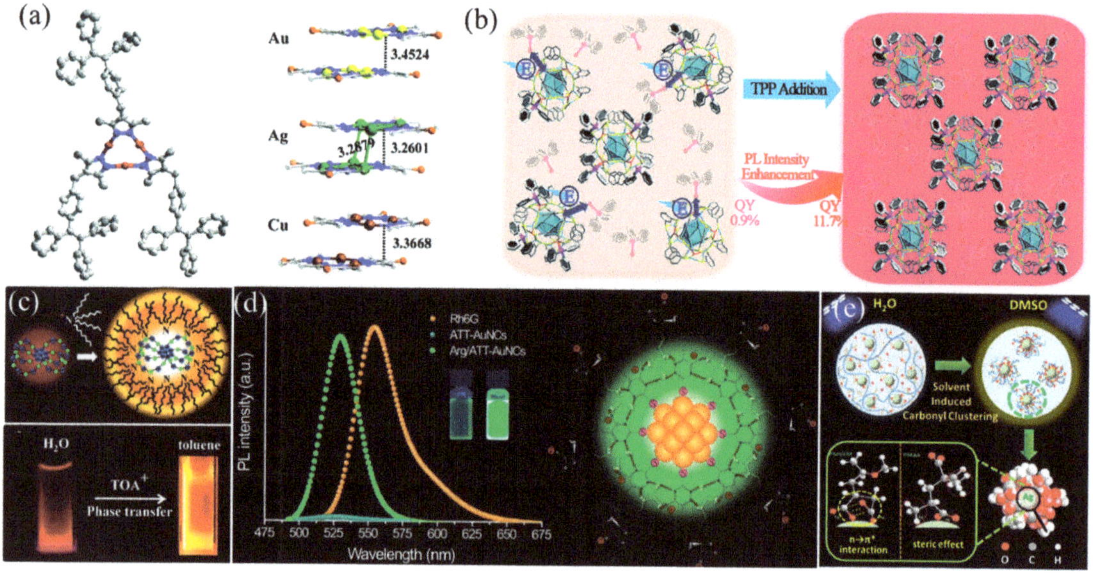

Figure 5. (**a**) Schematic representing the crystal structure of M_3 and interlayer stacking structures of Au_3, Ag_3, and Cu_3. Reprinted/adapted with permission from Ref. [43]. 2019, copyright Royal Society of Chemistry. (**b**) Mechanism model of TPP dissociation-aggregation process. Reprinted/adapted with permission from Ref. [46]. 2018, copyright Royal Society of Chemistry. (**c**) Mechanism model of binding TOA to $Au_{22}(SG)_{18}$ and a digital photograph of $Au_{22}(SG)_{18}$ and TOA-Au_{22}. Reprinted/adapted with permission from Ref. [10]. 2015, copyright American Chemical Society. (**d**) Mechanism model of the Arg-mediated ATT-AuNCs. Reprinted/adapted with permission from Ref. [47]. 2017, copyright American Chemical Society. (**e**) Mechanism model of the solvent-induced stage of Ag NCs. Reprinted/adapted with permission from Ref. [48]. 2017, copyright American Chemical Society.

2.6. Surface Constraint-Induced AIE

As a key component of AIE-type metal NCs, the restraining action of surface ligands can give a more rigidified molecular conformation, thereby preventing energy loss and maximizing emission efficiency. Thus, the better AIE performance of clusters could be triggered by adding mutually constrained ligands. A representative example is $Ag_{29}(BDT)_{12}(TPP)_4$ NC (BDT: 1,3-benzenedithiol; TPP: triphenylphosphine), reported by Zhu et al. in 2018, which exhibited obvious red fluorescence [46]. However, adding additional TPP in the DMF of $Ag_{29}(BDT)_{12}(TPP)_4$ can significantly increase the luminous intensity (about 13-fold, from 0.9% to 11.7%) (Figure 5b). The main reason for this phenomenon is that the TPP dissociation–aggregation stage on the cluster surface is prevented, due to the excess TPP ligands. The as-paired ligands/molecules rigidify the out-shell ligands and motifs to reduce nonradiative relaxation decay in LMCT and LMMCT processes. Based on the strong intermolecular and ion-pairing forces between glutathione (GSH) and T-tetraoctylammonium (TOA), the Lee group achieved rigidity of the $Au_{22}(SG)_{18}$ shell (Figure 5c) [10]. The measured QY value for the Au_{22}-TOA NCs was 62%, which is nine-fold higher than that of Au_{22} NCs (~7%). The better fluorescence behavior in Au_{22}-TOA is attributed to the suppressed nonradiative relaxations by TOA cations, which is demonstrated by the extended decay lifetime (from 380 ns to 2.44 μs). In 2017, the Chen group incorporated L-arginine (Arg) on 6-aza-2-thiothymine (ATT)-protected Au NCs to yield supramolecular host–guest assemblies [47]. As shown in Figure 5d, a strong hydrogen-bonding interaction was formed based on the guanidine group of Arg and ATT, keeping these clusters' surface rigid and suppressing the intramolecular rotation and vibration of ATT. As a result, the Arg/ATT-Au NCs show an excellent QY (65%), which is much better than that (1.8%) of ATT-Au NCs.

Endowing ligands with multiple interaction sites to anchor metal cores can effectively enhance the structural rigidity of metal NCs. Zhang et al. obtained two Ag(I) carbonyl clusters: PMVEM-Ag NCs and PMAA-Ag NCs (PMVEM: polymethyl vinyl ether-alt-maleic acid; PMAA: poly(methacrylic acid)), which showed distinct AIE photophysical behaviors [48]. Both PMVEM-Ag NCs and PMAA-Ag NCs show a short decay lifetime (1.0 ns) and poor QY values (1.0%) in water solution. The precise control of the aggregation degree is realized by adding dimethyl sulfoxide (DMSO) as a solvent. For PMVEM-Ag NCs, as DMSO is gradually added to water, the spectral peak gradually splits from a broad emission band (500 nm) to two peaks (460 and 530 nm). Moreover, the intensities of these two new peaks are enhanced by 3-fold and 54-fold, respectively. In particular, the 530 nm emission peak shows a long decay lifetime (97.1 μs) and higher QY (40%), which is different from the intrinsic fluorescence. Hydrogen bond evolution is the main factor affecting the aggregation of carbonyl groups. As shown in Figure 5e, in water, the interaction of carbonyl with a metal core is significantly weaker than that of water and carboxyl, which inhibits the aggregation of carbonyl. Meanwhile, in the DMSO solution, the strong hydrogen bond interactions are weakened, creating dense aggregation of the carbonyl groups on the surface of the metal nucleus, finally resulting in a strong electron delocalization effect. For PMAA-Ag NCs, DMSO fails to achieve delocalized electrons' conjugation between carbonyl groups because of the large steric hindrance effect of methyl groups in PMAA.

2.7. Mineral-Confined AIE

Minerals, with unique topology, uniform micro-pores, and abundant acid–base sites, can effectively anchor metal NCs' surface ligands to improve the AIE. Zeolites are an excellent candidate, because their spatial structure can be easily adjusted by changing coordination properties, total charge, and surface counter-ions. Inducing silver NCs into the rigid cavities of zeolites to trigger excellent luminescence properties has been an effective strategy. As an example, $(Ag_{0.5}Na_6)^+[Al_{6.5}Si_{17.5}O_{48}]$ (referred to as FAUY[$Ag_{0.5}$]) was prepared by calcining the combined parent zeolite and silver nitrate aqueous solution at 450 °C in the air [49]. The PLQY rises to near unity based on the regulation of non-framework metal cations, which is the highest QY of Ag clusters reported so far in the zeolite framework. $Ag_4(H_2O)_x^{2+}$ was presented by Marta and co-workers in 2018; its structure is shown in Figure 6a [50]. This structure produces long decay lifetime triplet emission from a superatom quantum system constructed by hybridized silver atoms and oxygen orbitals. Lievens et al. studied this material in detail, including the crystal structure and the ultrafast electron dynamics [51]. Temperature-induced emission tuning of silver-loaded zeolite from green to white was reported by the Li group in 2019 [52]. It was clear that the luminescence color was related to the Ag state: green emission comes from Ag_3^{n+}, and white emission comes from $(Ag_2)^+$ and $(Ag^+)_2$.

Layered double hydroxides (LDHs) with adjustable interlayer space can anchor guest molecules (such as metal NCs) in periodic long-range ordered arrays. The confinement of LDH can promote aggregation and restrict the movement of clusters, further enhancing the electron–hole pairs' recombination efficiency and fluorophore luminescence intensity. The first report of 2D LDH-confined gold NCs dates back to 2015, when $(\text{Au NCs/LDH})_n$ films were obtained by a layer-by-layer assembly process (Figure 6b) [53]. Much stronger QY (from 2.69 to 14.11%) and a longer average decay lifetime (from 1.84 to 14.67 μs) were observed for $(\text{Au NCs/LDH})_n$ complexes. In detail, the COO^- group in Au NCs was adsorbed by LDH nanosheets due to hydrogen bonds and host–guest electrostatic interactions, which inhibit the rotation and vibration of Au NCs, further resulting in boosted optical performance. In the following year, the same group constrained Au NCs using 2D exfoliated layered double hydroxide (ELDH) nanosheets to attain a QY of 19.05% [54]. Chromotropic acid (CTA) and LDH-confined Ag NCs (CTA-AgNCs/LDH) are another representative case of mineral-confined AIE of metal NCs, reported by Jin et al. in 2016 [55].

Based on the confinement effect of host–guest interactions, the CTA-AgNCs/LDH shows a high QY of 12.08% with an emission peak at 565 nm, which is much better than CTA-Ag NCs with a QY of 2.15% at 540 nm. Moreover, the Jia group observed a 14-fold QY improvement (from 0.11 to 14.27%) after achieving the assembly of Cu NCs and LDH [56]. The improved fluorescence intensity was attributed to the electron–hole being trapped in the quantum well structure of LDH.

2.8. MOF-Confined AIE

Metal-organic frameworks (MOFs) have proven to be an excellent carrier for metal NCs due to their ideal pore size, high loading capacity, abundant surface ligands, and cluster-based metal nodes. Metal NCs were rigidified in MOFs, which effectively limited nonradiative transition caused by intramolecular vibration and rotation, finally generating intense luminescence. A classical and interesting case presented by Zang et al. demonstrated the anchoring effects of MOF well [57]. By using 4,4′-bipyridine (bpy) to replace the CH_3CN ligands, $[(Ag_{12}(S^tBu)_8(CF_3COO)_4(bpy)_4)]_n$ (Ag_{12}bpy) with high stability (>1 year) and QY (12.1%) was constructed. From the perspective of structure, the bidentate ligands of bpy act as bridges to link Ag_{12} clusters to form a double-layer structure in the *a* and *b* directions. Moreover, the double-layer structure is stacked in the c direction, forming an ordered arrangement and a structurally rigidified composite framework (Figure 6c). The enhanced luminescence was achieved mainly through the following factors: (i) inhibited nonradiative relaxation decay; (ii) improved metal Au^I to ligand bpy charge transfer (MLCT), and (iii) improved ligand (S, O) to ligand (bpy) charge transfer (LLCT). Further, Zang introduced a $-NH_2$ group based on Ag_{12}bpy to obtain a fluorescence-phosphorescence dual-emission cluster [58]. The 456 nm blue emission originates from the amino group, and the yellow emission at 556 nm comes from the phenyl groups. The lone-pair electrons of $-NH_2$ strengthen spin–orbit coupling, resulting in the phosphorescence lifetime increasing from 0.2 μs to 3.12 ms. In 2019, the Tang group reported an Ag cluster (1⊃∩DMAC, DMAC = dimethylacetamide) from the self-assembly of the AIEgens 1,1,2,2-tetrakis(4-(pyridin-4-yl)phenyl)-ethene (tppe) and silver chalcogenolate cluster [59]. Interestingly, 1⊃∩DMAC showed a fluorescence color transition from blue (470 nm) to green (532 nm) when exposed to the atmosphere. Moreover, the fluorescence could return to blue immediately with treatment with DMAC. The free rotation of conjugated groups (e.g., phenyl and pyridyl) being restricted in the MOF is the fundamental reason for this reversible fluorescence change.

Among various MOFs, Zeolitic Imidazolate Framework-8 (ZIF-8) is considered a promising candidate for encapsulating metal NCs. Two representative examples are $Au_{25}(SG)_{18}$@ZIF-8 and $Au_{25}(SG)_{18}$/ZIF-8, reported by the Shi group in 2018 [60]. The $Au_{25}(SG)_{18}$@ZIF-8 was prepared by intercalating $Au_{25}(SG)_{18}$ into the ZIF-8, and $Au_{25}(SG)_{18}$/ZIF-8 was achieved by impregnating $Au_{25}(SG)_{18}$ NCs on the surface of ZIF-8 (Figure 6d). For $Au_{25}(SG)_{18}$@ZIF-8, the orientations of thiolate ligands were strictly restrained, thus showing similar luminescence as $Au_{25}(SG)_{18}$ solids. Concretely, it exhibited red emission centering at 680 nm and a weak infrared emission. For $Au_{25}(SG)_{18}$/ZIF-8, its fluorescence resembled $Au_{25}(SG)_{18}$ NCs′ solution, with a broad near-infrared emission (700–800 nm) accompanying a 665 nm red emissive. This behavior was attributed to the coordination effect between Zn^{2+} ions and surface carboxyl groups. Other representative examples of ZIF-8-based NCs have also been reported, such as Ag NCs-BSA@ZIF-8 [61], Ag NCs/ZIF-8 [62], Au NCs/ZIF-8 [63], etc.

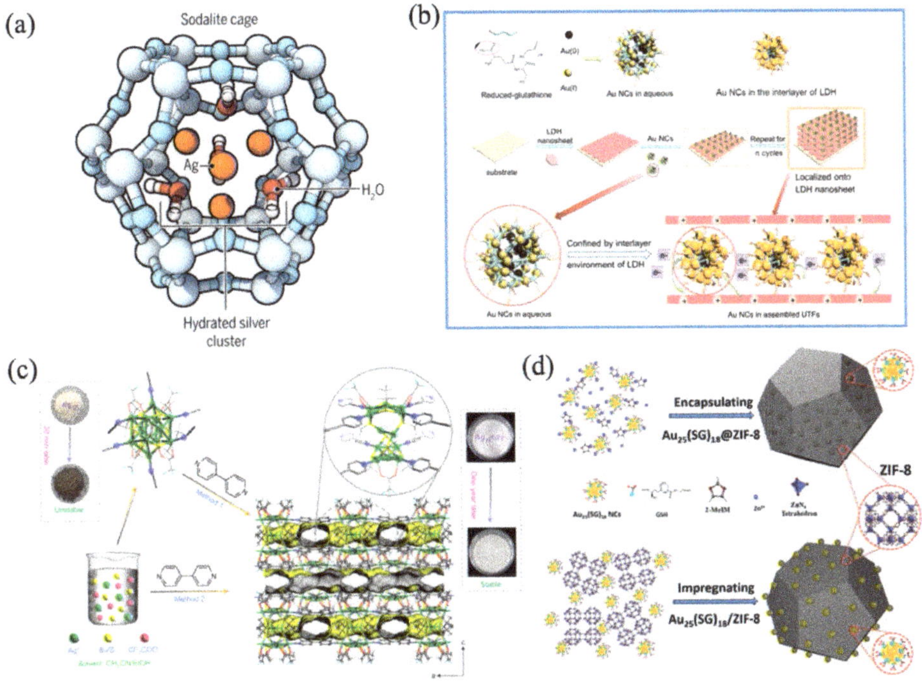

Figure 6. (**a**) Diagram of a hydrated Ag_4 NC-contained sodalite zeolite cage. Reprinted/adapted with permission from Ref. [50]. 2018, copyright American Association for the Advancement of Science. (**b**) Schematic representation of AuNCs' luminescence enhancement. Reprinted/adapted with permission from Ref. [53]. 2015, copyright Wiley-VCH. (**c**) Mechanism model of the preparation of Ag_{12}bpy NC-based metal-organic framework. Reprinted/adapted with permission from Ref. [57]. 2017, copyright Springer Nature. (**d**) Mechanism model of the synthesis processes for $Au_{25}(SG)_{18}$@ZIF-8 and $Au_{25}(SG)_{18}$/ZIF-8 nanocomposites. Reprinted/adapted with permission from Ref. [60]. 2017, copyright Wiley-VCH.

3. Applications of AIE-type metal NCs

AIE-type metal NCs retain both the excellent properties of metal NCs and AIE activity, attracting a large amount of research on cutting-edge applications. Over the past decade, plentiful sensory systems and optoelectronic devices have been studied based on AIE-type metal NCs, including bioimaging, photodynamic therapy, light-emitting diodes (LEDs), detection assays, circularly polarized luminescence (CPL), etc. In this section, we will briefly describe the recent developments in this field.

3.1. Bioimaging

The high emission efficiency of AIE-type metal NCs (especially Au NCs), combined with superior biocompatibility and little cytotoxicity, make them ideal candidates for bioimaging. Among them, peptides and alkaloid-capped Au NCs are promising [64–68]. For example, the Cheng group showed the potential in deep tissue visualization by employing NIR-II $Au_{25}(SG)_{18}$ (SG: glutathione) as translatable probes [69]. The $Au_{25}(SG)_{18}$ significantly combines with hydroxyapatite (HA) of the bone matrix and shows the bone structure in vivo in high resolution and contrast. Due to the ultrasmall size, these Au NCs could be rapidly excreted and hardly accumulate in the liver and spleen. Qu and coworkers designed nucleic acid-driven AIE-type Au NCs for telomerase detection with favorable sensitivity [70]. As shown in Figure 7a, strand A was modified on the surface of Au NCs through sulfhydryl groups and hybridized with telomerase substrate oligonucleotide (TS primer). In addition, the hairpin structure will be opened and connected to strand

B-modified Au NCs under the impact of telomerase, resulting in the aggregation of Au NCs and thus enhanced fluorescence. In this vein, significantly, the in situ visualization of telomerase activity in cells and in vivo can be achieved.

Based on pH sensitivity, Au NCs can be used to sense and monitor intracellular pH [71]. Cao et al. confined Au NCs in MOFs to hold aggregation and limit the rotation of ligands to enhance luminescence. Moreover, this complex was applied for real-time monitoring of drug release in vivo due to its pH-dependent luminescence (Figure 7b) [63]. Au NCs-MOF is stable in neutral and alkaline environments but quickly decomposes in acidic environments, resulting in reduced luminescence. Au NCs-MOF can transport drugs (e.g., CPT: camptothecin) into cancer cells. Drugs and Au NCs escaped from the MOF due to acidic environments (pH = 5–6) when CPT@Au NCs-MOF was close to cancer cells. Thus, drug release can be monitored in real time via Au NCs luminescence. Ligand-functionalized Au NCs can be used to target intracellular high reactive oxygen species (hROS, e.g., •OH, ClO^-, and $ONOO^-$) to prevent oxidative stress diseases [72]. Au NCs protected by quaternary ammonium with oligopeptides as a linker have been developed by Jiang et al. to be utilized for intracellular imaging (Figure 7c) [73]. The AuNCs became smaller, and their valence changed from Au(0) to Au(I) after the hROS effect, which caused the loss of fluorescence. To improve sensitivity and selectivity, the Quan group developed dual-emission ratiometric fluorescent probes, CNC@GNCs RFP, consisting of Au NCs, cellulose nanocrystals (CNCs), and nonluminous o-phenylenediamine (o-PD) [74]. Au NCs emit blue fluorescence, and o-PD can be oxidized through hROS to form 2, 3-diaminophenazine (o-PDox) to emit yellow fluorescence. Thus, based on the action of hROS, this probe can achieve an obvious color transition. Zebrafish experiments confirm the probe's potential for biological applications in both tissue penetrability and ROS responsiveness (Figure 7d–f).

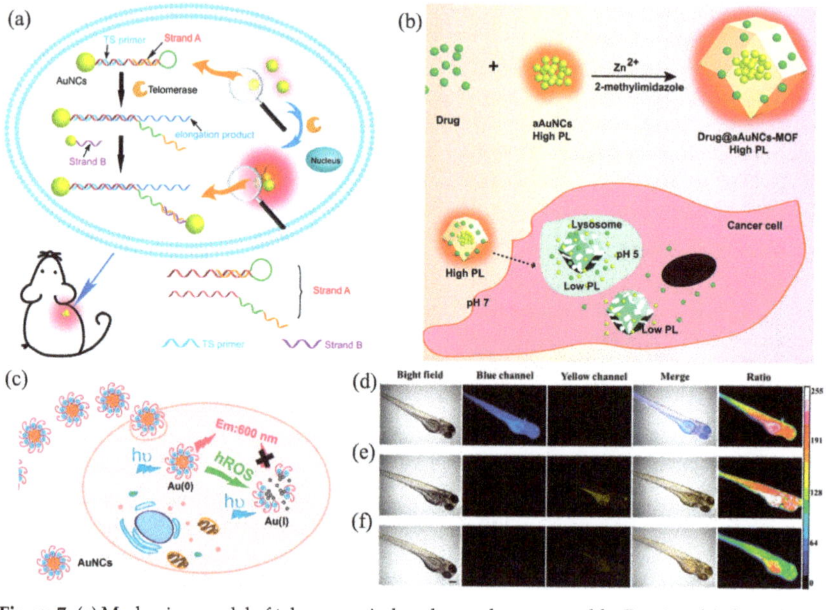

Figure 7. (**a**) Mechanism model of telomerase-induced nanocluster assembly. Reprinted/adapted with permission from Ref. [70]. 2021, copyright Royal Society of Chemistry. (**b**) Mechanism model of drugs encapsulated into Au NCs-MOF. Reprinted/adapted with permission from Ref. [63]. 2017, copyright Royal Society of Chemistry. (**c**) Mechanism model of the one-step preparation of red-emitting Au NCs. Reprinted/adapted with permission from Ref. [73]. 2018, copyright Wiley-VCH. (**d**–**f**) Selectivity of CNC@GNCs RFP for hROS in vivo of different Zebrafish. Reprinted/adapted with permission from Ref. [74]. 2022, copyright Elsevier.

3.2. Biological Diagnosis and Therapy

Besides bioimaging, AIE-type metal NCs are promising candidates for biological diagnosis and therapy, including antibacterial agents, cancer radiotherapy, and so on.

As early as 2013, the Xie group demonstrated that glutathione (GSH)-protected Ag NCs can generate a high concentration of reactive oxygen against the bacteria pseudomonas aeruginosa [75]. Furthermore, a higher antibacterial activity of Ag(I) than Ag(0) was revealed, and the corresponding Ag NCs have been applied against gram bacteria [76]. In 2016, antibiotic-grafting Ag NC technology was developed, raising the level of antibacterial ability (Figure 8a) [77]. D-Ag NCs can effectively destroy the bacterial membrane to form larger pores and continuously generate ROS, resulting in strong damage to bacterial DNA. There are also other antibacterial agents, such as Au NCs, which could interact with bacteria to cause an imbalance in bacterial cell metabolism, resulting in increased intracellular reactive oxygen species, and thereby killing bacteria [78–80]. Further research found that the more negatively charged Au NCs, the more ROS produced, allowing for a better bacterial killing efficiency (Figure 8b) [81].

The study of AIE-type metal NCs for the diagnosis and treatment of cancer is of great value for both basic research and practical applications [82,83]. GSH-$Au_{29-43}(SG)_{27-37}$ [12], GSH-Au_{25} [84], and GSH-$Au_{10-12}(SG)_{10-12}$ [85], for example, could increase tumor uptake and targeting specificity via an improved EPR effect. $Au_8(C_{21}H_{27}O_2)_8$ was developed by Zang et al. for a radiosensitizer (Figure 8c) [86]. $Au_8(C_{21}H_{27}O_2)_8$ produced ROS after X-ray exposure, causing irreversible cell apoptosis. Through rational structural design, AIE-type metal NCs can be used for photodynamic therapy (PDT) to achieve more effective anticancer treatments. Liu and coworkers have designed amine-terminated, PAMAM dendrimer-encapsulated Au NCs, which can consume H_2O_2 through catalase in the physiological pH range [87]. The possible mechanism was attributed to the fact that tertiary amines are easily protonated in acidic solutions, resulting in pre-adsorption of OH on the metal surface, thereby promoting catalase-like reactions. NIR-II-triggered PDT shows significantly increased tissue penetration, bypassing the limitations of conventional PDT. In the alkyl-thiolate Au NC@HSA/CAT system (HAS: serum albumin and CAT: catalase), the Au NCs can generate singlet oxygen to trigger PDT under 1064 nm laser excitation. Further, the HSA can improve the physiological stability of the nanoparticles. CAT is also used to enhance the efficacy of PDT by triggering the decomposition of tumor endogenous H_2O_2 into oxygen (Figure 8d–e) [88]. In addition, other biomedical applications of Au NCs have also been explored (e.g., for Parkinson's disease treatment). N-isobutyryl-L-cysteine (L-NIBC)-protected Au NCs significantly reversed dopaminergic (DA) neuron loss in substantia nigra and striatum, preventing the fibrillation of α-Synuclein (α-Syn) [89].

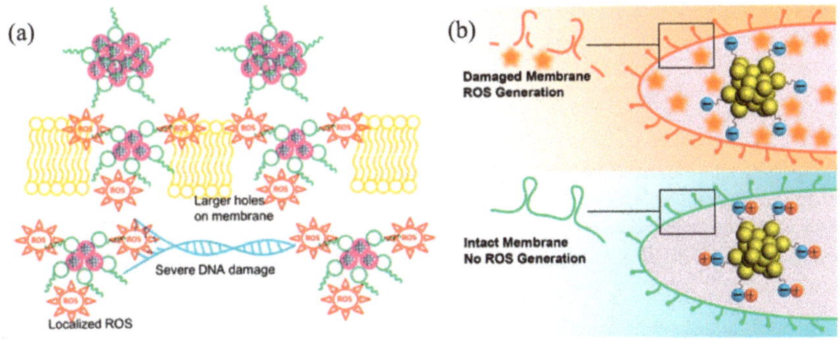

Figure 8. *Cont.*

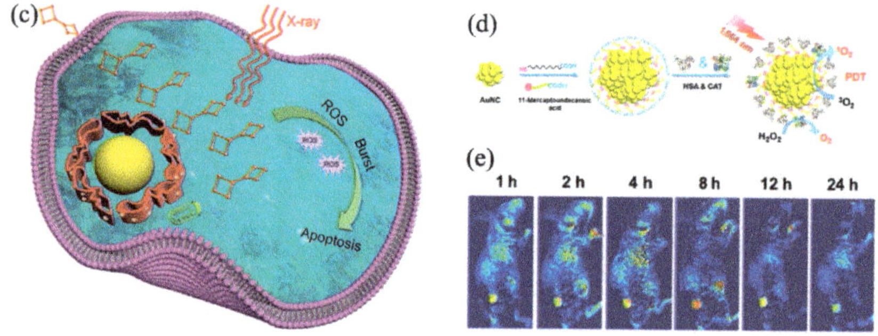

Figure 8. (**a**) Schematic illustration of D-Ag NCs damaging bacteria. Reprinted/adapted with permission from Ref. [77]. 2016, copyright American Chemical Society. (**b**) Influence model of surface properties of Au NCs on antibacterial activity. Reprinted/adapted with permission from Ref. [81]. 2018, copyright American Chemical Society. (**c**) Au_8NCs for cancer radiotherapy via the ROS burst. Reprinted/adapted with permission from Ref. [86]. 2019, copyright American Chemical Society. (**d**) Mechanism model of prepare Au NC@HSA/CAT nanoparticles. (**e**) In vivo fluorescence images of tumor-bearing nude mice taken at different time points. Reprinted/adapted with permission from Ref. [88]. 2018, copyright Springer Nature.

3.3. *Light-Emitting Diodes (LEDs)*

With the advantages of high luminescence efficiency, excellent durability, and low production costs, LEDs have led to innovations in the entire lighting industry. Normally, white light can be achieved by combining UV/n-UV/blue chips with polychromatic phosphors [90–93]. The AIE-type metal NCs are highly emissive in the solid state and are thus promising for LED phosphor applications [50,52]. Through ordered structure self-assembly technology, AIE NCs with emission colors can be easily designed and generated. A white device with color coordinates at (0.32, 0.36) was built by coating Cu and Au NCs on a 365 nm GaN LED chip [94]. Zn-coordinated glutathione-stabilized NCs (ZnGsH-Au NCs) have been reported for conversion layers. The device performance of its assembled LED showed (0.38, 0.38) of the CIE index and a CRI value of 75 [94]. CPL NCs can also be prepared for LED devices. For example, a circularly polarized light source was successfully realized based on copper(I)-iodide hybrid clusters [95]. Highly quantum-efficient (>95%) chiral silver clusters Ag_6L_6 were prepared into a WLED by combining 465 nm blue LEDs, which emitted nearly identical white emission as commercial lighting systems (blue chips + YAG: Ce^{3+}) [15].

Besides pc-WLED, the Zang group demonstrated that Au NCs (e.g., $(Au_4L_4)_n/(Au_4D_4)_n$) also have significant application potential in OLEDs [96]. $(Au_4L_4)_n/(Au_4D_4)_n$ was fabricated into multilayer OLED devices with the configuration poly(3,4-ethylenedioxythiophene): poly(styrenesulfonic acid) (PEDOT: PSS) (50 nm)/enantiomers (50 nm)/1,3,5-tris(N-phenyl-benzimidazol-2-yl)benzene (TPBi) (50 nm)/8-hydroxyquinolinolato-lithium (LiQ) (2 nm)/Al by a solution process. The OLEDs had a maximum EQE and dissymmetry factor of 1.5% and 1.12×10^{-2}, respectively. Koen Kenne's group reported ZEOLEDs based on silver-exchanged zeolites [97]. They showed that the key factors affecting the device performance were the concentration of metal NCs and the degree of zeolite anode contact. Last year, Li and coworkers introduced aggregation-enhanced $Cu_3(L_2)_3$ (L_2 = 4-hexyl-3,5-dimethyl-1Hpyrazole) acting as an emissive layer into OLEDs [29], where the OLED exhibited red emission centering at 610 nm, with a current efficiency (of 7.21 cd/A), a luminance (1200 cd/m^2), as well as EQE (4%).

3.4. *Detection Assays*

Because of the sensitivity of luminescence to the surrounding environment, AIE-type metal NCs can be used as fluorescent probes to detect metal ions, small organic molecules, polypeptides, amino acids, and proteins. For example, Au^+ ions have a high affinity for Hg^{2+} ions, which leads to the luminescence quenching of NCs [98]. Based on this principle, a series of clusters were prepared, such as lysozyme- [99], cytidine- [100], and protein-Au NCs [101] for the detection of Hg^{2+} ions, and the range of the detection limit was 0.5–60 nM. Similarly, metal NCs can detect other metal ions such as Cu^{2+}, Pb^{2+}, Ag^{+}, Fe^{3+}, and so on [102–105].

The luminescence of Au NCs heavily relies on the valence state of Au, thus enabling the detection of small molecules with redox properties. Chang et al. used cholesterol oxidase (ChOx) to reduce cholesterol to H_2O_2 and then quenched the luminescence of BSA-Au NCs, thereby allowing for the quantitative detection of cholesterol. This approach can detect cholesterol ranging from 1 to 100 μM, with a detection limit of 1.4 mM [106]. Glucose and doxycycline have also been detected through analogical means [107,108]. By impregnating the Au NCs into glycol-chitosan (GC) nanogel, Au NCs@GC was developed as an H_2S detector by Wang et al. in 2018 [109]. The confinement effect significantly improved sensitivity to aqueous sulfides. For amino acid and protein detection, metal NCs are suitable due to the interaction between the clusters and the detected object. The surface defects of Au NCs could be modified by cysteine to enhance luminescence, which allows for the quantitative detection of cysteine [89]. Based on the static quenching and inner filter effect (IFE), the fluorescent "on-off-on" of CQDs/Au NCs was achieved for detecting Cd^{2+} and L-ascorbic acid [110]. Protein enzymes can be detected through protein-Au NCs, because the luminescence signal intensity of Au NCs will change when enzymes degrade the protein templates [111,112].

3.5. Circularly Polarized Luminescence (CPL)

Polarization, as an intrinsic property of light, is a prerequisite for three-dimensional optical display, remote sensing, spin information optical communication, circular polarization tomography, and information encryption. Typically, CPL can be obtained when non-polarized light passes through a quarter-wave plate. However, energy loss and complex device structure restrict the popularization of this method. As an alternative efficient pathway, chiral luminescent materials were used to produce CPL directly [113]. The ideal CPL materials should have both a high quantum efficiency (QE) and a large g_{lum} value. The g_{lum} value means the luminescence dissymmetry factor, which can be defined by Equation (1) [114,115]:

$$g_{lum} = \frac{2(I_L - I_R)}{I_L + I_R} \tag{1}$$

where I_L and I_R represent the luminescence intensity of left- and right-CPL, respectively. A high QE indicates efficient energy conversion, whereas the performance of common organic compound-based emitters is always poor due to the ACQ effect [116,117]. Fortunately, the advent of AIE materials opens a window for CPL studies [118,119]. With the researchers' efforts, AIE-type metal NCs have become a member of the CPL materials family.

Recently, Yao et al. worked on a design for highly efficient chiral hybrid copper(I) halides NCs [95,120,121]. As shown in Figure 9a, R/S-$Cu_2I_2(BINAP)_2$ hybrid clusters with a layered conformation were designed and synthesized [121]. Benefiting from the biphosphine chelating coordination of chiral ligands, the stability of the chiral cluster was greatly improved and could be dissolved in polar solvents such as dimethyl sulfoxide (DMSO) to achieve solution processing. Further, with the assistance of polyvinylpyrrolidone (PVP), hexagonal flake-like microcrystals with a high g_{lum} (9.5×10^{-3}) were obtained in an ethanol solution based on intermolecular interactions (Figure 9b). On the basis above, Yao et al. synthesized high-performance CPL materials (PLQY: 32%, g_{lum}: 1.5×10^{-2}) with the biomimetic non-classical crystallization (BNCC) strategy [95]. The main crystallization process was revealed, involving nanocrystal nucleation, aggregation, oriented attachment, and mesoscopic transformation (Figure 9c). Moreover, electrostatic interactions and Van der Waals forces can induce these aggregated nanoparticles to assembly and chiral adjustment. Thus, chiral polycrystals with different morphologies can be achieved by adjusting the chiral center position, chain length, and concentration of ligands. In addition,

(R/S-MBA)$_2$CuCl$_2$ [122], (1-M/1-P)$_2$Cu$_2$I$_2$ [123], (R/S-MBA)$_4$Cu$_4$I$_4$ [124], and other chiral copper(I) halides also enrich the database of CPL materials.

Zang et al. incorporated (R/S)2,2-bis(diphenylphosphino)-1,1-binaphthyl on a BuSCu copper source to yield the CPL material R/S-Cu$_3$ [125]. In the aggregated state, R/S-Cu$_3$ displays notable circularly polarized luminescence (g_{lum}: 2×10^{-2}) due to the chiral structure and AIE feature. The luminescence comes from metal cluster-centered (MCC) and triplet metal-to-ligand charge-transfer (^{3}MLCT) processes and is limited by the intramolecular motion mechanism. Following that, in 2020, the authors further developed a copper cluster of [Cu$_{14}$(R/S-DPM)$_8$](PF$_6$)$_6$ (R/S-Cu$_{14}$) by modifying (R/S)-2-diphenyl-2-hydroxylmethylpyrrolidine-1-propyne ligands [126]. The R/S-Cu$_{14}$ dilute solution was non-luminescent and CPL inactive. However, by choosing n-hexane as the solvent, the copper clusters formed well-dispersed colloids of dense aggregates, showing strong red emission and CPL signals (g_{lum}: 1×10^{-2}), as shown in Figure 9d. This luminescence enhancement is attributed to electrostatic interaction, C−H···π, C−H···F, and O−H···F interactions of ligands, which could substantially restrict the intramolecular rotations and vibrations. Besides Cu NCs, ultrasmall Au NCs, (e.g., L/D-[Au$_4$(C$_9$H$_8$S$_2$N)$_4$], abbreviated as Au$_4$PL$_4$ and Au$_4$PD$_4$; and L/D-{[Au$_4$(C$_6$H$_{10}$S$_2$N)$_4$]$_3$}$_n$, abbreviated as (Au$_4$L$_4$)$_n$ and (Au$_4$D$_4$)$_n$, also have significant CPL characteristics [96]. Both Au$_4$PL$_4$/Au$_4$PD$_4$ and (Au$_4$L$_4$)$_n$/(Au$_4$D$_4$)$_n$ exhibited bright luminescence and were CPL active (g_{lum} = 7×10^{-3}) after adding poor solvent (H$_2$O), with the QE increasing by 6.5% and 4.7%, respectively (Figure 9e).

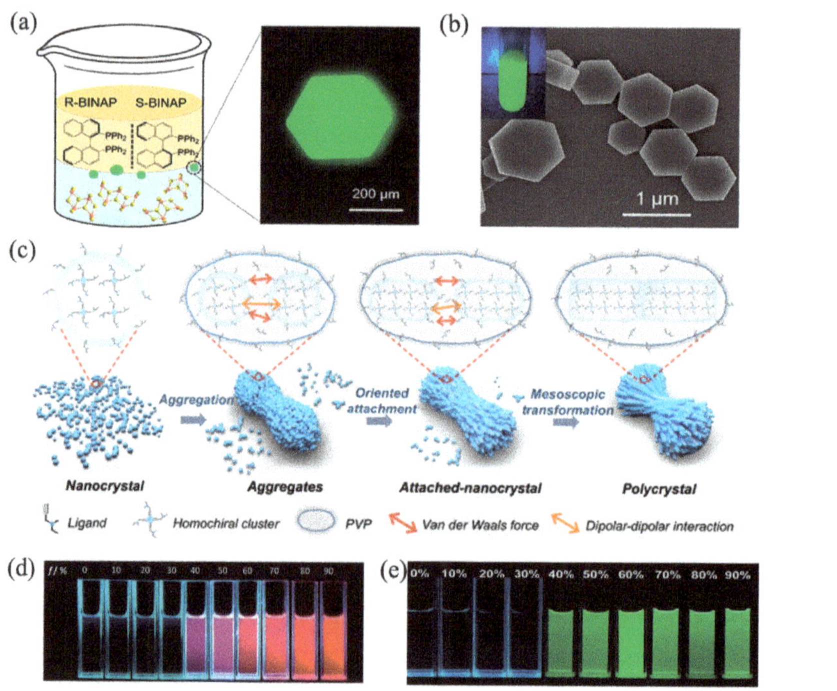

Figure 9. (**a**) Mechanism model of the single crystal growth process of R/S-Cu$_2$I$_2$(BINAP)$_2$ hybrid materials (left) and image of the R-Cu$_2$I$_2$(BINAP)$_2$ single crystal (right). (**b**) SEM image of R-Cu$_2$I$_2$(BINAP)$_2$ hexagon platelet-shaped microcrystals. Reprinted/adapted with permission from Ref. [121]. 2021, copyright American Chemical Society. (**c**) Mechanism model of the biomimetic non-classical crystallization process. Reprinted/adapted with permission from Ref. [95]. 2022, copyright Springer Nature. (**d**) Images of R-Cu$_{14}$ in different n-hexane solutions. Reprinted/adapted with permission from Ref. [126]. 2020, copyright Wiley-VCH. (**e**) Images of (Au$_4$L$_4$)$_n$ in dimethylformamide with 0–90% of H$_2$O. Reprinted/adapted with permission from Ref. [96]. 2020, copyright Springer Nature.

4. Conclusions and Outlook

Looking back, AIE metal NCs have taken the first step in terms of preparation techniques and practical applications. However, the current development of cluster chemistry has not yet reached an unambiguous agreement on the AIE fundamentals of metal NC luminescence. Moving forward, the research on AIE-type metal NCs will remain a scientific hotspot. Herein, we elaborated our perspectives, including but not limited to:

- Composition: Except for a few structural units for the synthesis of metal NCs, most organic ligands and noble metals (e.g., Au, Ag, and their alloys) are scarce and/or expensive. To accommodate large-scale production for commercial purposes, the development of low-cost metal NC materials is necessary. For instance, the AIE properties of copper, zinc, and other transition metal NCs are also worth designing for a breakthrough, because of their similar electronic configurations ($d^{10}s^{1}$) and wide range of oxidation states.
- Structure: Regarding the development of metal NCs' structural isomers, the following issues need to be addressed urgently in the future: (i) how to synthesize and stabilize novel structural isomers; (ii) how to rationally separate and structurally identify isomers. For this, DFT calculation, SXRD, ESI-MS, UV-vis-NIR absorption spectroscopy, and thin-layer chromatography techniques would help to address this issue; (iii) accurately mapping the internal relationship between structure and AIE properties at the molecular level, which will lay the foundation for prediction, clipping, and preparation of excellent AIE metal NCs.
- Property: The QY directly affects the future of metal NCs for application and commercialization. In particular, some red- and infrared-emitting metal NCs still have a large gap with commercial phosphors, which limits their application. Moreover, metal NCs easily decompose or lose their luminescent activity, exhibiting terrible thermo-stability and chemical stability under thermal and environmental stimuli. Some strategies, such as conferring strong chemical bond forces between the metal core and ligands, using MOF protection, etc., have improved stability. These measures are efficient to a certain extent but cannot solve all the problems of this issue.
- Mechanism: The current development of cluster chemistry has not yet reached an unambiguous agreement on the AIE fundamentals of metal NC luminescence, such as: (i) the correlation between metal NCs' structural characteristics and AIE behaviors; (ii) AIE concepts at the branch level in metal NCs such as aggregation-induced emission enhancement (AIEE) and clusterization-triggered emission (CTE); (iii) a clear identification of similar overlapping topics between AIE-type metal NCs and other metal counterparts. For example, the aggregation of metal nanoparticles supports the emission of adjacently located emitter moieties, which has been well-developed in systems such as cryosoret and soret colloid nano-assembly [127–130].
- Applications: To explore the application potential of AIE metal NCs in related biological and medical systems, the development of water-soluble AIE metal NCs with near-infrared emission or long afterglow luminescence is still in its infancy. In addition, there is an urgent need to develop metal NC-based optoelectronics devices to serve practical applications in various sectors. Last but not least, the preparation of high-quantum yield AIE metal NCs is necessary to develop their applications in the latest technologies (such as photonic crystal-coupled emission for biosensors).

The AIE mechanism has been widely accepted and continuously improved upon in the research community of luminescent metal NCs. However, quoting Laroche, to "have achieved unremittingly, indomitable than when subjected to failure is more important". AIE and/or metal NCs researchers still have a long way to go to witness the fundamentals and applications of AIE-type metal NCs. We are hopeful that this review will accelerate the advancement of such promising research, as well as provide models or springboards for many pending challenges in other cross-disciplines and research fields.

Author Contributions: Conceptualization, H.Z. and Z.W.; writing—T.L., H.Z. and Z.W.; funding acquisition, Z.W. All authors have read and agreed to the published version of the manuscript.

Funding: This work was supported by the financial support from the National Natural Science Foundation of China (12174151). This research received no external funding.

Institutional Review Board Statement: Not applicable.

Informed Consent Statement: Not applicable.

Data Availability Statement: The study did not report any data.

Acknowledgments: We gratefully acknowledge Ying Zhang from the Pediatric Respiratory Department, First Hospital of Jilin University for her significant discussion and contributions during writing and revision of this review.

Conflicts of Interest: The authors declare no conflict of interest.

References

1. Wu, Z.K.; Jin, R.C. On the Ligand's Role in the Fluorescence of Gold Nanoclusters. *Nano Lett.* **2010**, *10*, 2568–2573. [CrossRef]
2. Yuan, X.; Yu, Y.; Yao, Q.F.; Zhang, Q.B.; Xie, J.P. Fast Synthesis of Thiolated Au_{25} Nanoclusters via Protection-Deprotection Method. *J. Phys. Chem. Lett.* **2012**, *3*, 2310–2314. [CrossRef]
3. Wu, Z.N.; Yao, Q.F.; Zang, S.Q.; Xie, J.P. Directed Self-Assembly of Ultrasmall Metal Nanoclusters. *ACS Mater. Lett.* **2019**, *1*, 237–248. [CrossRef]
4. Zheng, J.; Lu, Z.; Wu, K.; Ning, G.H.; Li, D. Coinage-Metal-Based Cyclic Trinuclear Complexes with Metal-Metal Interactions: Theories to Experiments and Structures to Functions. *Chem. Rev.* **2020**, *120*, 9675–9742. [CrossRef]
5. Gan, Z.; Lin, Y.; Luo, L.; Han, G.; Liu, W.; Liu, Z.; Yao, C.; Weng, L.; Liao, L.; Chen, J.; et al. Fluorescent Gold Nanoclusters with Interlocked Staples and a Fully Thiolate-Bound Kernel. *Angew. Chem. Int. Ed. Engl.* **2016**, *55*, 11567–11571. [CrossRef]
6. Soldan, G.; Aljuhani, M.A.; Bootharaju, M.S.; AbdulHalim, L.G.; Parida, M.R.; Emwas, A.H.; Mohammed, O.F.; Bakr, O.M. Gold Doping of Silver Nanoclusters: A 26-Fold Enhancement in the Luminescence Quantum Yield. *Angew. Chem. Int. Ed.* **2016**, *55*, 5749–5753. [CrossRef]
7. Shang, L.; Xu, J.; Nienhaus, G.U. Recent Advances in Synthesizing Metal Nanocluster-based Nanocomposites for Application in Sensing, Imaging and Catalysis. *Nano Today* **2019**, *28*, 100767–100789. [CrossRef]
8. Kolay, S.; Bain, D.; Maity, S.; Devi, A.; Patra, A.; Antoine, R. Self-Assembled Metal Nanoclusters: Driving Forces and Structural Correlation with Optical Properties. *Nanomaterials* **2022**, *12*, 544. [CrossRef] [PubMed]
9. Wu, Z.N.; Liu, J.L.; Gao, Y.; Liu, H.W.; Li, T.T.; Zou, H.Y.; Wang, Z.G.; Zhang, K.; Wang, Y.; Zhang, H.; et al. Assembly-Induced Enhancement of Cu Nanoclusters Luminescence with Mechanochromic Property. *J. Am. Chem. Soc.* **2015**, *137*, 12906–12913. [CrossRef]
10. Pyo, K.; Thanthirige, V.D.; Kwak, K.; Pandurangan, P.; Ramakrishna, G.; Lee, D. Ultrabright Luminescence from Gold Nanoclusters: Rigidifying the Au(I)-Thiolate Shell. *J. Am. Chem. Soc.* **2015**, *137*, 8244–8250. [CrossRef]
11. Yu, Y.; Luo, Z.T.; Chevrier, D.M.; Leong, D.T.; Zhang, P.; Jiang, D.-e.; Xie, J.P. Identification of a Highly Luminescent $Au_{22}(SG)_{18}$ Nanocluster. *J. Am. Chem. Soc.* **2014**, *136*, 1246–1249. [CrossRef]
12. Zhang, X.D.; Luo, Z.; Chen, J.; Song, S.; Yuan, X.; Shen, X.; Wang, H.; Sun, Y.; Gao, K.; Zhang, L.; et al. Ultrasmall Glutathione-Protected Gold Nanoclusters as Next Generation Radiotherapy Sensitizers with High Tumor Uptake and High Renal Clearance. *Sci. Rep.* **2015**, *5*, 8669–8675. [CrossRef]
13. Zhang, H.; Liu, H.; Tian, Z.Q.; Lu, D.L.; Yu, Y.; Cestellos-Blanco, S.; Sakimoto, K.K.; Yang, P.D. Bacteria Photosensitized by Intracellular Gold Nanoclusters for Solar Fuel Production. *Nat. Nanotechnol.* **2018**, *13*, 900–905. [CrossRef]
14. Wu, Z.N.; Liu, H.W.; Li, T.T.; Liu, J.; Yin, J.; Mohammed, O.F.; Bakr, O.M.; Liu, Y.; Yang, B.; Zhang, H. Contribution of Metal Defects in the Assembly Induced Emission of Cu Nanoclusters. *J. Am. Chem. Soc.* **2017**, *139*, 4318–4321. [CrossRef]
15. Han, Z.; Dong, X.Y.; Luo, P.; Li, S.; Wang, Z.Y.; Zang, S.Q.; Mak, T.C.W. Ultrastable Atomically Precise Chiral Silver Clusters with More Than 95% Quantum Efficiency. *Sci. Adv.* **2020**, *6*, eaay0107–eaay0115. [CrossRef]
16. Chang, H.; Karan, N.S.; Shin, K.; Bootharaju, M.S.; Nah, S.; Chae, S.I.; Baek, W.; Lee, S.; Kim, J.; Son, Y.J.; et al. Highly Fluorescent Gold Cluster Assembly. *J. Am. Chem. Soc.* **2021**, *143*, 326–334. [CrossRef]
17. Luo, Z.T.; Yuan, X.; Yu, Y.; Zhang, Q.B.; Leong, D.T.; Lee, J.Y.; Xie, J.P. From Aggregation-Induced Emission of Au(I)−Thiolate Complexes to Ultrabright Au(0)@Au(I)−Thiolate Core−Shell Nanoclusters. *J. Am. Chem. Soc.* **2012**, *134*, 16662–16670. [CrossRef]
18. Wu, Z.N.; Yao, Q.F.; Zang, S.Q.; Xie, J.P. Aggregation-Induced Emission in Luminescent Metal Nanoclusters. *Natl. Sci. Rev.* **2021**, *8*, 208–210. [CrossRef]
19. Goswami, N.; Yao, Q.; Luo, Z.; Li, J.; Chen, T.; Xie, J. Luminescent Metal Nanoclusters with Aggregation-Induced Emission. *J. Phys. Chem. Lett.* **2016**, *7*, 962–975. [CrossRef]
20. Jin, Y.; Peng, Q.C.; Li, S.; Su, H.F.; Luo, P.; Yang, M.; Zhang, X.; Li, K.; Zang, S.Q.; Tang, B.Z.; et al. Aggregation-Induced Barrier to Oxygen-a new AIE Mechanism for Metal Clusters with Phosphorescence. *Natl. Sci. Rev.* **2022**, *9*, 216–224. [CrossRef]

21. Luo, J.; Xie, Z.; Lam, J.W.; Cheng, L.; Chen, H.; Qiu, C.; Kwok, H.S.; Zhan, X.; Liu, Y.; Zhu, D.; et al. Aggregation-induced Emission of 1-methyl-1,2,3,4,5-pentaphenylsilole. *Chem. Commun.* **2001**, *18*, 1740–1741. [CrossRef] [PubMed]
22. Leung, N.L.C.; Xie, N.; Yuan, W.; Liu, Y.; Wu, Q.; Peng, Q.; Miao, Q.; Lam, J.W.Y.; Tang, B.Z. Restriction of Intramolecular Motions: The General Mechanism behind Aggregation-Induced Emission. *Chem. Eur. J.* **2014**, *20*, 15349–15353. [CrossRef] [PubMed]
23. Wang, Y.; Li, H.; Wang, D.; Tang, B.Z. Inorganic–Organic Nanocomposites Based on Aggregation-Induced Emission Luminogens. *Adv. Funct. Mater.* **2020**, *31*, 2006952–2006969. [CrossRef]
24. Mei, J.; Hong, Y.; Lam, J.W.; Qin, A.; Tang, Y.; Tang, B.Z. Aggregation-induced Emission: The Whole is More Brilliant than the Parts. *Adv. Mater.* **2014**, *26*, 5429–5479. [CrossRef] [PubMed]
25. Zhang, H.; Zhao, Z.; Turley, A.T.; Wang, L.; McGonigal, P.R.; Tu, Y.; Li, Y.; Wang, Z.; Kwok, R.T.K.; Lam, J.W.Y.; et al. Aggregate Science: From Structures to Properties. *Adv. Mater.* **2020**, *32*, 2001457–2001471. [CrossRef]
26. Wu, Z.; Yao, Q.; Chai, O.J.H.; Ding, N.; Xu, W.; Zang, S.; Xie, J. Unraveling the Impact of Gold(I)-Thiolate Motifs on the Aggregation-Induced Emission of Gold Nanoclusters. *Angew. Chem. Int. Ed. Engl.* **2020**, *59*, 9934–9939. [CrossRef]
27. Maity, S.; Bain, D.; Patra, A. Engineering Atomically Precise Copper Nanoclusters with Aggregation Induced Emission. *J. Phys. Chem. C* **2019**, *123*, 2506–2515. [CrossRef]
28. Sugiuchi, M.; Zhang, M.; Hakoishi, Y.; Shichibu, Y.; Horimoto, N.N.; Yamauchi, Y.; Ishida, Y.; Konishi, K. Aggregation-Mode-Dependent Optical Properties of Cationic Gold Clusters: Formation of Ordered Assemblies in Solution and Unique Optical Responses. *J. Phys. Chem. C* **2020**, *124*, 16209–16215. [CrossRef]
29. Yang, H.; Zheng, J.; Xie, M.; Luo, D.; Tang, W.J.; Peng, S.K.; Cheng, G.; Zhang, X.J.; Zhou, X.P.; Che, C.M.; et al. Aggregation-Enhanced Emission in a Red Cu(I) Emitter with Quantum Yield >99%. *ACS Mater. Lett.* **2022**, *4*, 1921–1928. [CrossRef]
30. Wang, Z.P.; Zhu, Z.L.; Zhao, C.K.; Yao, Q.F.; Li, X.Y.; Liu, H.Y.; Du, F.L.; Yuan, X.; Xie, J.P. Silver Doping-Induced Luminescence Enhancement and Red-Shift of Gold Nanoclusters with Aggregation-Induced Emission. *Chem. Asian J.* **2019**, *14*, 765–769. [CrossRef]
31. Wang, S.X.; Meng, X.M.; Das, A.; Li, T.; Song, Y.B.; Cao, T.T.; Zhu, X.Y.; Zhu, M.Z.; Jin, R.C. A 200-fold Quantum Yield Boost in the Photoluminescence of Silver-Doped Ag_xAu_{25-x} Nanoclusters: The 13th Silver Atom Matters. *Angew. Chem. Int. Ed.* **2014**, *53*, 2376–2380. [CrossRef] [PubMed]
32. Kuppan, B.; Maitra, U. Instant Room Temperature Synthesis of Selfassembled Emission-tunable Gold Nanoclusters: Million-fold Emission Enhancement and Fluorimetric Detection of Zn^{2+}. *Nanoscale* **2017**, *9*, 15494–15504. [CrossRef] [PubMed]
33. Huang, K.Y.; Xiu, L.F.; Fang, X.Y.; Yang, M.R.; Noreldeen, H.A.A.; Chen, W.; Deng, H.H. Highly Efficient Luminescence from Charge-Transfer Gold Nanoclusters Enabled by Lewis Acid. *J. Phys. Chem. Lett.* **2022**, *13*, 9526–9533. [CrossRef] [PubMed]
34. Yuan, W.Z.; Shen, X.Y.; Zhao, H.; Lam, J.W.Y.; Tang, L.; Lu, P.; Wang, C.L.; Liu, Y.; Wang, Z.M.; Zheng, Q.; et al. Crystallization-Induced Phosphorescence of Pure Organic Luminogens at Room Temperature. *J. Phys. Chem. C* **2010**, *114*, 6090–6099. [CrossRef]
35. Chen, T.; Yang, S.; Chai, J.S.; Song, Y.B.; Fan, J.Q.; Rao, B.; Sheng, H.T.; Yu, H.Z.; Zhu, M.Z. Crystallization-Induced Emission Enhancement: A Novel Fluorescent Au-Ag Bimetallic Nanocluster with Precise Atomic Structure. *Sci. Adv.* **2017**, *3*, e1700956–e1700963. [CrossRef] [PubMed]
36. Khatun, E.; Bodiuzzaman, M.; Sugi, K.S.; Chakraborty, P.; Paramasivam, G.; Dar, W.A.; Ahuja, T.; Antharjanam, S.; Pradeep, T. Confining an Ag_{10} Core in an Ag_{12} Shell: A Four-Electron Superatom with Enhanced Photoluminescence upon Crystallization. *ACS Nano* **2019**, *13*, 5753–5759. [CrossRef]
37. Yang, M.; Chen, X.L.; Lu, C.Z. Efficiently Luminescent Copper(i) Iodide Complexes with Crystallization-Induced Emission Enhancement (CIEE). *Dalton Trans.* **2019**, *48*, 10790–10794. [CrossRef]
38. Nematulloev, S.; Huang, R.W.; Yin, J.; Shkurenko, A.; Dong, C.; Ghosh, A.; Alamer, B.; Naphade, R.; Hedhili, M.N.; Maity, P.; et al. $[Cu_{15}(PPh_3)_6(PET)_{13}]^{2+}$: A Copper Nanocluster with Crystallization Enhanced Photoluminescence. *Small* **2021**, *17*, e2006839–e2006844. [CrossRef]
39. Chakraborty, S.; Bain, D.; Maity, S.; Kolay, S.; Patra, A. Controlling Aggregation-Induced Emission in Bimetallic Gold–Copper Nanoclusters via Surface Motif Engineering. *J. Phys. Chem. C* **2022**, *126*, 2896–2904. [CrossRef]
40. Ding, W.C.; Liu, Y.; Li, Y.J.; Shi, Q.R.; Li, H.S.; Xia, H.B.; Wang, D.Y.; Tao, X.T. Water-soluble Gold Nanoclusters with pH-dependent Fluorescence and High Colloidal Stability Over a Wide pH Range via Co-reduction of Glutathione and Citrate. *RSC Adv.* **2014**, *4*, 22651–22659. [CrossRef]
41. Long, T.F.; Guo, Y.J.; Lin, M.; Yuan, M.K.; Liu, Z.D.; Huang, C.Z. Optically Active Red-emitting Cu Nanoclusters Originating from Complexation and Redox Reaction between Copper(ii) and d/l-penicillamine. *Nanoscale* **2016**, *8*, 9764–9770. [CrossRef] [PubMed]
42. Pan, S.; Liu, W.; Tang, J.; Yang, Y.; Feng, H.; Qian, Z.; Zhou, J. Hydrophobicity-guided Self-assembled Particles of Silver Nanoclusters with Aggregation-induced Emission and Their Use in Sensing and Bioimaging. *J. Mater. Chem. B* **2018**, *6*, 3927–3933. [CrossRef] [PubMed]
43. Tian, Y.; Wang, Z.Y.; Zang, S.Q.; Li, D.; Mak, T.C.W. Luminescent Cyclic Trinuclear Coinage Metal Complexes with Aggregation-Induced Emission (AIE) Performance. *Dalton Trans.* **2019**, *48*, 2275–2279. [CrossRef] [PubMed]
44. Li, H.; Luo, J.; Zhang, Z.Y.; Wei, R.J.; Xie, M.; Huang, Y.L.; Ning, G.H.; Li, D. Cyclic Trinuclear Copper(I) Complex Exhibiting Aggregation-Induced Emission: A Novel Fluorescent Probe for the Selective Detection of Gold(III) Ions. *Inorg. Chem.* **2022**, *61*, 414–421. [CrossRef] [PubMed]
45. Utrera-Melero, R.; Mevellec, J.Y.; Gautier, N.; Stephant, N.; Massuyeau, F.; Perruchas, S. Aggregation-Induced Emission Properties of Copper Iodide Clusters. *Chem. Asian J.* **2019**, *14*, 3166–3172. [CrossRef] [PubMed]

46. Kang, X.; Wang, S.; Zhu, M. Observation of a New Type of Aggregation-Induced Emission in Nanoclusters. *Chem. Sci.* **2018**, *9*, 3062–3068. [CrossRef]
47. Deng, H.H.; Shi, X.Q.; Wang, F.F.; Peng, H.P.; Liu, A.L.; Xia, X.H.; Chen, W. Fabrication of Water-Soluble, Green-Emitting Gold Nanoclusters with a 65% Photoluminescence Quantum Yield via Host–Guest Recognition. *Chem. Mater.* **2017**, *29*, 1362–1369. [CrossRef]
48. Yang, T.; Dai, S.; Yang, S.; Chen, L.; Liu, P.; Dong, K.; Zhou, J.; Chen, Y.; Pan, H.; Zhang, S.; et al. Interfacial Clustering-Triggered Fluorescence-Phosphorescence Dual Solvoluminescence of Metal Nanoclusters. *J. Phys. Chem. Lett.* **2017**, *8*, 3980–3985. [CrossRef]
49. Fenwick, O.; Coutino-Gonzalez, E.; Grandjean, D.; Baekelant, W.; Richard, F.; Bonacchi, S.; De Vos, D.; Lievens, P.; Roeffaers, M.; Hofkens, J.; et al. Tuning the Energetics and Tailoring the Optical Properties of Silver Clusters Confined in Zeolites. *Nat. Mater.* **2016**, *15*, 1017–1022. [CrossRef]
50. Quintanilla, M.; Liz-Marzan, L.M. Caged Clusters Shine Brighter. *Science* **2018**, *361*, 645–646. [CrossRef]
51. Grandjean, D.; Coutiño-Gonzalez, E.; Ngo Tuan Cuong, E.F.; Baekelant, W.; Aghakhani, S.; Schlexer, P.; D'Acapito, F.; Banerjee, D.; Roeffaers, M.B.J.; Nguyen, M.T.; et al. Origin of the Bright Photoluminescence of Few-atom Silver Clusters Confined in LTA Zeolites. *Science* **2018**, *361*, 686–690. [CrossRef]
52. Yao, D.C.; Xu, S.; Wang, Y.G.; Li, H.R. White-Emitting Phosphors with High Color-Rendering Index Based on Silver Cluster-Loaded Zeolites and Their Application to Near-UV LED-based White LEDs. *Mater. Chem. Front.* **2019**, *3*, 1080–1084. [CrossRef]
53. Tian, R.; Zhang, S.T.; Li, M.W.; Zhou, Y.Q.; Lu, B.; Yan, D.P.; Wei, M.; Evans, D.G.; Duan, X. Localization of Au Nanoclusters on Layered Double Hydroxides Nanosheets: Confinement-Induced Emission Enhancement and Temperature-Responsive Luminescence. *Adv. Funct. Mater.* **2015**, *25*, 5006–5015. [CrossRef]
54. Tian, R.; Yan, D.; Li, C.; Xu, S.; Liang, R.; Guo, L.; Wei, M.; Evans, D.G.; Duan, X. Surface-Confined Fluorescence Enhancement of Au Nanoclusters Anchoring to a Two-dimensional Ultrathin Nanosheet toward Bioimaging. *Nanoscale* **2016**, *8*, 9815–9821. [CrossRef]
55. Ren, H.; Li, M.Z.; Fu, Y.Y.; Jin, L. Silver Nanoclusters Functionalized by Chromotropic Acid and Layered Double Hydroxides for the Turn-on Detection of Melamine. *J. Mater. Chem. C* **2016**, *4*, 6104–6109. [CrossRef]
56. Yang, J.L.; Song, N.Z.; Jia, Q. Investigation of the Surface Confinement Effect of Copper Nanoclusters: Construction of an Ultrasensitive Fluorescence Turn-on Bio-enzyme Sensing Platform. *Nanoscale* **2019**, *11*, 21927–21933. [CrossRef]
57. Huang, R.W.; Wei, Y.S.; Dong, X.Y.; Wu, X.H.; Du, C.X.; Zang, S.Q.; Mak, T.C.W. Hypersensitive Dual-function Luminescence Switching of a Silver-Chalcogenolate Cluster-Based Metal-Organic Framework. *Nat. Chem.* **2017**, *9*, 689–697. [CrossRef]
58. Dong, X.Y.; Si, Y.; Yang, J.S.; Zhang, C.; Han, Z.; Luo, P.; Wang, Z.Y.; Zang, S.Q.; Mak, T.C.W. Ligand Engineering to Achieve Enhanced Ratiometric Oxygen Sensing in a Silver Cluster-Based Metal-Organic Framework. *Nat. Commun.* **2020**, *11*, 3678–3686. [CrossRef] [PubMed]
59. Wu, X.H.; Luo, P.; Wei, Z.; Li, Y.Y.; Huang, R.W.; Dong, X.Y.; Li, K.; Zang, S.Q.; Tang, B.Z. Guest-Triggered Aggregation-Induced Emission in Silver Chalcogenolate Cluster Metal-Organic Frameworks. *Adv. Sci.* **2019**, *6*, 1801304–1801310. [CrossRef]
60. Luo, Y.C.; Fan, S.Y.; Yu, W.Q.; Wu, Z.L.; Cullen, D.A.; Liang, C.L.; Shi, J.Y.; Su, C.Y. Fabrication of $Au_{25}(SG)_{18}$-ZIF-8 Nanocomposites: A Facile Strategy to Position $Au_{25}(SG)_{18}$ Nanoclusters Inside and Outside ZIF-8. *Adv. Mater.* **2018**, *30*, 1704576–1704584. [CrossRef]
61. Fan, C.; Lv, X.X.; Liu, F.J.; Feng, L.P.; Liu, M.; Cai, Y.Y.; Liu, H.; Wang, J.Y.; Yang, Y.L.; Wang, H. Silver Nanoclusters Encapsulated into Metal-Organic Frameworks with Enhanced Fluorescence and Specific Ion Accumulation toward the Microdot Array-Based Fluorimetric Analysis of Copper in Blood. *ACS Sens.* **2018**, *3*, 441–450. [CrossRef]
62. Feng, L.P.; Liu, M.; Liu, H.; Fan, C.; Cai, Y.Y.; Chen, L.J.; Zhao, M.L.; Chu, S.; Wang, H. High-Throughput and Sensitive Fluorimetric Strategy for MicroRNAs in Blood Using Wettable Microwells Array and Silver Nanoclusters with Red Fluorescence Enhanced by Metal Organic Frameworks. *ACS Appl. Mater. Interfaces* **2018**, *10*, 23647–23656. [CrossRef]
63. Cao, F.F.; Ju, E.G.; Liu, C.Q.; Li, W.; Zhang, Y.; Dong, K.; Liu, Z.; Ren, J.S.; Qu, X.G. Encapsulation of Aggregated Gold Nanoclusters in a Metal-organic Framework for Real-time Monitoring of Drug Release. *Nanoscale* **2017**, *9*, 4128–4134. [CrossRef] [PubMed]
64. Shang, L.; Azadfar, N.; Stockmar, F.; Send, W.; Trouillet, V.; Bruns, M.; Gerthsen, D.; Nienhaus, G.U. One-pot Synthesis of Near-infrared Fluorescent Gold Clusters for Cellular Fluorescence Lifetime Imaging. *Small* **2011**, *7*, 2614–2620. [CrossRef] [PubMed]
65. Shang, L.; Dorlich, R.M.; Brandholt, S.; Schneider, R.; Trouillet, V.; Bruns, M.; Gerthsen, D.; Nienhaus, G.U. Facile Preparation of Water-Soluble Fluorescent Gold Nanoclusters for Cellular Imaging Applications. *Nanoscale* **2011**, *3*, 2009–2014. [CrossRef]
66. Qiao, J.; Mu, X.Y.; Qi, L.; Deng, J.J.; Mao, L.Q. Folic Acid-Functionalized Fluorescent Gold Nanoclusters with Polymers as Linkers for Cancer Cell Imaging. *Chem. Commun.* **2013**, *49*, 8030–8032. [CrossRef] [PubMed]
67. Tao, Y.; Li, Z.H.; Ju, E.G.; Ren, J.S.; Qu, X.G. Polycations-Functionalized Water-Soluble Gold Nanoclusters: A Potential Platform for Simultaneous Enhanced Gene Delivery and Cell Imaging. *Nanoscale* **2013**, *5*, 6154–6160. [CrossRef] [PubMed]
68. Yu, Z.X.; Musnier, B.; Wegner, K.D.; Henry, M.; Chovelon, B.; Desroches-Castan, A.; Fertin, A.; Resch-Genger, U.; Bailly, S.; Coll, J.L.; et al. High-Resolution Shortwave Infrared Imaging of Vascular Disorders Using Gold Nanoclusters. *ACS Nano* **2020**, *14*, 4973–4981. [CrossRef] [PubMed]
69. Li, D.; Liu, Q.; Qi, Q.; Shi, H.; Hsu, E.C.; Chen, W.; Yuan, W.; Wu, Y.; Lin, S.; Zeng, Y.; et al. Gold Nanoclusters for NIR-II Fluorescence Imaging of Bones. *Small* **2020**, *16*, e2003851–e2003860. [CrossRef] [PubMed]

70. Ran, X.; Wang, Z.Z.; Pu, F.; Ju, E.G.; Ren, J.S.; Qu, X.G. Nucleic Acid-Driven Aggregation-Induced Emission of Au Nanoclusters for Visualizing Telomerase Activity in Living Cells and in Vivo. *Mater. Horiz.* **2021**, *8*, 1769–1775. [CrossRef]
71. Wang, J.X.; Goswami, N.; Shu, T.; Su, L.; Zhang, X.J. pH-Responsive Aggregation-Induced Emission of Au Nanoclusters and Crystallization of the Au(i)–thiolate Shell. *Mater. Chem. Front.* **2018**, *2*, 923–928. [CrossRef]
72. Wang, J.L.; Zhang, G.; Li, Q.W.; Jiang, H.; Liu, C.Y.; Amatore, C.; Wang, X.M. In Vivo Self-bio-imaging of Tumors Through in Situ Biosynthesized Fluorescent Gold Nanoclusters. *Sci. Rep.* **2013**, *3*, 1157–1162. [CrossRef] [PubMed]
73. Xie, Y.Z.Y.; Xianyu, Y.L.; Wang, N.X.; Yan, Z.Y.; Liu, Y.; Zhu, K.; Hatzakis, N.S.; Jiang, X.Y. Functionalized Gold Nanoclusters Identify Highly Reactive Oxygen Species in Living Organisms. *Adv. Funct. Mater.* **2018**, *28*, 1702026–1702032. [CrossRef]
74. Quan, Z.Y.; Xue, F.; Li, H.Y.; Chen, Z.P.; Wang, L.; Zhu, H.X.; Pang, C.L.; He, H. A Bioinspired Ratiometric Fluorescence Probe Based on Cellulose Nanocrystal-Stabilized Gold Nanoclusters for Live-Cell and Zebrafish Imaging of Highly Reactive Oxygen species. *Chem. Eng. J.* **2022**, *431*, 133954–133964. [CrossRef]
75. Yuan, X.; Setyawati, M.I.; Tan, A.S.; Ong, C.N.; Leong, D.T.; Xie, J.P. Highly Luminescent Silver Nanoclusters with Tunable Emissions: Cyclic Reduction–Decomposition Synthesis and Antimicrobial Properties. *NPG Asia Mater.* **2013**, *5*, e39–e46. [CrossRef]
76. Yuan, X.; Setyawati, M.I.; Leong, D.T.; Xie, J.P. Ultrasmall Ag^+-Rich Nanoclusters as Highly Efficient Nanoreservoirs for Bacterial Killing. *Nano Res.* **2013**, *7*, 301–307. [CrossRef]
77. Zheng, K.Y.; Setyawati, M.I.; Lim, T.P.; Leong, D.T.; Xie, J.P. Antimicrobial Cluster Bombs: Silver Nanoclusters Packed with Daptomycin. *ACS Nano* **2016**, *10*, 7934–7942. [CrossRef]
78. Zheng, K.Y.; Setyawati, M.I.; Leong, D.T.; Xie, J.P. Antimicrobial Gold Nanoclusters. *ACS Nano* **2017**, *11*, 6904–6910. [CrossRef]
79. Zheng, K.Y.; Setyawati, M.I.; Leong, D.T.; Xie, J.P. Antimicrobial Silver Nanomaterials. *Coord. Chem. Rev.* **2018**, *357*, 1–17. [CrossRef]
80. Chang, T.K.; Cheng, T.M.; Chu, H.L.; Tan, S.H.; Kuo, J.C.; Hsu, P.H.; Su, C.Y.; Chen, H.M.; Lee, C.M.; Kuo, T.R. Metabolic Mechanism Investigation of Antibacterial Active Cysteine-Conjugated Gold Nanoclusters in Escherichia Coli. *ACS Sustain. Chem. Eng.* **2019**, *7*, 15479–15486. [CrossRef]
81. Zheng, K.; Setyawati, M.I.; Leong, D.T.; Xie, J.P. Surface Ligand Chemistry of Gold Nanoclusters Determines Their Antimicrobial Ability. *Chem. Mater.* **2018**, *30*, 2800–2808. [CrossRef]
82. Goswami, N.; Luo, Z.T.; Yuan, X.; Leong, D.T.; Xie, J.P. Engineering Gold-Based Radiosensitizers for Cancer Radiotherapy. *Mater. Horiz.* **2017**, *4*, 817–831. [CrossRef]
83. Zhang, X.D.; Wu, D.; Shen, X.; Liu, P.X.; Fan, F.Y.; Fan, S.J. In Vivo Renal Clearance, Biodistribution, Toxicity of Gold Nanoclusters. *Biomaterials* **2012**, *33*, 4628–4638. [CrossRef]
84. Zhang, X.D.; Chen, J.; Luo, Z.T.; Wu, D.; Shen, X.; Song, S.S.; Sun, Y.M.; Liu, P.X.; Zhao, J.; Huo, S.S.; et al. Enhanced Tumor Accumulation of Sub-2 nm Gold Nanoclusters for Cancer Radiation Therapy. *Adv. Healthc. Mater.* **2014**, *3*, 133–141. [CrossRef]
85. Zhang, X.D.; Luo, Z.T.; Chen, J.; Shen, X.; Song, S.S.; Sun, Y.M.; Fan, S.J.; Fan, F.Y.; Leong, D.T.; Xie, J.P. Ultrasmall $Au_{10-12}(SG)_{10-12}$ Nanomolecules for High Tumor Specificity and Cancer Radiotherapy. *Adv. Mater.* **2014**, *26*, 4565–4568. [CrossRef]
86. Jia, T.T.; Yang, G.; Mo, S.J.; Wang, Z.Y.; Li, B.J.; Ma, W.; Guo, Y.X.; Chen, X.; Zhao, X.; Liu, J.Q.; et al. Atomically Precise Gold-Levonorgestrel Nanocluster as a Radiosensitizer for Enhanced Cancer Therapy. *ACS Nano* **2019**, *13*, 8320–8328. [CrossRef]
87. Liu, C.P.; Wu, T.H.; Liu, C.Y.; Chen, K.C.; Chen, Y.X.; Chen, G.S.; Lin, S.Y. Self-Supplying O_2 through the Catalase-Like Activity of Gold Nanoclusters for Photodynamic Therapy against Hypoxic Cancer Cells. *Small* **2017**, *13*, 1700278–1700286. [CrossRef]
88. Chen, Q.; Chen, J.W.; Yang, Z.J.; Zhang, L.; Dong, Z.L.; Liu, Z. NIR-II Light Activated Photodynamic Therapy with Protein-Capped Gold Nanoclusters. *Nano Res.* **2018**, *11*, 5657–5669. [CrossRef]
89. Gao, G.B.; Chen, R.; He, M.; Li, J.; Li, J.; Wang, L.Y.; Sun, T.L. Gold Nanoclusters for Parkinson's Disease Treatment. *Biomaterials* **2019**, *194*, 36–46. [CrossRef]
90. Dang, P.P.; Liu, D.J.; Li, G.G.; Al Kheraif, A.A.; Lin, J. Recent Advances in Bismuth Ion-Doped Phosphor Materials: Structure Design, Tunable Photoluminescence Properties, and Application in White LEDs. *Adv. Opt. Mater.* **2020**, *8*, 1901993–1902025. [CrossRef]
91. Xia, Z.G.; Meijerink, A. Ce^{3+}-Doped Garnet Phosphors: Composition Modification, Luminescence Properties and Applications. *Chem. Soc. Rev.* **2017**, *46*, 275–299. [CrossRef]
92. Zhu, H.F.; Feng, S.Q.; Kong, Z.H.; Huang, X.; Peng, L.; Wang, J.; Wong, W.-Y.; Zhou, Z.; Xia, M. Bi^{3+} Occupancy Rearrangement in $K_{2-x}A_xMgGeO_4$ Phosphor to Achieve Ultra-broad-band White Emission Based on Alkali Metal Substitution Engineering. *Appl. Surf. Sci.* **2021**, *563*, 150252–150262. [CrossRef]
93. Zhou, Z.; Zhu, H.F.; Huang, X.; She, Y.L.; Zhong, Y.; Wang, J.; Liu, M.; Li, W.; Xia, M. Anti-thermal-quenching, Color-tunable and Ultra-narrow-band Cyan Green-emitting Phosphor for W-LEDs with Enhanced Color Rendering. *Chem. Eng. J.* **2022**, *433*, 134079–134089. [CrossRef]
94. Huang, H.Y.; Cai, K.B.; Talite, M.J.; Chou, W.C.; Chen, P.W.; Yuan, C.T. Coordination-Induced Emission Enhancement in Gold-Nanoclusters with Solid-State Quantum Yields up to 40% for Eco-friendly, Low-reabsorption Nano-phosphors. *Sci. Rep.* **2019**, *9*, 4053–4063. [CrossRef]
95. Feng, L.Z.; Wang, J.J.; Ma, T.; Yin, Y.C.; Song, K.H.; Li, Z.D.; Zhou, M.M.; Jin, S.; Zhuang, T.; Fan, F.J.; et al. Biomimetic Non-Classical Crystallization Drives Hierarchical Structuring of Efficient Circularly Polarized Phosphors. *Nat. Commun.* **2022**, *13*, 3339–3349. [CrossRef]

96. Han, Z.; Zhao, X.L.; Peng, P.; Li, S.; Zhang, C.; Cao, M.; Li, K.; Wang, Z.Y.; Zang, S.Q. Intercluster Aurophilicity-Driven Aggregation Lighting Circularly Polarized Luminescence of Chiral Gold Clusters. *Nano Res.* **2020**, *13*, 3248–3252. [CrossRef]
97. Kennes, K.; Martin, C.; Baekelant, W.; Coutino-Gonzalez, E.; Fron, E.; Roeffaers, M.B.J.; Hofkens, J.; Van der Auweraer, M. Silver Zeolite Composite-Based LEDs: Origin of Electroluminescence and Charge Transport. *ACS Appl. Mater. Interfaces* **2019**, *11*, 12179–12183. [CrossRef]
98. Xie, J.P.; Zheng, Y.G.; Ying, J.Y. Highly Selective and Ultrasensitive Detection of Hg^{2+} Based on Fluorescence Quenching of Au Nanoclusters by Hg^{2+}-Au^{+} Interactions. *Chem. Commun.* **2010**, *46*, 961–963. [CrossRef]
99. Wei, H.; Wang, Z.D.; Yang, L.M.; Tian, S.L.; Hou, C.J.; Lu, Y. Lysozyme-Stabilized Gold Fluorescent Cluster: Synthesis and Application as Hg^{2+} sensor. *Analyst* **2010**, *135*, 1406–1410. [CrossRef]
100. Zhang, Y.Y.; Jiang, H.; Wang, X.M. Cytidine-Stabilized Gold Nanocluster as a Fluorescence Turn-On and Turn-Off Probe for Dual Functional Detection of Ag^{+} and Hg^{2+}. *Anal. Chim. Acta* **2015**, *870*, 1–7. [CrossRef]
101. Hofmann, C.M.; Essner, J.B.; Baker, G.A.; Baker, S.N. Protein-Templated Gold Nanoclusters Sequestered within Sol-Gel Thin Films for the Selective and Ratiometric Luminescence Recognition of Hg^{2+}. *Nanoscale* **2014**, *6*, 5425–5431. [CrossRef] [PubMed]
102. Babaee, E.; Barati, A.; Gholivand, M.B.; Taherpour, A.A.; Zolfaghar, N.; Shamsipur, M. Determination of Hg^{2+} and Cu^{2+} Ions by Dual-Emissive Ag/Au Nanocluster/Carbon Dots Nanohybrids: Switching the Selectivity by pH Adjustment. *J. Hazard. Mater.* **2019**, *367*, 437–446. [CrossRef] [PubMed]
103. Bian, R.X.; Wu, X.T.; Chai, F.; Li, L.; Zhang, L.Y.; Wang, T.T.; Wang, C.G.; Su, Z.M. Facile Preparation of Fluorescent Au Nanoclusters-Based Test Papers for Recyclable Detection of Hg^{2+} and Pb^{2+}. *Sens. Actuators B* **2017**, *241*, 592–600. [CrossRef]
104. Huang, H.; Li, H.; Feng, J.J.; Wang, A.J. One-Step Green Synthesis of Fluorescent Bimetallic Au/Ag Nanoclusters for Temperature Sensing and in Vitro Detection of Fe^{3+}. *Sens. Actuators B* **2016**, *223*, 550–556. [CrossRef]
105. Das, N.K.; Ghosh, S.; Priya, A.; Datta, S.; Mukherjee, S. Luminescent Copper Nanoclusters as a Specific Cell-Imaging Probe and a Selective Metal Ion Sensor. *J. Phys. Chem. C* **2015**, *119*, 24657–24664. [CrossRef]
106. Chang, H.C.; Ho, J.A. Gold Nanocluster-Assisted Fluorescent Detection for Hydrogen Peroxide and Cholesterol Based on the Inner Filter Effect of Gold Nanoparticles. *Anal. Chem.* **2015**, *87*, 10362–10367. [CrossRef]
107. Yang, K.C.; Wang, S.S.; Wang, Y.; Miao, H.Y.; Yang, X.M. Dual-Channel Probe of Carbon Dots Cooperating with Gold Nanoclusters Employed for Assaying Multiple Targets. *Biosens. Bioelectron.* **2017**, *91*, 566–573. [CrossRef]
108. Zhao, Q.; Chen, S.N.; Huang, H.W.; Zhang, L.Y.; Wang, L.Q.; Liu, F.; Chen, J.P.; Zeng, Y.L.; Chu, P.K. Colorimetric and Ultra-Sensitive Fluorescence Resonance Energy Transfer Determination of H_2O_2 and Glucose by Multi-Functional Au Nanoclusters. *Analyst* **2014**, *139*, 1498–1503. [CrossRef]
109. Bai, X.; Xu, S.Y.; Wang, L.Y. Full-Range pH Stable Au-Clusters in Nanogel for Confinement-Enhanced Emission and Improved Sulfide Sensing in Living Cells. *Anal. Chem.* **2018**, *90*, 3270–3275. [CrossRef]
110. Niu, W.J.; Shan, D.; Zhu, R.H.; Deng, S.Y.; Cosnier, S.G.; Zhang, X.J. Dumbbell-Shaped Carbon Quantum Dots/AuNCs Nanohybrid as an Efficient Ratiometric Fluorescent Probe for Sensing Cadmium (II) ions and L-ascorbic Acid. *Carbon* **2016**, *96*, 1034–1042. [CrossRef]
111. Wang, Y.; Wang, Y.; Zhou, F.; Kim, P.; Xia, Y. Protein-protected Au clusters as a new class of nanoscale biosensor for label-free fluorescence detection of proteases. *Small* **2012**, *8*, 3769–3773. [CrossRef] [PubMed]
112. Chen, W.Y.; Chen, L.Y.; Ou, C.M.; Huang, C.C.; Wei, S.C.; Chang, H.T. Synthesis of fluorescent gold nanodot-liposome hybrids for detection of phospholipase C and its inhibitor. *Anal. Chem.* **2013**, *85*, 8834–8840. [CrossRef] [PubMed]
113. Zhang, D.W.; Li, M.; Chen, C.F. Recent Advances in Circularly Polarized Electroluminescence Based on Organic Light-Emitting Diodes. *Chem. Soc. Rev.* **2020**, *49*, 1331–1343. [CrossRef] [PubMed]
114. Richardson, F.S.; Riehl, J.P. Circularly Polarized Luminescence Spectroscopy. *Chem. Rev.* **1977**, *77*, 773–792. [CrossRef]
115. Riehl, J.P.; Richardson, F.S. Circularly Polarized Luminescence Spectroscopy. *Chem. Rev.* **1986**, *86*, 1–16. [CrossRef]
116. Kwok, R.T.; Leung, C.W.; Lam, J.W.; Tang, B.Z. Biosensing by Luminogens with Aggregation-Induced Emission Characteristics. *Chem. Soc. Rev.* **2015**, *44*, 4228–4238. [CrossRef]
117. Gong, Z.L.; Zhu, X.F.; Zhou, Z.H.; Zhang, S.W.; Yang, D.; Zhao, B.; Zhang, Y.P.; Deng, J.P.; Cheng, Y.X.; Zheng, Y.X.; et al. Frontiers in Circularly Polarized Luminescence: Molecular Design, Self-assembly, Nanomaterials, and Applications. *Sci. China Chem.* **2021**, *64*, 2060–2104. [CrossRef]
118. Roose, J.; Tang, B.Z.; Wong, K.S. Circularly-Polarized Luminescence (CPL) from Chiral AIE Molecules and Macrostructures. *Small* **2016**, *12*, 6495–6512. [CrossRef]
119. Yang, J.; Fang, M.M.; Li, Z. Organic Luminescent Materials: The Concentration on Aggregates from Aggregation-Induced Emission. *Aggregate* **2020**, *1*, 6–18. [CrossRef]
120. Wang, J.J.; Mao, X.; Yang, J.N.; Yin, Y.C.; Yao, J.S.; Feng, L.Z.; Zhu, F.; Ma, C.; Yang, C.; Zou, G.; et al. Bright and Near-Unity Polarized Light Emission Enabled by Highly Luminescent Cu_2I_2-Dimer Cluster-Based Hybrid Materials. *Nano Lett.* **2021**, *21*, 4115–4121. [CrossRef]
121. Wang, J.J.; Zhou, H.T.; Yang, J.N.; Feng, L.Z.; Yao, J.S.; Song, K.H.; Zhou, M.M.; Jin, S.; Zhang, G.; Yao, H.B. Chiral Phosphine-Copper Iodide Hybrid Cluster Assemblies for Circularly Polarized Luminescence. *J. Am. Chem. Soc.* **2021**, *143*, 10860–10864. [CrossRef] [PubMed]

122. Ge, F.; Li, B.H.; Cheng, P.; Li, G.; Ren, Z.; Xu, J.; Bu, X.H. Chiral Hybrid Copper(I) Halides for High Efficiency Second Harmonic Generation with a Broadband Transparency Window. *Angew. Chem. Int. Ed.* **2022**, *61*, e202115024–e202115030. [CrossRef] [PubMed]
123. Yu, M.X.; Liu, C.P.; Zhao, Y.F.; Li, S.C.; Yu, Y.L.; Lv, J.Q.; Chen, L.; Jiang, F.L.; Hong, M.C. White-Light Emission and Circularly Polarized Luminescence from a Chiral Copper(I) Coordination Polymer through Symmetry-Breaking Crystallization. *Angew. Chem. Int. Ed. Engl.* **2022**, *61*, e202201590–e202201595.
124. Yao, L.; Niu, G.D.; Li, J.Z.; Gao, L.; Luo, X.F.; Xia, B.; Liu, Y.H.; Du, P.P.; Li, D.H.; Chen, C.; et al. Circularly Polarized Luminescence from Chiral Tetranuclear Copper(I) Iodide Clusters. *J. Phys. Chem. Lett.* **2020**, *11*, 1255–1260. [CrossRef] [PubMed]
125. Kong, Y.J.; Yan, Z.P.; Li, S.; Su, H.F.; Li, K.; Zheng, Y.X.; Zang, S.Q. Photoresponsive Propeller-like Chiral AIE Copper(I) Clusters. *Angew. Chem. Int. Ed.* **2020**, *59*, 5336–5340. [CrossRef]
126. Zhang, M.M.; Dong, X.Y.; Wang, Z.Y.; Li, H.Y.; Li, S.J.; Zhao, X.; Zang, S.Q. AIE Triggers the Circularly Polarized Luminescence of Atomically Precise Enantiomeric Copper(I) Alkynyl Clusters. *Angew. Chem. Int. Ed. Engl.* **2020**, *59*, 10052–10058. [CrossRef]
127. Bhaskar, S.; Das, P.; Moronshing, M.; Rai, A.; Subramaniam, C.; Bhaktha, S.B.N.; Ramamurthy, S.S. Photoplasmonic Assembly of Dielectric-metal, Nd_2O_3-Gold Soret Nanointerfaces for Dequenching the Luminophore Emission. *Nanophotonics* **2021**, *10*, 3417–3431. [CrossRef]
128. Bhaskar, S.; Jha, P.; Subramaniam, C.; Ramamurthy, S.S. Multifunctional Hybrid Soret Nanoarchitectures for Mobile Phone-based Picomolar Cu^{2+} ion Sensing and Dye Degradation Applications. *Phys. E* **2021**, *132*, 114764–114771. [CrossRef]
129. Rai, A.; Bhaskar, S.; Ganesh, K.M.; Ramamurthy, S.S. Hottest Hotspots from the Coldest Cold: Welcome to Nano 4.0. *ACS Appl. Nano Mater.* **2022**, *5*, 12245–12264. [CrossRef]
130. Li, H.; Hu, X.; Hong, W.; Cai, F.; Tang, Q.; Zhao, B.; Zhang, D.; Cheng, P. Photonic Crystal Coupled Plasmonic Nanoparticle Array for Resonant Enhancement of Light Harvesting and Power Conversion. *Phys. Chem. Chem. Phys.* **2012**, *14*, 14334–14339. [CrossRef]

nanomaterials

MDPI

Review

Ultrasmall-in-Nano: Why Size Matters

Ryan D. Mellor and Ijeoma F. Uchegbu *

School of Pharmacy, University College London (UCL), 29–39 Brunswick Square, London WC1N 1AX, UK; ryan.mellor.16@ucl.ac.uk
* Correspondence: ijeoma.uchegbu@ucl.ac.uk

Abstract: Gold nanoparticles (AuNPs) are continuing to gain popularity in the field of nanotechnology. New methods are continuously being developed to tune the particles' physicochemical properties, resulting in control over their biological fate and applicability to in vivo diagnostics and therapy. This review focuses on the effects of varying particle size on optical properties, opsonization, cellular internalization, renal clearance, biodistribution, tumor accumulation, and toxicity. We review the common methods of synthesizing ultrasmall AuNPs, as well as the emerging constructs termed ultrasmall-in-nano—an approach which promises to provide the desirable properties from both ends of the AuNP size range. We review the various applications and outcomes of ultrasmall-in-nano constructs in vitro and in vivo.

Keywords: ultrasmall-in-nano; gold nanoparticles; clearance; biodistribution; tumor accumulation; toxicity

1. Introduction

Colloidal gold is the subject of ever-growing interest in the field of nanotechnology. This is due to its versatility and tunability in terms of size, shape, and surface chemistry. With a rigorous understanding of the properties of gold nanoparticles (AuNPs) comes the ability to exploit them for a plethora of therapeutic and diagnostic applications.

Almost any material will display three distinct size-dependent ranges of properties, in their atomic-, nano-, and bulk-scale [1]. Thus, most materials can feasibly exist as a 'nanomaterial' between 1 and 1000 nanometers; however, to be of any practical use, its properties must be precisely and reproducibly manipulated at scale and, under this criteria, AuNPs excel. Various methods (chemical and physical) have been developed to accurately control AuNP's size (from 1 to 330 nm), shape (spheres, rods, stars, plates, cubes, cages, and shells), surface chemistry, and optical-electronic properties. Furthermore, AuNPs are inert, non-toxic, and can be made to be stable in a range of solvents and pH values, properties which are desirable from a biological standpoint [2].

With progress into the tunability of almost every aspect of AuNPs comes the opportunity to investigate how varying each one of these properties independently will affect the physicochemical properties and biological outcome of the particles. One of the easiest and most effective ways to control the properties of AuNPs is by varying the size. There are advantages and disadvantages for AuNPs in both the ultrasmall (<5 nm) and the nano (5–1000 nm) size range, in terms of the optical properties, cellular uptake, opsonization, toxicity, biodistribution, tumor accumulation, and excretability. This work will outline some of the methods implemented to synthesize ultrasmall AuNPs, the trends observed with varying AuNP size, and finally, the approaches and applications of ultrasmall-in-nano—a new construct which is able to combine the advantages from both ends of the size range.

2. Effects of Varying Size

2.1. Effect on Optical Properties

Varying the size of AuNPs is one of the most straightforward and efficient ways to alter the optical properties and enhance scattering [3]. AuNPs exhibit surface plasmon resonance (SPR), meaning that, at a particular range of wavelengths, dependent on the particles' size and shape, they will display increased absorbance. This is a distinctly nano property, not observed at either the atomic or bulk scale of gold [4]. This SPR band can be exploited to create diagnostic techniques such as surfaced enhanced Raman spectroscopy (SERS) [5] and spatially offset Raman spectroscopy (SORS) [6], and for therapeutic techniques including photothermal therapy (PTT) [7] and photodynamic therapy (PDT) [8]. For diagnostic and therapeutic applications, the SPR band should reside in a region known as the phototherapeutic window—650–850 nm, a range of low absorbance by biomolecules in human tissue, resulting in the high depth penetration of the incident laser light [9]. Unfortunately, particles which exhibit this SPR band tend to be in the size range of 100–200 nm [10], a size that would not be sufficiently excreted by the kidneys.

Generally, larger particles will have a more redshifted SPR peak compared to their smaller counterparts [11]. Spheres in the ultrasmall size range (~5 nm) will exhibit SPR peaks in the range of 515–520 nm and this will bathochromically shift to over 570 nm with an increasing particle diameter beyond 100 nm [12]. This bathochromic shift is accompanied by a broadening of the resonance [3,11].

It is also reported that SERS intensity increases with particle size [11], this is of huge significance for theranostic applications of AuNPs as a stronger SERS intensity results in particles which can be detected at lower concentrations and greater depth. Further investigation is required to determine the interplay between particle size, laser wavelength, and SERS intensity [11].

2.2. Effect on Opsonization

Nanoparticles, upon introduction to the bloodstream, collide with proteins, some of which will bind to the particle, forming a biological identity known as a protein corona. A fraction of these proteins will be opsonins, such as complement proteins [13] and antibodies [14], which tag the particle for uptake by phagocytes and elimination from the body. This is a vital aspect of the immune system and it is responsible for the removal of, among other things, pathogens, diseased cells and protein aggregates [15]; however, this phagocytosis leads to the undesirable reduced circulation time of nanoparticles. One of the most successful approaches to avoid opsonization is by altering the nanoparticle's surface chemistry, for example, by PEGylation [16,17]; however, the particle's size can also have a dramatic effect.

It has been shown that, for nanoparticles and proteins of similar sizes, their attraction resulting from van der Walls potential scales with the particles radii [18]. Therefore, it follows that larger particles will experience a greater degree of opsonization, as has been shown experimentally for particles with diameters of 7–22 nm [19]. The increased opsonization can be explained in part by the increased surface area; however, this is not a complete explanation, as it has been shown that the density of adsorbed proteins also increases [20]. This trend would not be expected to hold indefinitely, since there reaches a point where, due to the decreasing curvature, the particle's surface is indistinguishable from being flat from the perspective of a protein. Indeed, that size limit has been shown to be on the order of 50 nm, after which the protein binding diminishes with increasing size [21].

2.3. Effect on Cellular Internalization

It is usually desirable to either suppress or promote cellular internalization, depending on the theranostic application. Suppression is generally advantageous if the application does not require activity within the cell, for example, imaging or PTT; this allows efficient elimination of the nanoparticle after it has served its purpose [22]. Conversely, the promotion of cellular internalization also has its applications, for example, the delivery of cargo which acts intracellularly. However, when a system is designed to promote initialization, efforts should be made to minimize systemic toxicity arising from non-specific internalization, and this may be achieved by coating the particle with a targeting moiety [23].

Nanoparticles rely heavily on receptor-mediated endocytosis for cellular internalization [24,25]. This results in the rate of internalization, with regards to nanoparticle size, being governed by two opposing phenomena. Larger particles can bind to many cell-surface receptors simultaneously, which leads a decrease in the Gibbs free energy, resulting in the membrane wrapping around the particle. On the one hand, a smaller particle is only able to interact with a few receptors at a time; this means that there is no risk of a localized receptor shortage which would act to reduce the overall rate of particle uptake. Similarly, larger particles require a greater surface area of the cell membrane to envelope the particle, whereas smaller particles require a greater surface curvature of the cell membrane, neither of which are favorable due to either kinetics or energetics, respectively. Several studies, both theoretical [25–27] and experimental [28,29], have found the sweet spot for efficient internalization to be around 25–50 nm. However, this range is heavily dependent on the particles' surface chemistry [30] and cell-line [31] being tested. More generally, 10–100 nm is considered the optimal range for favoring internalization, where particles outside of this range will suffer from one of the aforementioned phenomena.

2.4. *Effect on Renal Clearance*

The renal clearance of injected agents is desirable to avoid hazards resulting from the accumulation and/or decomposition of said agents. In order for efficient renal clearance, particles must poses a hydrodynamic diameter below that of the kidney filtration threshold (KFT) of ~5.5 nm [32,33], a limit set by the size of the glomerular pores that filter blood plasma. Particles of up to 8 nm may be filtered by the kidneys provided so that the particle surface is positively charged; this is due to favorable interactions with the filtration barrier, which are not present for neutral or negatively charged particles [34]. Particles that are not able to be filtered by the kidneys will rely on the slower hepatobiliary pathway for clearance, if they are to be excreted at all.

It is noteworthy that there exists a Goldilocks zone for the size of renally clearable particles of approximately 1–5 nm; this is assumed to be due to fact that particles below 1 nm can enter the ~1 nm pores of the glomerular glycocalyx. This is evidenced by the exponential decrease in rate of glomerular filtration with increasing particle size for atomically precise AuNPs below 1 nm, where the size is inferred from the particles' mass as determined by electrospray ionization mass spectroscopy [35].

Multiple studies [36,37] have compared particles within the size range 1–5 nm to particles larger than 5 nm and the findings are in almost unanimous agreement. That is, larger (>5 nm) particles show low to undetectable levels in the urine as they cannot pass through the glomerular pores, and significant levels accumulating in the liver and excreted via the hepatobiliary pathway. Meanwhile, smaller (<5 nm) particles show higher rates of excretion via both the renal and hepatobiliary pathways, where more than 50% of the injected dose (%ID) is cleared in hours [38], rather than weeks [39] to months [40], for larger particles. Notably, it is the hydrodynamic diameter, and not that of the solid core, which determines the fate of a nanoparticle. This is nicely demonstrated by varying the molecular weight, and therefore the thickness, of a PEG coating while maintaining a gold core size of 2.5 nm; the <5 nm hydrodynamic diameter particles showed preferential renal clearance, while the >5 nm hydrodynamic diameter particles showed a decreased rate that would be expected of larger particles [41]. The correlation of size vs. rate of excretion has been shown to be exponential in nature [42], where particles of 2, 6, and 13 nm demonstrated renal clearance efficiencies of 50, 4, and 0.5%, respectively, 24 h post-injection.

2.5. Effect on Biodistribution

The biodistribution of nanoparticles is heavily impacted by not only the particle size, but also the shape and surface chemistry; therefore, it is important to bear in mind that the trends observed when changing any one of these parameters can be enhanced or countered by changing one of the others. That said, all else being equal, there are significant correlations between biodistribution and size. Figure 1 summarizes some of the ways in which biodistribution is affected by the particles' properties.

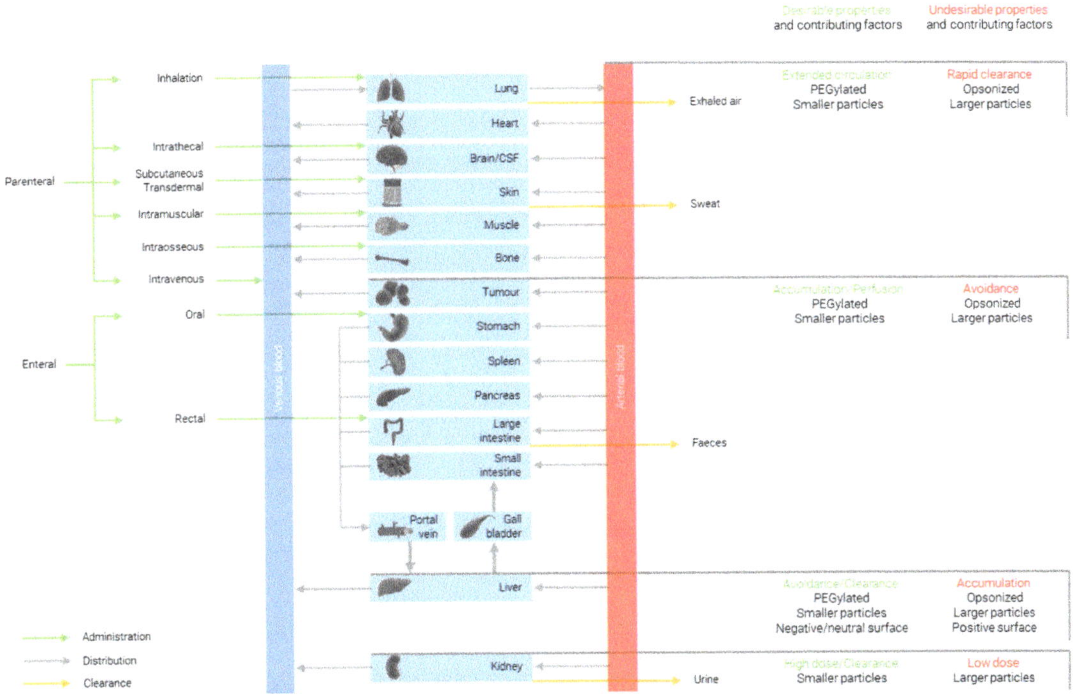

Figure 1. Approaches for modulating biodistribution of nanoparticles.

Generally, it is shown that smaller particles exhibit a more widespread distribution than larger ones. One study compared the biodistribution of 10, 50, 100, and 250 nm particles 24 h after intravenous injection [43]. They found detectable levels of the largest particles (100 and 250 nm) in the blood, liver, and spleen, with negligible quantities (0.1 %ID) in the kidneys; smaller 50 nm particles were further detected in the lungs and heart; and only the smallest 10 nm particles were detected in the remaining tissues in the testis, thymus, and brain, with a larger quantity (1 %ID) present in the kidneys. Across all sizes tested, the highest organ accumulation was found to be in the liver. A similar study, comparing 15, 50, 100, and 200 nm particles, found almost identical results, with the exception that 50 nm particles were additionally detected in the brain [44]; this may be attributed to the different animal models and AuNP preparations used in the two studies [45].

2.6. Effect on Tumor Accumulation

For any cancer theranostic, it is necessary to achieve some degree of targeting. In the case of diagnostic applications, a construct must be able to accumulate at the tumor in order to distinguish it from healthy tissue and identify the tumor. For therapy, tumor accumulation allows tissue damage to be focused on the diseased tissue to the greatest extent possible, maximizing efficacy and minimizing adverse side effects.

One popular approach is to functionalize the surface of constructs with tumor-targeting antibodies, aptamers, peptides, or small molecules; however, this is outside of the scope of this review; we refer the reader to the review "Active targeting of gold nanoparticles as cancer therapeutics" [46] for an overview of the subject. Within the scope is the opportunity for size-based passive targeting. Nanoparticles need to fall within a particular size rage (40–400 nm [47]) in order to take advantage of the enhanced permeability and retention (EPR) effect. The EPR effect leads to passive preferential accumulation in tumor tissue due to a combination of the leaky vasculature, compared to the continuous endothelial junctions of healthy tissue, and reduced lymphatic drainage that would normally help to carry away cytotoxic compounds [48]. For passive targeting to be effective, the therapeutic must remain in circulation for as long as possible. To accomplish an extended half-life, the NPs may be functionalized to tune the particle's size, surface charge, hydrophobicity, and surface chemistry in order to reduce the renal and phagocytic clearance [49].

Size plays a key role in a particle's ability to preferentially accumulate at a tumor. Small particles show more widespread distribution amongst all tissues, both healthy and diseased; they can diffuse more freely into the tumor from the vasculature [50], but this also means that they diffuse out again at a greater rate. Larger particles, close to or exceeding 1 μm, are unable to pass through the tumor fenestrae, with pores in the order of a few hundred nanometers [51]; they are also less able to diffuse into solid tumors, which may not be problematic if the intended application is diagnostic, where locating at the tumor boundary is sufficient.

As discussed in the previous section, smaller particles exhibit extended circulation times and this tends to lead to increased tumor accumulation compared to larger particles; ultrasmall AuNPs have been shown to have a particularly high tumor accumulation when compared to >10 nm AuNPs [52].

2.7. *Effect on Toxicity*

The correlation between size and toxicity of AuNPs is not easy to discern from the literature, due to inconsistent synthesis methods, capping ligands, cell/animal models, dosages, and routes of administration [53]. Toxicity has been reported for ultrasmall (<2 nm) AuNPs, and when the bare gold surface is accessible [54]. The toxicity of ultrasmall AuNPs may be attributed to the more widespread biodistribution and longer circulation times when compared to larger particles which rapidly accumulate in the liver.

Most studies report AuNPs as being non-toxic, and where toxicity is observed, it is generally attributed to physicochemical properties derived from the capping ligand as opposed to the gold core itself.

On the other hand, one study looked at citrate-capped particles ranging from 3 to 100 nm, performing an in vitro MTT ((3-(4,5-dimethylthiazol-2-yl)-2,5-diphenyltetrazolium bromide) assay against Hela cells to assess the particles and found no cytotoxicity for any size at any concentration (up to 0.4 mM). The study also determined the average lifespan (L_{50}) of BALB/c mice dosed intraperitoneally with 8 mg/kg/week for each particle size and no toxicity or lethality was observed for small particles (3, and 5 nm) or for large particles (50, and 100 nm); however, intermediate particles (8, 12, 17, and 37 nm) all showed an L_{50} of less than 21 days. The authors suggest that the zone of toxicity is attributed to particles being small enough to enter cells but large enough to avoid initiating a specific immune response [55].

3. Methods to Synthesize Ultrasmall AuNPs

A plethora of methods have been developed to precisely control the size, shape, and surface chemistry of AuNPs. These range from green synthesis methods, where the AuNPs are produced either by microorganisms or plant extracts [56–58], to physical methods such as laser ablation [59–61], thermal decomposition [62,63], and mechanical milling [64], and finally, chemical synthesis methods. Chemical synthesis methods are perhaps the most widely employed methods, owing to the vast array of physicochemical properties which can be achieved and the specificity with which they can be obtained. Many approaches exist to chemically synthesize AuNPs; however, they all proceed via essentially the same steps (Figure 2):

- **Reduction** of Au^{3+}—afforded by a gold salt, usually $HAuCl_4$—to atomic Au^0; this process is rapid and continues until the concentration of gold atoms in solution reaches supersaturation.
- **Nucleation** of gold atoms into gold clusters; the number of nucleation sites determines the number concentration of AuNPs, i.e., for a fixed mass concentration more nucleation events results in smaller particles and vice versa.
- **Growth** via coalescence of gold clusters and diffusion of remaining soluble gold atoms onto the surface of gold agglomerates.

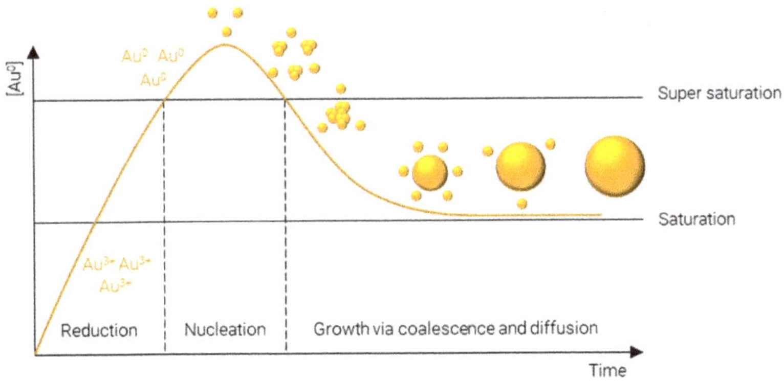

Figure 2. LaMer model of metal nanoparticle formation.

The following sections will outline some of the most common methods used to synthesize ultrasmall AuNPs; they have been divided into the four main methods used in the literature—Turkevich/Frens, reduction by sodium borohydride, Brust–Schiffrin, and seeded growth. Where reagent names have been abbreviated, the meaning can be found in Table 1.

Table 1. Abbreviations for reagents use in AuNP synthesis.

Abbreviation	Meaning
BDAC	Benzyldimethylhexadecylammonium chloride
CTAB	Cetyltrimethylammonium bromide
CTAC	Cetyltrimethylammonium chloride
GSH	Glutathione
$HAuCl_4$	Chloroauric acid
HQL	8-hydroxyquinoline
MPA	Mercaptopropionic acid
$NaBH_4$	Sodium borohydride
NaI	Sodium iodide
ODA	Octadecylamine
PVP	Polyvinylpyrrolidone
TOAB	Tetraoctylammonium bromide

Turkevich/Frens (Figure 3A) synthesis is the classical method of producing AuNPs. It was one of the first systematic approaches to the size-controlled synthesis of AuNPs and is still popular today, owing largely to its simplicity and reliability. The method was pioneered Turkevich et al. in 1951 [65], producing 15–24 nm AuNPs, and later refined by Frens in 1973 [66], extending the size range to 16–147 nm. In this synthesis, citrate is used as both reducing agent and capping agent; however, citrate is not a strong enough reducing agent to rapidly generate atomic gold at room temperature; therefore, the synthesis is carried out at elevated temperature, typically boiling. The AuNPs' size is controlled predominantly by the ratio of citrate:Au, where more citrate results in more rapid nucleation and, subsequently, smaller particles. Particle size and distribution may also be controlled by pH [67], temperature [68], and order of reagent addition [69]. While AuNPs with an average diameter of 4 nm have been reportedly synthesized by the Turkevich method with minor modifications [70], it is far more common for particles to be larger than 10 nm in diameter.

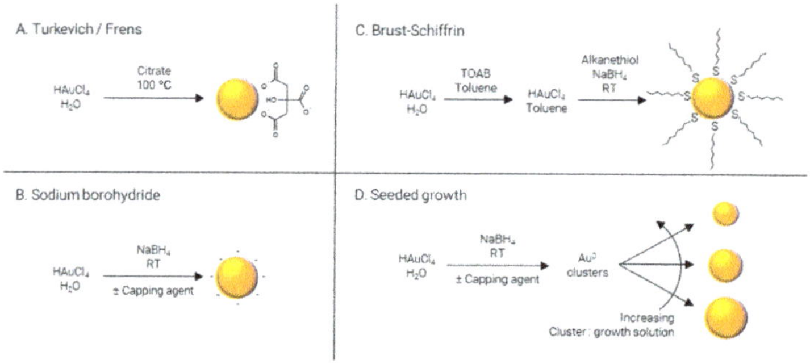

Figure 3. Methods of AuNP synthesis.

Sodium borohydride (Figure 3B) is often implemented as a strong reducing agent in the synthesis of AuNPs enabling the reaction to be performed at room temperature and allowing for rapid nucleation and formation of smaller AuNPs, frequently sub 5 nm. Like the Turkevich method, citrate may be included; however, when $NaBH_4$ is used as a reducing agent, citrate severs solely as a capping agent [71]. Alternatively, citrate can be replaced by other hydrophilic capping agents such as alginate [72] or chitosan [73]. The synthesis may also be performed in non-polar solvents such as chloroform, implementing hydrophobic capping agents such as CTAB [74] and ODA [75,76]. Finally, capping agents may be omitted entirely to produce "bare" AuNPs [77].

Brust–Schiffrin (Figure 3C) synthesis is a two-phase approach to produce alkanethiol-capped AuNPs which are soluble in hydrophobic solvents. TOAB is employed to transfer $AuCl_4^-$ from the aqueous phase to an organic phase, typically toluene; $NaBH_4$ is used to reduce the gold salt in the presence of a capping agent, traditionally an alkanethiol. The capping agent first used, and still commonly used today, is the alkanethiol dodecanethiol [78]; however, this may be replaced with other alkanethiols such as pentanethiol [79] or hexanethiol [80], surfactants such as CTAB or CTAC [81], or even ionizable molecules for the synthesis of water soluble AuNPs, for example via the use of MPA [82].

Seeded growth (Figure 3D) is synthetic process of first producing Au^0 clusters, often via $NaBH_4$, although Turkevich/Frens particles may also be used as seeds, which are then introduced as presynthesized nuclei into a growth solution. Essentially, the particle number concentration of the resulting solution can be finely controlled by varying the number of nuclei introduced and the final particle size is regulated by the gold concentration in the growth solution. This approach is not particularly well suited to the formation of ultrasmall AuNPs and is more commonly employed for the preparation of particles over a large size range [83,84]. As well as being applicable over large size ranges, seeded growth is also capable of producing a wide variety of shapes by using different shape directing agents, such as CTAC for spheres [85] and cubes [86], CTAC/NaI for triangles [87], CTAB/CTAC/HQL for bipyramids/javelins [88], PVP for stars [89], and BDAC/CTAB for rods [90].

The methods outlined in the previous sections are summarized in Table 2.

Table 2. Methods of AuNP Synthesis. Focusing mainly on ultrasmall spheres, with several prominent examples of methods for synthesizing larger or non-spherical particles.

Method of Synthesis	Size Range	Shape	Surface Chemistry	Polarity	Solvent	Ref.
Turkevich	15–24 nm	Sphere	Citrate	Hydrophilic	H_2O	[65]
Frens	16–147 nm	Sphere	Citrate	Hydrophilic	H_2O	[66]
Turkevich/Frens	4 nm	Sphere	Citrate	Hydrophilic	H_2O	[70]
Sodium borohydride	3–5 nm	Sphere	Citrate	Hydrophilic	H_2O	[71]
Sodium borohydride	3.3–12	Sphere	Alginate	Hydrophilic	H_2O	[72]
Sodium borohydride	3.5–14 nm	Sphere	Chitosan	Hydrophilic	H_2O	[73]
Sodium borohydride	3–14 nm	Sphere	CTAB	Hydrophobic	$CHCl_3$	[74]
Sodium borohydride	4.7 nm	Sphere	ODA	Hydrophobic	$CHCl_3$	[75]
Sodium borohydride	3 nm	Sphere	ODA	Hydrophobic	$CHCl_3$	[76]
Sodium borohydride	3–5 nm	Sphere	Bare	Hydrophilic	H_2O	[77]
Turkevich/Frenz—modified	3.6–13 nm	Sphere	Citrate/tannic acid	Hydrophilic	H_2O	[91]
Turkevich/Frenz—modified	3.5–15 nm	Sphere	PDEAEM	Hydrophilic	H_2O	[92]
Turkevich/Frenz—modified	2–330 nm	Sphere	Citrate	Hydrophilic	H_2O	[93]
Brust-Schiffrin	1–3 nm	Sphere	Dodecanethiol	Hydrophobic	Toluene	[78]
Brust-Schiffrin	5 nm	Sphere	Pentanethiol	Hydrophobic	Toluene	[79]
Brust-Schiffrin	2 nm	Sphere	Hexanethiol	Hydrophobic	Toluene	[80]
Brust-Schiffrin	10 nm	Sphere	CTAB/CTAC	Hydrophobic	Toluene	[81]
Brust-Schiffrin	3 nm	Sphere	MPA	Variable	Toluene/H_2O	[82]
Seeded growth	8.4–180.5 nm	Sphere	Citrate	Hydrophilic	H_2O	[83]
Seeded growth	15–300 nm	Sphere	Citrate	Hydrophilic	H_2O	[84]
Seeded growth	5–150	Sphere	CTAC	Hydrophilic	H_2O	[85]
Seeded growth	60 nm	Triangle	CTAC/NaI	Hydrophilic	H_2O	[87]
Seeded growth	76 nm	Cube	CTAC	Hydrophilic	H_2O	[86]
Seeded growth	40–300 nm	Bipyramid/Javelin	CTAB/CTAC/HQL	Hydrophilic	H_2O	[88]
Seeded growth	45–116 nm	Star	PVP	Hydrophilic	DMF	[89]
Seeded growth	10–100 nm	Rod	BDAC/CTAB	Hydrophilic	H_2O	[90]
Other—GSH reduction	2.5 nm	Sphere	GSH	Hydrophilic	H_2O	[94]
Other—GSH reduction	2.3 nm	Sphere	GSH/cysteamine	Hydrophilic	H_2O	[95]
Other—HEPES reduction	23 nm	Star	HEPES	Hydrophilic	H_2O	[96]
Other—TBAB reduction	2–7 nm	Sphere	Oleylamine	Hydrophobic	DCM	[97]
Other—TBAB reduction	3–10 nm	Sphere	Oleylamine	Hydrophobic	Hexane	[98]
Other—thermal reduction	2 nm	Sphere	PEG	Hydrophilic	H_2O	[99]
Other—mechanochemical	1–4 nm	Sphere	Various	Various	None	[64]

4. Ultrasmall-in-Nano—Approaches and Applications

One approach which has been gaining popularity in the past few years is referred to as an ultrasmall-in-nano [30] construct: 'ultrasmall' represents the sub-5 nm gold cores, and 'nano' represents the 100–500 nm clusters of said AuNPs. While gold itself is inert and biocompatible, problems arise when considering persistence of the gold in the body [30]. The body's main mechanism for excretion of compounds from circulation, specifically, via the renal pathway, is not efficient at removing particles larger than 5 nm in diameter [100]. This is mainly due to the functional pore size of the glomerular capillary wall of 4.5–5 nm [101] and larger particles will instead rely on elimination via the hepatobiliary pathway. Consequently, particles below this threshold are desirable to avoid the potential retention of gold in the body. However, several desirable properties result from particles only in the 100–500 nm range. These properties include increased circulation times [102], superior accumulation in tumor tissue due to the enhanced permeability and retention effect [103], and specifically in the case of AuNPs strong absorbance in the phototherapeutic window [104]. A work around for these diametrically opposed concepts is to have a system which can be converted from the latter to the former after its function has been served, i.e., from larger (100–500 nm) nanoparticles to ultrasmall (sub-5 nm) particles.

To display the desired optical properties, specifically, a bathochromic shift in absorbance upon clustering, the ultrasmall particles must come into proximity of one another. The effect of interparticle electromagnetic coupling is proportional to the inverse of interparticle distance [105], suggesting that the most prominent bathochromic shift is obtained when the ultrasmall particles are as close as possible without touching to avoid irreversible fusion, as would be observed with other aggregation methods, for example, upon the addition of glucosamine phosphate to small AuNPs [106].

Many approaches have been reported for forming ultrasmall-in-nano constructs; while they all follow what is essentially the same schematic, depicted in Figure 4, and have some common characteristics, they differ in the approach used to cluster the ultrasmalls and in the application of the nano construct. Some of these approaches are outlined in the following sections and summarized in Table 3.

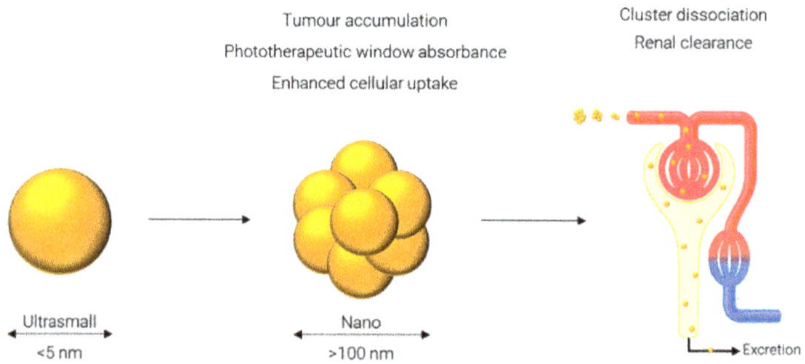

Figure 4. Generalized principle of the ultrasmall-in-nano approach.

Table 3. Methods of generating ultrasmall-in-nano constructs.

Ultrasmall (Surface Chemistry, and Size)	Nano (Clustering Principle, and Size)	SPR	Reversible	Refs.
ODA 4.67 ± 1.74 nm	Crosslinking with EGBMA 254–278 nm	710 nm	Yes	[75]
NA 2–8 nm	Coating of DSPC: cholesterol liposomes 100–120 nm	760 nm	Yes	[107]
Tannic acid and/or citrate 3, 5, and 13 nm	Single stranded DNA-coated AuNPs + complementary linker 50–150 nm	Nr	Yes	[108]
PSS ~3 nm (varies with article)	Ionic interactions with PL ~100 nm (varies with article)	530 nm	Yes	[109–114]
AcetalDextran-pMBA-AuNPs 2.1 ± 0.5 nm	Encapsulation in PEG-PCL 111.1 ± 38 nm	Nr	Nr	[115]
11-MUA or GSH 2–5 nm	Encapsulation in PCPP 40–500 nm	>650 nm	Yes	[116]
GSH ~2 nm	Encapsulation in PAA HCl 120 nm	Nr	Nr	[117]
Citrate/lysine 4.1 ± 0.8 nm	Interaction with PLA(2K)-PEG(10K)-PLA(2K) 83.0 ± 4.6 nm	Broad, NIR absorbance	Yes	[118]
NA 6.1 ± 1.8 nm	Self-assembly with PCL-PHEMA and $PMEO_2MA$ 300 nm	800 nm	Nr	[119]

Nr: Nor reported.

4.1. Approaches to Clustering

4.1.1. Small Molecule Crosslinking

Mellor et al. [75] showed that using a labile dithiol molecule—ethylene glycol bis-mercaptoacetate (EGBMA)—they were able cluster octadecylamine (ODA) AuNPs (sub-5 nm) and demonstrated that, under physiological conditions, the clusters can revert to ultrasmall AuNPs. The nano constructs exhibited strong absorbance in the phototherapeutic window and were labelled with a Raman reporter, allowing for specific detection in a biological matrix, demonstrating its feasibility for future theranostic applications.

4.1.2. Coating of Liposomes

Rengan et al. [107] prepared a formulation that can be described as ultrasmall-on-nano, by first forming 1,2-distearoyl-sn-glycero-3-phosphocholine (DSPC): cholesterol liposomes in the nano range, and subsequently coating them with ultrasmall AuNPs (2–8 nm). They demonstrate the ability of these particles, following intratumoral injection, to kill cancer cells using PTT with complete ablation of the tumor mass following 750 nm laser illumination. The particles are shown to be degraded in hepatocytes and cleared through the hepato-biliary and renal routes. On days 1, 7, and 14, the %ID detected in the liver was 52%, 9.8%, and 3%, respectively, and in the kidney was 2.7%, 0.25%, and 0.22%, respectively. This demonstrated a significant reduction in gold levels in just 14 days, and the authors hypothesized that renal excretion would be increased when the constructs are subjected to both photothermal and enzymatic degradation.

4.1.3. DNA Assembly

Chou et al. [108] designed a core-satellite architecture whereby AuNPs (13 nm cores, and 3 or 5 nm satellites) coated with thiolated, single-stranded DNA were then assembled using complementary sequence linker DNA. They demonstrated that, by careful selection of the DNA sequence and capping ligand, they could tune a number of properties: the constructs' ability to encapsulate small molecules, their propensity for cellular uptake, elimination, and tumor-targeting. In addition, in this study, gold content in urine was quantified 48 h after systemic injection of cores, satellites, and core-satellite constructs. They found the highest levels of renal excretion for the smallest particles tested—15 %ID for 3 nm satellites with a 1 kDa PEG coating—and decreasing levels with increasing particle size or PEG molecular weight. Similarly, for the core-satellite constructs, they found that the urine levels were proportional to the size of the satellite tested, and lower than that of the respective satellite alone. This finding suggests, as would be expected, that only the satellites are being excreted and not the 13 nm cores.

4.1.4. Encapsulation/Ionic Interaction

Voliani and their team demonstrated, in multiple articles [109–114], their ultrasmall-in-nano construct, referred to as 'passion fruit'-like, formed by ionic interaction between poly(l-lysine) (PL) and poly(sodium 4-styrene sulfonate) (PSS)-coated ultrasmall AuNPs (~3 nm). They studied their tumor-targeting ability following modification with a peptide [109]; biocompatibility and excretion in murine models where they showed excretion from both renal and biliary pathways [110]; pharmacokinetics following inhalation [111], showing accumulation in the lungs, translocation to secondary organs, and almost complete excretion within 10 days; and the suitability of their ultrasmall-in-nano construct for PTT applications [114]. The researchers monitored the biodistribution and excretion for 10 days following intravenous injection through the tail vein. They observed that the gold concentration in the liver decreased over the course of time, and that gold was detected in both urine and faces for the duration of the experiment for a cumulative excretion of ~16 %ID in 10 days.

Higbee-Dempsey et al. [115] prepared ultrasmall p-MBA-AuNPs coated with thiolated dextran (2.1 nm); the dextran was made to be hydrophobic by the covalent incorporation of acetyl groups. The hydrophobic AcetalDextran-pMBA-AuNPs were mixed with poly(ethylene glycol)-block-poly(ε-caprolactone) (PEG-PCL), resulting in micellization and dense packing of the ultrasmall cores. Acetyl groups were shown to be cleaved in an acidic environment, resulting in a hydrophilic polymer, demicellization, and dispersion of gold cores. The group showed clearance of gold from mice organs over 3 months, following a single bolus injection via the tail vein. They report that levels in the liver and spleen reduced by 86% and 72%, respectively, and that gold is detected in urine and faeces at decreasing levels over the time course; however, the reported levels represent samples collected on the day of sacrifice, and do not represent the cumulative excretions between time points.

Cheheltani et al. [116] encapsulated ultrasmall GSH-coated AuNPs (2–5 nm) in a biodegradable poly di(carboxylatophenoxy)phosphazene (PCPP) polymer; they were able to control the size of the particles, and therefore, the position of the SPR peak in the NIR region, by varying the amount of polyethylene glycol-polylysine block co-polymer in the formulation. They demonstrate the potential of these constructs as computed tomography (CT) and photoacoustic (PA) imaging contrast agents both in vitro and in vivo. They also demonstrate the ability of the clusters to degrade upon incubation in serum.

Yahia-Ammar et al. [117] first synthesized ultrasmall GSH-coated AuNPs (~2 nm); these particles were then encapsulated by the addition of poly(allyl amine hydrochloride) (PAA HCl) polymer. They show that the clustered particles exhibit enhanced fluorescence compared to the ultrasmall particles, with quantum yield increasing from 7 to 25%. They also demonstrated the increased cellular uptake of these particles and the application of this to enhanced cellular delivery of both peptides and antibodies.

Tam et al. [118] clustered ultrasmall lysine/citrate-capped AuNPs (4.1 nm) via ionic interaction with biodegradable triblock copolymer of polylactic acid and polyethylene glycol (PLA(2K)-PEG(10K)-PLA(2K)). Upon cluster formation, the absorbance maximum shifted from 520 nm into the NIR region with a fairly constant absorbance of 700 to 900 nm. Limited degradation was observed after 4 weeks at neutral pH, whereas almost total degradation was observed after just 1 week at pH 5, a finding attributed to the stability of PLA at neutral pH. Nanocluster degradation was confirmed in vitro by TEM and scattering spectra from hyperspectral images of treated and untreated murine macrophage cells over 168 h.

Deng et al. [119] worked with 6.1 nm particles, which is on the upper limit of what may be considered ultrasmall, as acknowledged by the researchers; however, the approach may be transferable to smaller particles to ensure excretability. Their system is based on the self-assembly of AuNPs within a novel comb-like amphipathic polymer composed of hydrophobic poly(ε-caprolactone)/poly(2-hydroxyethyl methacrylate) (PCL-PHEMA) and hydrophilic poly(2-(2-methoxyethoxy) ethyl methacrylate) ($PMEO_2MA$). The group tested the potential of the particles for PTT by exposing solutions of various concentrations to 808 nm NIR laser at a power density of 1.5 W cm^{-2}; at the highest concentration (0.4 mg mL^{-1}), they observed heating up to 71 °C after laser irradiation for 5 min. They also loaded DOX into the particles, which was released following laser irradiation to demonstrate the combined chemotherapeutic and phototherapeutic abilities of the construct. The constructs were shown to be suitable as contrast agents for both CT and PAT imaging. The researchers monitored the levels of the gold in the tumor, vital organs (heart, liver, spleen, lung, and kidney), and vital metabolic products (bile, urine, and faces) for 7 days following intratumoral injection. They found that levels in the tumor decreased over the course of time; the levels in the vital organs increased until day 2, but then had decreased by day 7, and the levels in metabolic products increased over the course of time. The increasing renal excretion over the course of the experiment, with 4 %ID detected in the urine on day 7, suggests disassembly of the constructs back to ultrasmall AuNPs, and that 6 nm AuNPs are still renally excreted, if only to a lower extent than smaller particles.

4.2. *Accomplishments of Ultrasmall-in-Nano Constructs*

It is clear that ultrasmall-in-nano provides a viable route to produce constructs with the properties of particles larger than their constituent cores; in various studies, they have been shown to have absorbance in the phototherapeutic window [75,107,116,118,119], to be SERS active [75], to be suitable as PTT agents [107,114,119], and to function as PA and CT contrast agents [116], amongst other key characteristics.

What has also been demonstrated, to a lesser extent, is the excretion of ultrasmall particles following administration of nano constructs. Where in vivo studies have been performed to monitor retention and/or excretion, they have demonstrated quite varied results, suggesting that the excretion rate is highly dependent on the approach to clustering, as well as the size of the ultrasmall particle. Rengan et al. [103] showed fairly quick decrease in levels of gold detected in the liver of 52% to 3%ID over 14 days, whereas Higbee-Dempsey et al. [115] reported ~85% elimination from the liver over three months. It is promising that studies such as those performed by Chou et al. [108] demonstrated that the rate of excretion of ultrasmall-in-nano constructs is proportional to the sizes of the constituent ultrasmall particles. This highlights the necessity for the entirety of the construct to be comprised of ultrasmall particles, as the 13 nm cores do not appear to have been excreted.

Several studies, where excretion has not yet been demonstrated experimentally, have reported that their ultrasmall-in-nano constructs are able to liberate ultrasmall particles [75,117,119]. This, combined with the data of the aforementioned in vivo studies, suggests that there are several emerging constructs with the potential to deliver the capabilities of nanoparticles, while retaining the excretable nature of ultrasmall particles.

5. Conclusions and Future Perspectives

Optical properties, opsonization, cellular internalization, renal clearance, biodistribution, tumor accumulation, and toxicity all display size-dependent relationships with gold nanoparticles. In some regards, smaller particles are favored, for example for increased renal clearance; in others its larger particles, such as to display strong absorbance in the phototherapeutic window. Clearly, it is not possible to have a single static construct which displays all these properties; however, ultrasmall-in-nano promises a dynamic structure to provide the advantages of the small and the large.

Where the properties of the larger particles are disadvantageous, work-arounds have been proposed which usually entail the modification of the particles' surface, for example, PEGylation to reduce the propensity for opsonization, or conjugation to a targeting moiety to increase tumor accumulation.

It is noteworthy that the examples of ultrasmall-in-nano reviewed here were all from the last decade or so; this emphasizes how novel this work is and the interest its subject matter has gained in a relatively short time.

Ultrasmall-in-nano constructs have already been shown to deliver a lot of what the concept promises, from phototherapeutic absorbance and SERS activity to declustering and excretion. Future works will likely focus on combining many of these properties into a single construct and demonstrating complete excretion in a reasonable timeframe.

Author Contributions: Writing—original draft preparation, R.D.M.; writing—review and editing, R.D.M. and I.F.U. All authors have read and agreed to the published version of the manuscript.

Funding: This research was funded by UK Engineering and Physical Sciences Research Council, grant number EP/R020965/1.

Institutional Review Board Statement: Not applicable.

Informed Consent Statement: Not applicable.

Data Availability Statement: Not applicable.

Conflicts of Interest: The authors declare no conflict of interest.

References

1. Roduner, E. Size matters: Why nanomaterials are different. *Chem. Soc. Rev.* **2006**, *35*, 583–592. [CrossRef] [PubMed]
2. Chan, W.C.W. (Ed.) *Bio-Applications of Nanoparticles*; Advances in Experimental Medicine and Biology; Springer: Berlin/Heidelberg, Germany; Landes Bioscience: Austin, TX, USA, 2007; ISBN 978-0-387-76712-3.
3. Lee, K.-S.; El-Sayed, M.A. Dependence of the Enhanced Optical Scattering Efficiency Relative to That of Absorption for Gold Metal Nanorods on Aspect Ratio, Size, End-Cap Shape, and Medium Refractive Index. *J. Phys. Chem. B* **2005**, *109*, 20331–20338. [CrossRef] [PubMed]
4. Mohapatra, S.; Ranjan, S.; Dasgupta, N.; Kumar, R.; Thomas, S. *Characterization and Biology of Nanomaterials for Drug Delivery: Nanoscience and Nanotechnology in Drug Delivery*; Elsevier: Amsterdam, The Netherlands, 2018; ISBN 978-0-12-814032-1.
5. Langer, J.; Jimenez de Aberasturi, D.; Aizpurua, J.; Alvarez-Puebla, R.A.; Auguié, B.; Baumberg, J.J.; Bazan, G.C.; Bell, S.E.J.; Boisen, A.; Brolo, A.G.; et al. Present and Future of Surface-Enhanced Raman Scattering. *ACS Nano* **2020**, *14*, 28–117. [CrossRef] [PubMed]
6. Mosca, S.; Conti, C.; Stone, N.; Matousek, P. Spatially offset Raman spectroscopy. *Nat. Rev. Methods Primer* **2021**, *1*, 21. [CrossRef]
7. Wei, W.; Zhang, X.; Zhang, S.; Wei, G.; Su, Z. Biomedical and bioactive engineered nanomaterials for targeted tumor photothermal therapy: A review. *Mater. Sci. Eng. C* **2019**, *104*, 109891. [CrossRef]
8. Chen, J.; Fan, T.; Xie, Z.; Zeng, Q.; Xue, P.; Zheng, T.; Chen, Y.; Luo, X.; Zhang, H. Advances in nanomaterials for photodynamic therapy applications: Status and challenges. *Biomaterials* **2020**, *237*, 119827. [CrossRef]
9. Diwu, Z.; William Lown, J. Phototherapeutic potential of alternative photosensitizers to porphyrins. *Pharmacol. Ther.* **1994**, *63*, 1–35. [CrossRef]
10. Niu, J.; Zhu, T.; Liu, Z. One-step seed-mediated growth of 30–150 nm quasispherical gold nanoparticles with 2-mercaptosuccinic acid as a new reducing agent. *Nanotechnology* **2007**, *18*, 325607. [CrossRef]
11. Njoki, P.N.; Lim, I.-I.S.; Mott, D.; Park, H.-Y.; Khan, B.; Mishra, S.; Sujakumar, R.; Luo, J.; Zhong, C.-J. Size Correlation of Optical and Spectroscopic Properties for Gold Nanoparticles. *J. Phys. Chem. C* **2007**, *111*, 14664–14669. [CrossRef]
12. Haiss, W.; Thanh, N.T.K.; Aveyard, J.; Fernig, D.G. Determination of Size and Concentration of Gold Nanoparticles from UV−Vis Spectra. *Anal. Chem.* **2007**, *79*, 4215–4221. [CrossRef]
13. Merle, N.S.; Noé, R.; Halbwachs-Mecarelli, L.; Fremeaux-Bacchi, V.; Roumenina, L.T. Complement System Part II: Role in Immunity. *Front. Immunol.* **2015**, *6*, 257. [CrossRef] [PubMed]
14. Chiu, M.L.; Goulet, D.R.; Teplyakov, A.; Gilliland, G.L. Antibody Structure and Function: The Basis for Engineering Therapeutics. *Antibodies* **2019**, *8*, 55. [CrossRef] [PubMed]
15. Cockram, T.O.J.; Dundee, J.M.; Popescu, A.S.; Brown, G.C. The Phagocytic Code Regulating Phagocytosis of Mammalian Cells. *Front. Immunol.* **2021**, *12*, 629979. [CrossRef] [PubMed]
16. Petros, R.A.; DeSimone, J.M. Strategies in the design of nanoparticles for therapeutic applications. *Nat. Rev. Drug Discov.* **2010**, *9*, 615–627. [CrossRef] [PubMed]
17. Yoo, J.-W.; Chambers, E.; Mitragotri, S. Factors that Control the Circulation Time of Nanoparticles in Blood: Challenges, Solutions and Future Prospects. *Curr. Pharm. Des.* **2010**, *16*, 2298–2307. [CrossRef]
18. Lane, L.A.; Qian, X.; Smith, A.M.; Nie, S. Physical Chemistry of Nanomedicine: Understanding the Complex Behaviors of Nanoparticles in Vivo. *Annu. Rev. Phys. Chem.* **2015**, *66*, 521–547. [CrossRef]
19. Deng, Z.J.; Liang, M.; Toth, I.; Monteiro, M.J.; Minchin, R.F. Molecular Interaction of Poly(acrylic acid) Gold Nanoparticles with Human Fibrinogen. *ACS Nano* **2012**, *6*, 8962–8969. [CrossRef]
20. Kaur, K.; Forrest, J.A. Influence of particle size on the binding activity of proteins adsorbed onto gold nanoparticles. *Langmuir* **2012**, *28*, 2736–2744. [CrossRef]
21. Lacerda, S.H.D.P.; Park, J.J.; Meuse, C.; Pristinski, D.; Becker, M.L.; Karim, A.; Douglas, J.F. Interaction of Gold Nanoparticles with Common Human Blood Proteins. *ACS Nano* **2010**, *4*, 365–379. [CrossRef]
22. Yu, M.; Zheng, J. Clearance Pathways and Tumor Targeting of Imaging Nanoparticles. *ACS Nano* **2015**, *9*, 6655–6674. [CrossRef]
23. Ruoslahti, E.; Bhatia, S.N.; Sailor, M.J. Targeting of drugs and nanoparticles to tumors. *J. Cell Biol.* **2010**, *188*, 759–768. [CrossRef] [PubMed]
24. Jiang, W.; Kim, B.Y.S.; Rutka, J.T.; Chan, W.C.W. Nanoparticle-mediated cellular response is size-dependent. *Nat. Nanotechnol.* **2008**, *3*, 145–150. [CrossRef]
25. Gao, H.; Shi, W.; Freund, L.B. Mechanics of receptor-mediated endocytosis. *Proc. Natl. Acad. Sci. USA* **2005**, *102*, 9469–9474. [CrossRef] [PubMed]
26. Yuan, H.; Li, J.; Bao, G.; Zhang, S. Variable nanoparticle-cell adhesion strength regulates cellular uptake. *Phys. Rev. Lett.* **2010**, *105*, 138101. [CrossRef] [PubMed]
27. Decuzzi, P.; Ferrari, M. The role of specific and non-specific interactions in receptor-mediated endocytosis of nanoparticles. *Biomaterials* **2007**, *28*, 2915–2922. [CrossRef]
28. Chithrani, B.D.; Ghazani, A.A.; Chan, W.C.W. Determining the Size and Shape Dependence of Gold Nanoparticle Uptake into Mammalian Cells. *Nano Lett.* **2006**, *6*, 662–668. [CrossRef] [PubMed]
29. Liu, X.; Huang, N.; Li, H.; Jin, Q.; Ji, J. Surface and size effects on cell interaction of gold nanoparticles with both phagocytic and nonphagocytic cells. *Langmuir* **2013**, *29*, 9138–9148. [CrossRef]

30. Li, B.; Lane, L.A. Probing the biological obstacles of nanomedicine with gold nanoparticles. *Wiley Interdiscip. Rev. Nanomed. Nanobiotechnol.* **2019**, *11*, e1542. [CrossRef]
31. Albanese, A.; Tang, P.S.; Chan, W.C.W. The Effect of Nanoparticle Size, Shape, and Surface Chemistry on Biological Systems. *Annu. Rev. Biomed. Eng.* **2012**, *14*, 1–16. [CrossRef]
32. Longmire, M.; Choyke, P.L.; Kobayashi, H. Clearance Properties of Nano-sized Particles and Molecules as Imaging Agents: Considerations and Caveats. *Nanomedicine* **2008**, *3*, 703–717. [CrossRef]
33. Liu, J.; Yu, M.; Zhou, C.; Zheng, J. Renal clearable inorganic nanoparticles: A new frontier of bionanotechnology. *Mater. Today* **2013**, *16*, 477–486. [CrossRef]
34. Adhipandito, C.F.; Cheung, S.-H.; Lin, Y.-H.; Wu, S.-H. Atypical Renal Clearance of Nanoparticles Larger Than the Kidney Filtration Threshold. *Int. J. Mol. Sci.* **2021**, *22*, 11182. [CrossRef] [PubMed]
35. Du, B.; Jiang, X.; Das, A.; Zhou, Q.; Yu, M.; Jin, R.; Zheng, J. Glomerular Barrier Behaves As an Atomically Precise Bandpass Filter in a Sub-nanometre Regime. *Nat. Nanotechnol.* **2017**, *12*, 1096–1102. [CrossRef] [PubMed]
36. Semmler-Behnke, M.; Kreyling, W.G.; Lipka, J.; Fertsch, S.; Wenk, A.; Takenaka, S.; Schmid, G.; Brandau, W. Biodistribution of 1.4- and 18-nm gold particles in rats. *Small* **2008**, *4*, 2108–2111. [CrossRef]
37. Balogh, L.; Nigavekar, S.S.; Nair, B.M.; Lesniak, W.; Zhang, C.; Sung, L.Y.; Kariapper, M.S.T.; El-Jawahri, A.; Llanes, M.; Bolton, B.; et al. Significant effect of size on the in vivo biodistribution of gold composite nanodevices in mouse tumor models. *Nanomed. Nanotechnol. Biol. Med.* **2007**, *3*, 281–296. [CrossRef]
38. Hainfeld, J.F.; Slatkin, D.N.; Focella, T.M.; Smilowitz, H.M. Gold nanoparticles: A new X-ray contrast agent. *Br. J. Radiol.* **2006**, *79*, 248–253. [CrossRef]
39. Renaud, G.; Hamilton, R.L.; Havel, R.J. Hepatic metabolism of colloidal gold-low-density lipoprotein complexes in the rat: Evidence for bulk excretion of lysosomal contents into bile. *Hepatology* **1989**, *9*, 380–392. [CrossRef]
40. Sadauskas, E.; Danscher, G.; Stoltenberg, M.; Vogel, U.; Larsen, A.; Wallin, H. Protracted elimination of gold nanoparticles from mouse liver. *Nanomed. Nanotechnol. Biol. Med.* **2009**, *5*, 162–169. [CrossRef]
41. Zhao, Y.; Sultan, D.; Detering, L.; Luehmann, H.; Liu, Y. Facile synthesis, pharmacokinetic and systemic clearance evaluation, and positron emission tomography cancer imaging of ^{64}Cu–Au alloy nanoclusters. *Nanoscale* **2014**, *6*, 13501–13509. [CrossRef]
42. Zhou, C.; Long, M.; Qin, Y.; Sun, X.; Zheng, J. Luminescent Gold Nanoparticles with Efficient Renal Clearance. *Angew. Chem. Int. Ed.* **2011**, *50*, 3168–3172. [CrossRef]
43. De Jong, W.H.; Hagens, W.I.; Krystek, P.; Burger, M.C.; Sips, A.J.A.M.; Geertsma, R.E. Particle size-dependent organ distribution of gold nanoparticles after intravenous administration. *Biomaterials* **2008**, *29*, 1912–1919. [CrossRef] [PubMed]
44. Sonavane, G.; Tomoda, K.; Makino, K. Biodistribution of colloidal gold nanoparticles after intravenous administration: Effect of particle size. *Colloids Surf. B Biointerfaces* **2008**, *66*, 274–280. [CrossRef] [PubMed]
45. Khlebtsov, N.; Dykman, L. Biodistribution and toxicity of engineered gold nanoparticles: A review of in vitro and in vivo studies. *Chem. Soc. Rev.* **2011**, *40*, 1647–1671. [CrossRef] [PubMed]
46. Goddard, Z.R.; Marín, M.J.; Russell, D.A.; Searcey, M. Active targeting of gold nanoparticles as cancer therapeutics. *Chem. Soc. Rev.* **2020**, *49*, 8774–8789. [CrossRef] [PubMed]
47. Subhan, M.A.; Yalamarty, S.S.K.; Filipczak, N.; Parveen, F.; Torchilin, V.P. Recent Advances in Tumor Targeting via EPR Effect for Cancer Treatment. *J. Pers. Med.* **2021**, *11*, 571. [CrossRef]
48. Greish, K. Enhanced Permeability and Retention (EPR) Effect for Anticancer Nanomedicine Drug Targeting. In *Cancer Nanotechnology: Methods and Protocols*; Grobmyer, S.R., Moudgil, B.M., Eds.; Methods in Molecular Biology; Humana Press: Totowa, NJ, USA, 2010; pp. 25–37. ISBN 978-1-60761-609-2.
49. Blanco, E.; Shen, H.; Ferrari, M. Principles of nanoparticle design for overcoming biological barriers to drug delivery. *Nat. Biotechnol.* **2015**, *33*, 941–951. [CrossRef]
50. Perrault, S.D.; Walkey, C.; Jennings, T.; Fischer, H.C.; Chan, W.C.W. Mediating Tumor Targeting Efficiency of Nanoparticles Through Design. *Nano Lett.* **2009**, *9*, 1909–1915. [CrossRef]
51. Jain, R.K.; Stylianopoulos, T. Delivering nanomedicine to solid tumors. *Nat. Rev. Clin. Oncol.* **2010**, *7*, 653–664. [CrossRef]
52. Huang, K.; Ma, H.; Liu, J.; Huo, S.; Kumar, A.; Wei, T.; Zhang, X.; Jin, S.; Gan, Y.; Wang, P.C.; et al. Size-dependent localization and penetration of ultrasmall gold nanoparticles in cancer cells, multicellular spheroids, and tumors in vivo. *ACS Nano* **2012**, *6*, 4483–4493. [CrossRef]
53. Sani, A.; Cao, C.; Cui, D. Toxicity of gold nanoparticles (AuNPs): A review. *Biochem. Biophys. Rep.* **2021**, *26*, 100991. [CrossRef]
54. Schmid, G.; Kreyling, W.G.; Simon, U. Toxic effects and biodistribution of ultrasmall gold nanoparticles. *Arch. Toxicol.* **2017**, *91*, 3011–3037. [CrossRef] [PubMed]
55. Chen, Y.-S.; Hung, Y.-C.; Liau, I.; Huang, G.S. Assessment of the In Vivo Toxicity of Gold Nanoparticles. *Nanoscale Res. Lett.* **2009**, *4*, 858–864. [CrossRef] [PubMed]
56. Bharadwaj, K.K.; Rabha, B.; Pati, S.; Sarkar, T.; Choudhury, B.K.; Barman, A.; Bhattacharjya, D.; Srivastava, A.; Baishya, D.; Edinur, H.A.; et al. Green Synthesis of Gold Nanoparticles Using Plant Extracts as Beneficial Prospect for Cancer Theranostics. *Molecules* **2021**, *26*, 6389. [CrossRef] [PubMed]
57. Lee, K.X.; Shameli, K.; Miyake, M.; Kuwano, N.; Khairudin, N.B.B.A.; Mohamad, S.E.B.; Yew, Y.P. Green Synthesis of Gold Nanoparticles Using Aqueous Extract of Garcinia mangostana Fruit Peels. *J. Nanomater.* **2016**, *2016*, e8489094. [CrossRef]

58. Esther, J.; Sridevi, V. Synthesis and characterization of chitosan-stabilized gold nanoparticles through a facile and green approach. *Gold Bull.* **2017**, *50*, 1–5. [CrossRef]
59. Mafuné, F.; Kohno, J.; Takeda, Y.; Kondow, T.; Sawabe, H. Formation of Gold Nanoparticles by Laser Ablation in Aqueous Solution of Surfactant. *J. Phys. Chem. B* **2001**, *105*, 5114–5120. [CrossRef]
60. Lévy, A.; De Anda Villa, M.; Laurens, G.; Blanchet, V.; Bozek, J.; Gaudin, J.; Lamour, E.; Macé, S.; Mignon, P.; Milosavljević, A.R.; et al. Surface Chemistry of Gold Nanoparticles Produced by Laser Ablation in Pure and Saline Water. *Langmuir* **2021**, *37*, 5783–5794. [CrossRef]
61. Amendola, V.; Polizzi, S.; Meneghetti, M. Laser Ablation Synthesis of Gold Nanoparticles in Organic Solvents. *J. Phys. Chem. B* **2006**, *110*, 7232–7237. [CrossRef]
62. Cho, S.P.; Jang, S.; Jo, H.N.; Lee, S.A.; Bae, S.; Lee, S.H.; Hwang, J.; Joh, H.I.; Wang, G.; Kim, T.W. One step synthesis of Au nanoparticle-cyclized polyacrylonitrile composite films and their use in organic nano-floating gate memory applications. *J. Mater. Chem. C* **2016**, *4*, 1511–1516. [CrossRef]
63. Bakrania, S.D.; Rathore, G.K.; Wooldridge, M.S. An investigation of the thermal decomposition of gold acetate. *J. Therm. Anal. Calorim.* **2009**, *95*, 117–122. [CrossRef]
64. Rak, M.J.; Saadé, N.K.; Friščić, T.; Moores, A. Mechanosynthesis of ultra-small monodisperse amine-stabilized gold nanoparticles with controllable size. *Green Chem.* **2014**, *16*, 86–89. [CrossRef]
65. Turkevich, J.; Stevenson, P.C.; Hillier, J. A study of the nucleation and growth processes in the synthesis of colloidal gold. *Discuss. Faraday Soc.* **1951**, *11*, 55–75. [CrossRef]
66. Frens, G. Controlled Nucleation for the Regulation of the Particle Size in Monodisperse Gold Suspensions. *Nat. Phys. Sci.* **1973**, *241*, 20–22. [CrossRef]
67. Tyagi, H.; Kushwaha, A.; Kumar, A.; Aslam, M. A Facile pH Controlled Citrate-Based Reduction Method for Gold Nanoparticle Synthesis at Room Temperature. *Nanoscale Res. Lett.* **2016**, *11*, 362. [CrossRef]
68. Tran, M.; DePenning, R.; Turner, M.; Padalkar, S. Effect of citrate ratio and temperature on gold nanoparticle size and morphology. *Mater. Res. Express* **2016**, *3*, 105027. [CrossRef]
69. Ojea-Jiménez, I.; Bastús, N.G.; Puntes, V. Influence of the Sequence of the Reagents Addition in the Citrate-Mediated Synthesis of Gold Nanoparticles. *J. Phys. Chem. C* **2011**, *115*, 15752–15757. [CrossRef]
70. Yang, Y.; Shi, J.; Chen, H.; Dai, S.; Liu, Y. Enhanced off-resonance optical nonlinearities of Au@CdS core-shell nanoparticles embedded in $BaTiO_3$ thin films. *Chem. Phys. Lett.* **2003**, *370*, 1–6. [CrossRef]
71. Al-Johani, H.; Abou-Hamad, E.; Jedidi, A.; Widdifield, C.M.; Viger-Gravel, J.; Sangaru, S.S.; Gajan, D.; Anjum, D.H.; Ould-Chikh, S.; Hedhili, M.N.; et al. The structure and binding mode of citrate in the stabilization of gold nanoparticles. *Nat. Chem.* **2017**, *9*, 890–895. [CrossRef]
72. Beishenaliev, A.; Faruqu, F.N.; Leo, B.F.; Lit, L.C.; Loke, Y.L.; Chang, C.-C.; Teo, Y.Y.; Chik, Z.; Foo, Y.Y.; Chung, L.Y.; et al. Facile synthesis of biocompatible sub-5 nm alginate-stabilised gold nanoparticles with sonosensitising properties. *Colloids Surf. A Physicochem. Eng. Asp.* **2021**, *627*, 127141. [CrossRef]
73. Abrica-González, P.; Zamora-Justo, J.A.; Sotelo-López, A.; Vázquez-Martínez, G.R.; Balderas-López, J.A.; Muñoz-Diosdado, A.; Ibáñez-Hernández, M. Gold nanoparticles with chitosan, N-acylated chitosan, and chitosan oligosaccharide as DNA carriers. *Nanoscale Res. Lett.* **2019**, *14*, 258. [CrossRef] [PubMed]
74. Fan, C.; Jiang, L. Preparation of Hydrophobic Nanometer Gold Particles and Their Optical Absorption in Chloroform. *Langmuir* **1997**, *13*, 3059–3062. [CrossRef]
75. Mellor, R.D.; Schätzlein, A.G.; Uchegbu, I.F. Development of Bio-Functionalized, Raman Responsive, and Potentially Excretable Gold Nanoclusters. *Nanomaterials* **2021**, *11*, 2181. [CrossRef]
76. Chen, X.Y.; Li, J.R.; Jiang, L. Two-dimensional arrangement of octadecylamine-functionalized gold nanoparticles using the LB technique. *Nanotechnology* **2000**, *11*, 108–111. [CrossRef]
77. Shellaiah, M.; Simon, T.; Sun, K.W.; Ko, F.-H. Simple bare gold nanoparticles for rapid colorimetric detection of Cr^{3+} ions in aqueous medium with real sample applications. *Sens. Actuators B Chem.* **2016**, *226*, 44–51. [CrossRef]
78. Brust, M.; Walker, M.; Bethell, D.; Schiffrin, D.J.; Whyman, R. Synthesis of thiol-derivatised gold nanoparticles in a two-phase Liquid–Liquid system. *J. Chem. Soc. Chem. Commun.* **1994**, 801–802. [CrossRef]
79. Briñas, R.P.; Maetani, M.; Barchi, J.J. A survey of place-exchange reaction for the preparation of water-soluble gold nanoparticles. *J. Colloid Interface Sci.* **2013**, *392*, 415–421. [CrossRef]
80. Dichello, G.A.; Fukuda, T.; Maekawa, T.; Whitby, R.L.D.; Mikhalovsky, S.V.; Alavijeh, M.; Pannala, A.S.; Sarker, D.K. Preparation of liposomes containing small gold nanoparticles using electrostatic interactions. *Eur. J. Pharm. Sci.* **2017**, *105*, 55–63. [CrossRef]
81. Praharaj, S.; Panigrahi, S.; Basu, S.; Pande, S.; Jana, S.; Ghosh, S.K.; Pal, T. Effect of bromide and chloride ions for the dissolution of colloidal gold. *J. Photochem. Photobiol. A Chem.* **2007**, *187*, 196–201. [CrossRef]
82. Kuroda, Y.; Fukumoto, K.; Kuroda, K. Uniform and high dispersion of gold nanoparticles on imogolite nanotubes and assembly into morphologically controlled materials. *Appl. Clay Sci.* **2012**, *55*, 10–17. [CrossRef]
83. Bastús, N.G.; Comenge, J.; Puntes, V. Kinetically Controlled Seeded Growth Synthesis of Citrate-Stabilized Gold Nanoparticles of up to 200 nm: Size Focusing versus Ostwald Ripening. *Langmuir* **2011**, *27*, 11098–11105. [CrossRef]
84. Ziegler, C.; Eychmüller, A. Seeded Growth Synthesis of Uniform Gold Nanoparticles with Diameters of 15−300 nm. *J. Phys. Chem. C* **2011**, *115*, 4502–4506. [CrossRef]

85. Zheng, Y.; Zhong, X.; Li, Z.; Xia, Y. Successive, Seed-Mediated Growth for the Synthesis of Single-Crystal Gold Nanospheres with Uniform Diameters Controlled in the Range of 5–150 nm. *Part. Part. Syst. Charact.* **2014**, *31*, 266–273. [CrossRef]
86. Oh, J.-H.; Sa, Y.-J.; Joo, S.-H.; Lee, J.-S. Assembling Gold Nanocubes Into a Nanoporous Gold Material. *Bull. Korean Chem. Soc.* **2012**, *33*, 1777–1780. [CrossRef]
87. Bhattarai, S.R.; Derry, P.J.; Aziz, K.; Singh, P.K.; Khoo, A.M.; Chadha, A.S.; Liopo, A.; Zubarev, E.R.; Krishnan, S. Gold nanotriangles: Scale up and X-ray radiosensitization effects in mice. *Nanoscale* **2017**, *9*, 5085–5093. [CrossRef] [PubMed]
88. Chateau, D.; Liotta, A.; Vadcard, F.; Navarro, J.R.G.; Chaput, F.; Lermé, J.; Lerouge, F.; Parola, S. From gold nanobipyramids to nanojavelins for a precise tuning of the plasmon resonance to the infrared wavelengths: Experimental and theoretical aspects. *Nanoscale* **2015**, *7*, 1934–1943. [CrossRef]
89. Khoury, C.G.; Vo-Dinh, T. Gold Nanostars For Surface-Enhanced Raman Scattering: Synthesis, Characterization and Optimization. *J. Phys. Chem. C* **2008**, *112*, 18849–18859. [CrossRef]
90. Nikoobakht, B.; El-Sayed, M.A. Preparation and Growth Mechanism of Gold Nanorods (NRs) Using Seed-Mediated Growth Method. *Chem. Mater.* **2003**, *15*, 1957–1962. [CrossRef]
91. Piella, J.; Bastús, N.G.; Puntes, V. Size-Controlled Synthesis of Sub-10-nanometer Citrate-Stabilized Gold Nanoparticles and Related Optical Properties. *Chem. Mater.* **2016**, *28*, 1066–1075. [CrossRef]
92. Cortez-Lemus, N.A.; Licea-Claverie, A.; Paraguay-Delgado, F.; Alonso-Nuñez, G. Gold Nanoparticles Size Design and Control by Poly(*N*,*N*′-diethylaminoethyl methacrylate). *J. Nanomater.* **2015**, *2015*, e273814. [CrossRef]
93. Xia, H.; Xiahou, Y.; Zhang, P.; Ding, W.; Wang, D. Revitalizing the Frens Method To Synthesize Uniform, Quasi-Spherical Gold Nanoparticles with Deliberately Regulated Sizes from 2 to 330 nm. *Langmuir* **2016**, *32*, 5870–5880. [CrossRef]
94. Liu, J.; Yu, M.; Zhou, C.; Yang, S.; Ning, X.; Zheng, J. Passive Tumor Targeting of Renal-Clearable Luminescent Gold Nanoparticles: Long Tumor Retention and Fast Normal Tissue Clearance. *J. Am. Chem. Soc.* **2013**, *135*, 4978–4981. [CrossRef]
95. Yu, M.; Zhou, C.; Liu, L.; Zhang, S.; Sun, S.; Hankins, J.D.; Sun, X.; Zheng, J. Interactions of Renal-Clearable Gold Nanoparticles with Tumor Microenvironments: Vasculature and Acidity Effects. *Angew. Chem. Int. Ed.* **2017**, *56*, 4314–4319. [CrossRef]
96. Xi, W.; Haes, A.J. Elucidation of HEPES Affinity to and Structure on Gold Nanostars. *J. Am. Chem. Soc.* **2019**, *141*, 4034–4042. [CrossRef] [PubMed]
97. Yang, Y.; Serrano, L.A.; Guldin, S. A Versatile AuNP Synthetic Platform for Decoupled Control of Size and Surface Composition. *Langmuir* **2018**, *34*, 6820–6826. [CrossRef]
98. Wu, B.-H.; Yang, H.-Y.; Huang, H.-Q.; Chen, G.-X.; Zheng, N.-F. Solvent effect on the synthesis of monodisperse amine-capped Au nanoparticles. *Chin. Chem. Lett.* **2013**, *24*, 457–462. [CrossRef]
99. Liu, J.; Yu, M.; Ning, X.; Zhou, C.; Yang, S.; Zheng, J. PEGylation and Zwitterionization: Pros and Cons in the Renal Clearance and Tumor Targeting of Near-IR-Emitting Gold Nanoparticles. *Angew. Chem. Int. Ed.* **2013**, *52*, 12572–12576. [CrossRef] [PubMed]
100. Choi, H.S.; Liu, W.; Misra, P.; Tanaka, E.; Zimmer, J.P.; Ipe, B.I.; Bawendi, M.G.; Frangioni, J.V. Renal clearance of quantum dots. *Nat. Biotechnol.* **2007**, *25*, 1165–1170. [CrossRef] [PubMed]
101. *Advances in Nanotheranostics, I.*; Springer: New York, NY, USA, 2015; ISBN 978-3-662-48542-2.
102. Patra, J.K.; Das, G.; Fraceto, L.F.; Campos, E.V.R.; del Pilar Rodriguez-Torres, M.; Acosta-Torres, L.S.; Diaz-Torres, L.A.; Grillo, R.; Swamy, M.K.; Sharma, S.; et al. Nano based drug delivery systems: Recent developments and future prospects. *J. Nanobiotechnol.* **2018**, *16*, 71. [CrossRef]
103. Duan, X.; Li, Y. Physicochemical Characteristics of Nanoparticles Affect Circulation, Biodistribution, Cellular Internalization, and Trafficking. *Small* **2013**, *9*, 1521–1532. [CrossRef]
104. Kim, H.; Lee, D. Near-Infrared-Responsive Cancer Photothermal and Photodynamic Therapy Using Gold Nanoparticles. *Polymers* **2018**, *10*, 961. [CrossRef]
105. Le, K.Q.; Alù, A.; Bai, J. Multiple Fano interferences in a plasmonic metamolecule consisting of asymmetric metallic nanodimers. *J. Appl. Phys.* **2015**, *117*, 023118. [CrossRef]
106. Martínez, Á.; Lyu, Y.; Mancin, F.; Scrimin, P. Glucosamine Phosphate Induces AuNPs Aggregation and Fusion into Easily Functionalizable Nanowires. *Nanomaterials* **2019**, *9*, 622. [CrossRef] [PubMed]
107. Rengan, A.K.; Bukhari, A.B.; Pradhan, A.; Malhotra, R.; Banerjee, R.; Srivastava, R.; De, A. In Vivo Analysis of Biodegradable Liposome Gold Nanoparticles as Efficient Agents for Photothermal Therapy of Cancer. *Nano Lett.* **2015**, *15*, 842–848. [CrossRef] [PubMed]
108. Chou, L.Y.T.; Zagorovsky, K.; Chan, W.C.W. DNA assembly of nanoparticle superstructures for controlled biological delivery and elimination. *Nat. Nanotechnol.* **2014**, *9*, 148–155. [CrossRef] [PubMed]
109. Mapanao, A.K.; Santi, M.; Faraci, P.; Cappello, V.; Cassano, D.; Voliani, V. Endogenously Triggerable Ultrasmall-in-Nano Architectures: Targeting Assessment on 3D Pancreatic Carcinoma Spheroids. *ACS Omega* **2018**, *3*, 11796–11801. [CrossRef] [PubMed]
110. Cassano, D.; Summa, M.; Pocoví-Martínez, S.; Mapanao, A.-K.; Catelani, T.; Bertorelli, R.; Voliani, V. Biodegradable Ultrasmall-in-Nano Gold Architectures: Mid-Period In Vivo Distribution and Excretion Assessment. *Part. Part. Syst. Charact.* **2019**, *36*, 1800464. [CrossRef]
111. Katrina Mapanao, A.; Giannone, G.; Summa, M.; Laura Ermini, M.; Zamborlin, A.; Santi, M.; Cassano, D.; Bertorelli, R.; Voliani, V. Biokinetics and clearance of inhaled gold ultrasmall-in-nano architectures. *Nanoscale Adv.* **2020**, *2*, 3815–3820. [CrossRef]

112. Santi, M.; Mapanao, A.K.; Cassano, D.; Vlamidis, Y.; Cappello, V.; Voliani, V. Endogenously-Activated Ultrasmall-in-Nano Therapeutics: Assessment on 3D Head and Neck Squamous Cell Carcinomas. *Cancers* **2020**, *12*, 1063. [CrossRef]
113. Mapanao, A.K.; Santi, M.; Voliani, V. Combined chemo-photothermal treatment of three-dimensional head and neck squamous cell carcinomas by gold nano-architectures. *J. Colloid Interface Sci.* **2021**, *582*, 1003–1011. [CrossRef]
114. Cassano, D.; Santi, M.; D'Autilia, F.; Mapanao, A.K.; Luin, S.; Voliani, V. Photothermal effect by NIR-responsive excretable ultrasmall-in-nano architectures. *Mater. Horiz.* **2019**, *6*, 531–537. [CrossRef]
115. Higbee-Dempsey, E.M.; Amirshaghaghi, A.; Case, M.J.; Bouché, M.; Kim, J.; Cormode, D.P.; Tsourkas, A. Biodegradable Gold Nanoclusters with Improved Excretion Due to pH-Triggered Hydrophobic-to-Hydrophilic Transition. *J. Am. Chem. Soc.* **2020**, *142*, 7783–7794. [CrossRef]
116. Cheheltani, R.; Ezzibdeh, R.M.; Chhour, P.; Pulaparthi, K.; Kim, J.; Jurcova, M.; Hsu, J.C.; Blundell, C.; Litt, H.I.; Ferrari, V.A.; et al. Tunable, biodegradable gold nanoparticles as contrast agents for computed tomography and photoacoustic imaging. *Biomaterials* **2016**, *102*, 87–97. [CrossRef] [PubMed]
117. Yahia-Ammar, A.; Sierra, D.; Mérola, F.; Hildebrandt, N.; Le Guével, X. Self-Assembled Gold Nanoclusters for Bright Fluorescence Imaging and Enhanced Drug Delivery. *ACS Nano* **2016**, *10*, 2591–2599. [CrossRef] [PubMed]
118. Tam, J.M.; Tam, J.O.; Murthy, A.; Ingram, D.R.; Ma, L.L.; Travis, K.; Johnston, K.P.; Sokolov, K.V. Controlled Assembly of Biodegradable Plasmonic Nanoclusters for Near-Infrared Imaging and Therapeutic Applications. *ACS Nano* **2010**, *4*, 2178–2184. [CrossRef] [PubMed]
119. Deng, H.; Dai, F.; Ma, G.; Zhang, X. Theranostic Gold Nanomicelles made from Biocompatible Comb-like Polymers for Thermochemotherapy and Multifunctional Imaging with Rapid Clearance. *Adv. Mater.* **2015**, *27*, 3645–3653. [CrossRef] [PubMed]

nanomaterials

MDPI

Review

Self-Assembled Metal Nanoclusters: Driving Forces and Structural Correlation with Optical Properties

Sarita Kolay [1], Dipankar Bain [2], Subarna Maity [1], Aarti Devi [2], Amitava Patra [1,2,*] and Rodolphe Antoine [3,*]

[1] School of Materials Sciences, Indian Association for the Cultivation of Science, Kolkata 700032, India; mssk2256@iacs.res.in (S.K.); stsm2@iacs.res.in (S.M.)
[2] Energy and Environment Unit, Institute of Nano Science and Technology, Knowledge City, Sector 81, Mohali 140306, India; dipankar.ra202102@inst.ac.in (D.B.); aarti.ph20257@inst.ac.in (A.D.)
[3] CNRS, Institut Lumière Matière UMR 5306, Univ Lyon, Université Claude Bernard Lyon 1, F-69100 Villeurbanne, France
* Correspondence: msap@iacs.res.in (A.P.); rodolphe.antoine@univ-lyon1.fr (R.A.)

Abstract: Studies on self-assembly of metal nanoclusters (MNCs) are an emerging field of research owing to their significant optical properties and potential applications in many areas. Fabricating the desired self-assembly structure for specific implementation has always been challenging in nanotechnology. The building blocks organize themselves into a hierarchical structure with a high order of directional control in the self-assembly process. An overview of the recent achievements in the self-assembly chemistry of MNCs is summarized in this review article. Here, we investigate the underlying mechanism for the self-assembly structures, and analysis reveals that van der Waals forces, electrostatic interaction, metallophilic interaction, and amphiphilicity are the crucial parameters. In addition, we discuss the principles of template-mediated interaction and the effect of external stimuli on assembly formation in detail. We also focus on the structural correlation of the assemblies with their photophysical properties. A deep perception of the self-assembly mechanism and the degree of interactions on the excited state dynamics is provided for the future synthesis of customizable MNCs with promising applications.

Keywords: self-assembly; metal nanoclusters; nanoscale forces; structural correlation; optical property

1. Introduction

The study of self-assembly of nanomaterials has been an efficient and powerful strategy in nanotechnology for decades and is still relevant today. Self-assembly is a flexible process where the building blocks spontaneously assemble themselves into a highly ordered large structure through direct interaction and with the help of an external influence [1–4]. A comprehensive understanding of the thermodynamic forces that drive the self-assembly process is a prerequisite for designing and controlling the morphology of the assembled structure for ideal applications [5–7]. The assembly formation offers an enhanced collective property for a wide range of applications [8–12]. The self-assembled architectures with different morphologies have unique directing forces with exclusive photophysical properties.

However, the structural correlation with the optical properties is limited in the nanomaterials, especially for the nanoparticles (NPs) due to their polydisperse nature and the inherent difficulty in crystallization. Herein, a subclass of NPs, MNCs grabbed attention in self-assembly research due to their atomically precise composition and well-defined structure [13–18]. MNCs are ultra-small particles consisting of a controlled aggregation of metal atoms protected by a shell of organic ligands. Such MNCs are usually obtained at the atomic precision (i.e., with a defined number of metal atoms (nM) and ligand molecules (mL) leading to formula (M_nL_m)) and display molecular-like properties [15,19–21]. The geometry of the clusters must be determined by quantum chemistry methods that often use group theory, and the optic response is described in terms of molecular transitions whose positions and intensities are predicted by sophisticated calculations of quantum mechanics [22–24]. Among these molecular-like properties, luminescence has attracted tremendous research interest [25–29]. However, compared with conventional luminescent materials (e.g., organic dyes and quantum dots), the quantum yield (QY) of MNCs is relatively low [30]. Constructing hierarchical self-assembly or inducing aggregation of ligated MNCs has opened the way to boost the luminescence of MNCs [31]. However, uncontrolled aggregation during self-assembly may usually occur due to MNCs' large specific surface area. On the other hand, the richness of surface ligands in MNCs and the versatile surface functionalities allow for elaborating strategies involving manipulating the driving forces guiding the intercluster interactions. These attractive forces act as a glue between surface ligands of different NCs and can be triggered by external stimuli such as metal ions, pH, macromolecules, solvents, light, etc. MNCs are increasingly used as versatile building blocks to fabricate nanoarchitectures via a self-assembly process. The assembled geometry can be tuned with specific size and composition by controlling the driving forces, templates, or external stimuli [32–35]. Modulating the nanoscale forces such as dipolar interactions, van der Waals forces, electrostatic interaction, and metallophilic interaction, assembled architectures can be tailored. The correlation between the ligand structure and intra-cluster ligand interaction can play a significant role in fabricating ordered assemblies [36]. The hierarchical assembly of MNCs with atomic precision can be significantly modulated by tailoring the surface ligands [37]. The diverse surface chemistry of the ligands with different moieties provides a versatile weapon for controlling the assembly geometry as well as their photophysical properties [38,39]. A deep understanding of the control of morphology is required to customize the architectures for ideal applications. Hierarchically assembled structures consisting of MNCs are important in nanotechnology because of their collective properties and potential applications in catalysis, electronics, sensors, and storage devices [40–44]. Although most MNCs show weak luminescence, their assembly can enhance the luminescence property based on aggregation-induced emission strategy [30,45–49]. The metal–metal, metal–ligand and ligand–ligand interactions control the relaxation dynamics. The rotational and vibrational motions of the ligands are restricted during assembly formation, and thus the non-radiative channels for relaxation decrease, which enhances the luminescence.

There are many recent reviews on the synthetic strategies to produce self-assembly of NCs and their applications [50–59]. However, the structural correlation of the MNCs with their optical properties remains unknown, which is also important to design the self-assembled structures with customizable properties. In this review, we focus particularly on the self-assembled structure–optical properties relationship and the driving forces behind assembly formation. The current review article is thus intended to give insights into the fundamental driving forces leading to the self-assembly of metal NCs (copper, gold, and silver) and their impact on photophysical properties. Specifically, we discuss the fundamentals of leading interactions in self-assembled MNCs. The template effect controls the DNA-based self-assembly of MNCs. The use of external stimuli, e.g., light and temperature, controls the assembly process. Versatility in enhanced photoluminescence of self-assembly of MNCs, e.g., QY enhancement and tuning emission color, will be highlighted. Finally, we outline some perspectives on the development of this area.

2. On the Structure–Optical Properties Relationship of Metal Nanocluster Assembly

The origin of the photoluminescence (PL), in particular in the near-infrared, from thiolate-protected gold nanoclusters remains elusive. Indeed, it is still a major challenge for researchers to map out a definitive relationship between the atomic structure and the PL property and understand how the metal core (through excitations via the Au(0) kernel) and Au(I)–S surface (through charge transfer excitations) contribute to the PL of Au NCs. The lowest excited states in absorption spectra usually belong to the "core" in nature. In the case of the highest excited states, both "interface-like" and "core-like" (or a combination of both) are involved [60]. The characteristic of such excited states is more "interface-like" (or also called "oligomer band" or "Au−S band"), for which contribution from the Au-S interface in orbitals is more pronounced. Further, electron-rich donor groups from surface ligands may contribute to "surface-like" excited states. In addition, multiple energy transfers associated with intersystem crossings (ISC reinforced) allow an overall boost in PL emission and longer PL lifetimes [61]. The following experimental and theoretical findings were assembled from the literature to derive the energy diagram in Figure 1.

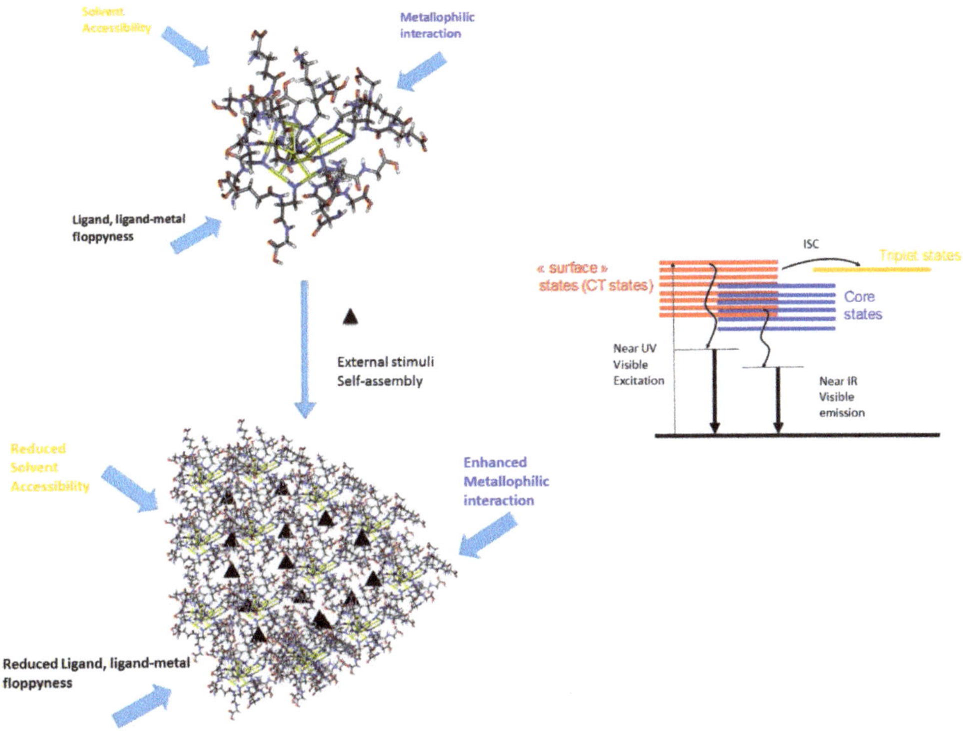

Figure 1. Schematic illustration of the critical parameters affecting the PL properties of MNCs (exemplified with $Au_{10}SG_{10}$) and their self-assembly. Proposed mechanism for PL (photoluminescence) in MNCs (metal nanoclusters).

For thiolated metal NCs, many parameters affect how the energy flows following photoexcitation. The main parameters are molecular floppiness, solvent accessibility, and metallophilic interactions, as exemplified with the smallest $Au_{10}SG_{10}$ NCs (with SG as glutathione ligand, see Figure 1). Generally speaking, loose and floppy ligand molecules such as glutathione will exhibit several rotational and vibrational degrees of freedom and seriously lower the fluorescence intensity. Solvents, in particular solvent accessibility to the metal core or metal-sulfur interface, may play a role in the deexcitation pathway of NCs. Suppose MNCs are in a high vibrational level of the triplet state. In that case, energy can be lost through collisions with solvent molecules (vibrational relaxation), leaving it at the lowest vibrational level of the triplet state. It can then again undergo intersystem crossing to a high vibrational level of the electronic singlet ground state, where it returns to the lowest vibrational level through vibrational relaxation. Solvent accessibility has been found to play a critical role in the luminescence properties of ligated gold NCs. It is strongly dependent on the nature of ligands and the size of MNCs [62,63]. Aurophilic interactions between interlocked Au(I)-thiolate ring structures for $Au_{10}SG_{10}$ catenane NCs [64,65] cause the enhancement of the luminescence for Au(I) oligomers. Usually, the surface ligands are the components that direct aggregation/assembly. Thus, when the assembly of MNCs is produced, there will be a substantial restriction of vibrational and rotational motions of the outer surface ligands of individual MNCs [66]. Building a massive assembly of NCs onto higher-order hybrid superstructures will lead to strong confinement on many length scales. This confinement will also reduce the solvent accessibility, increasing the aggregation-induced emission (AIE) process of luminescent MNC-based assemblies. Finally, intercluster metallophilic interactions also control the aggregations of MNCs and the AIE process. The impacts of the different forces on the AIE process of MNC-based assemblies will be discussed in the following section.

3. Nanoscale Forces on Assembly

Surface interactions can play a vital role in the self-assembly process of MNCs. The structural control and morphological evolution of the directed self-assembly can be controlled by the nanoscale forces such as dipolar attraction, van der Waals interactions, electrostatic interactions, π–π stacking, metallophilic interactions, etc. This section will highlight a couple of examples in which intercluster interactions are the driving force to form assemblies.

3.1. Dipolar Interaction

Numerous building blocks can self-assemble into ordered structures driven by the entropy factor. Ordering the noninteracting hard-sphere results in an entropy increment by having more available free space in the ordered structure compared to the disordered state. Theoretical calculations and simulations predict the most stable ordering of the assembled systems. Self-assembly of particles strongly depends on the particle size and the behavior of the particles based on coulombic, charge–dipole, and dipole–dipole interactions. The dipole moment can originate for the nanocrystals due to the noncentrosymmetric distribution of polar facets or asymmetric lattice truncations [67]. In the case of MNCs, the dipolar attraction between the individual NCs plays a substantial role in structural change. For example, dipolar attraction and van der Waals attraction control the 2D self-assembly of $Au_{15}DT_{15}$ into mono, few, and multi-layered sheets [68]. Au_{15} NCs involve dipole–dipole attraction and van der Waals attractions when the solvent is nonpolar. Dipole–dipole attraction arranges the 1D orientation of Au_{15} NCs. The permanent dipole moment (μ) of $Au_{15}DT_{15}$ was calculated to be 13.27 D, and the energy of dipolar attraction between two NCs was 3.8 kJ/mol, calculated using the formula [68]

$$E = -\frac{\mu^2}{2\pi\varepsilon_0 r(r^2 - d_{NC}^2)} \tag{1}$$

where ε_0 is 8.85×10^{-12} $C^2J^{-1}m^{-1}$, r is the center to center interdipolar separation, and d_{NC} is the diameter of NCs. The 3.8 kJ/mol dipolar energy is higher than the regular dipole–dipole attraction (1.5 kJ/mol). The self-assembly of Au_{15} NCs induced by anisotropic dipole–dipole attraction directs the linear alignment along the x-direction (Figure 2). Further, the inter-NC distance in the self-assembly can be modulated with different solvents resulting in a change in inter-NC dipolar attraction [38]. The dipole moment of Cu_6 is 0.87 D, corresponding to an energy of 0.02 kJ/mol. Cu_3 shows a dipole moment of 3.11 D and more energy (0.64 kJ/mol). The combined effect of theoretical and experimental results of dipolar interactions plays a vital role in the self-assembly process and can control the morphology. The advantage of these interactions over van der Waals is that they only provide internal interaction and thus stabilize the assembled structure.

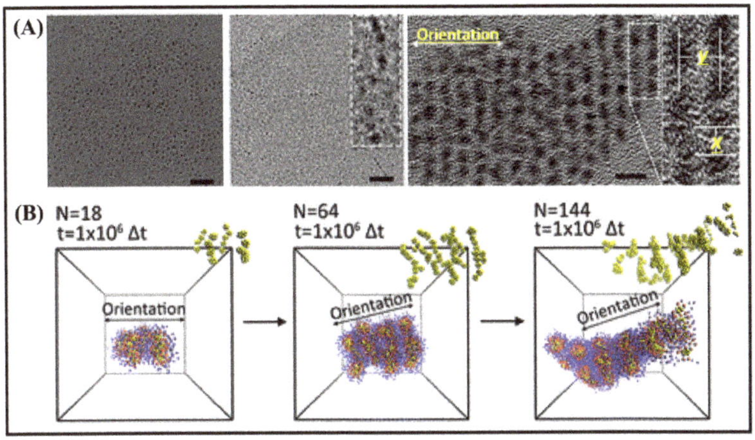

Figure 2. (**A**) TEM images of the morphological evolution of the self-assembled Au15 NCs (a–c), after 1 h at 25 °C (a), 24 h (b) and at 90 °C for 5 min (c). The scale bar is 10 nm (a,b) and 5 nm (c). (**B**) Simulation results of the dipole-induced linear arrangement of Au_{15} NCs. It was adapted from Ref. [68], Copyright 2015 American Chemical Society.

3.2. Van der Waals Interactions

Directed self-assembly can be achieved by controlling the van der Waals (vdW) attractions among the NCs. In the MNC assembly, van der Waals forces between the capping ligands of NCs play a pivotal role in forming an ordered structure. The vdW forces are the most ubiquitous nanoscale interaction, originating due to the electromagnetic fluctuations that arise from the continuous movements of opposite charges within the atoms, molecules, and bulk materials. The vdW interactions between the metal cores can be calculated using the formula [69]

$$u_{\text{vdW}}(r) = -C_{\text{vdW}} / r^6 \tag{2}$$

where r is the distance between the atomic or molecular center, and C_{vdw} is a constant characterizing the interacting species and the surrounding medium. The forces are proportional to $\frac{1}{r^6}$. It is clear that in MNC assemblies, the strength of the van der Waals force is comparable to the strength of other long-range interactions, such as electrostatic interactions. Another approach to estimating the vdW interactions is the additive Hamaker integral approximation [70]. Here, the interacting particles are the sphere, and the dispersion interaction is only present within the particles.

$$U_{vdW} = \frac{A}{3}\left[\frac{a_1a_2}{r^2-(a_1+a_2)^2} + \frac{a_1a_2}{r^2-(a_1-a_2)^2} + \frac{1}{2}\ln\left(\frac{r^2-(a_1+a_2)^2}{r^2-(a_1-a_2)^2}\right)\right] \tag{3}$$

where A is the Hamaker coefficient and can be calculated using $A = C_{\text{vdW}}\pi^2 / v_1v_2$ (v_i is the molar volume of material i) [69].

The morphology of self-assembled Cu NCs capped by 1-dodecanthiol (DT) can be tuned from wire to ribbon (Figure 3A) [71]. The van der Waals attraction between the capping ligands results in assembled nanowires having a diameter of 26 nm (wire-26). The composition of Cu NCs in wire-26 was revealed to be $Cu_{12}DT_8Ac_4$. The vdW attraction between two $Cu_{12}DT_8Ac_4$ NCs can be calculated using Equation (3) and was found to be 4.1 K_BT (K_B is the Boltzmann constant and T is the absolute temperature), which directs the self-assembly formation of the hydrophobic structures. Here, the use of temperature accelerates the self-assembly formation, whereas the low-temperature kinetically controlled process gives rise to wires like morphology. Prolonged annealing of these wires like geometry leads to reorganization into flower like structures. The thickness can be modulated by annealing pre-assembled wire-26 at 120 °C for 30 min. The thickness is reduced to 1.3 nm, indicating the formation of ribbon-1.3, and the composition changes to Cu_8DT_8. The addition of DT facilitates the conversion of the wire into a ribbon by increasing the extent of compactness and arrangement order of the self-assembly. Here, the driving force is mainly vdW attraction, which enables the reorganization of the self-assembled NCs, and the energy is calculated to be 3.8 K_BT between the Cu_8DT_8 NCs. The flexibility of the alkyl chains is more dynamic at high temperature and thus redirected into a compact and ordered structure. As both systems are composed of a self-assembly strategy, the ribbon and wire are highly luminescent. The driving force behind the 1D orientation of self-assembled Cu NCs into nanowires is firmly dipolar attraction [71].

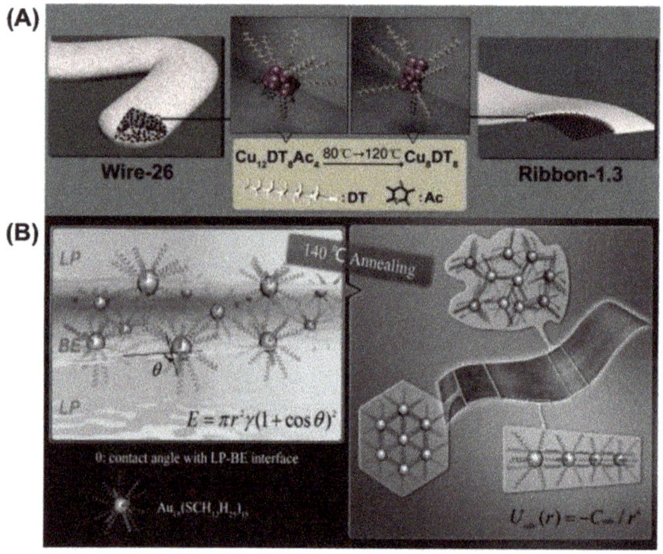

Figure 3. Schematic representation of the morphological evolution of (**A**) the self-assembled Cu NCs from wire-26 into ribbon 1.3. Adapted with permission from Ref. [71]. Copyright 2014 Wiley-VCH Verlag GmbH & Co. KGaA, Weinheim. (**B**) Au_{15} NCs into single-cluster-thick nanosheets. Reprinted from Ref. [72]. with permission. Copyright 2013 Wiley-VCH Verlag GmbH & Co. KGaA, Weinheim.

We will discuss another example where Au NCs protected with hydrophobic alkyl thiol are self-assembled into single-cluster-thick sheets through vdW and intercluster hydrophobic–hydrophobic attraction [72]. The self-assembly and reorganization are due to the slight difference between two miscible solvents, BE and LP (dibenzyl ether and liquid paraffin, respectively) (Figure 3B). Solvent evaporation is enhanced after annealing at 140 °C, and Au_{15} clusters aggregate [72]. The elevated temperature to enhance the mobility of the DT alkyl chain and the hydrophobic attraction between two NCs is also found to be 7 K_BT. This strong interaction leads to the formation of the aggregate into nanosheets. The interfacial energy at the LP-BE surface also drives the reorganization process to form single-cluster-thick sheets. Therefore, by choosing stabilizing ligands as the capping ligand for MNCs or appropriate solvents, van der Waals interactions can be controlled to guide the self-assembly processes.

3.3. Electrostatic Interaction

The electrostatic interactions between the NCs can primarily control the formation of directed self-assembly. Electrostatic interaction is almost as ubiquitous as van der Waals interactions. The unique property is that electrostatic interaction can be attractive, repulsive or directional, whereas van der Waals interactions are attractive in nature. The charge of the overall NCs or alteration of the surface capping groups introduces variation in self-assembly morphology [73]. Several electrostatic interactions such as C–H···π interactions, agostic interactions, H– bonding, and π–π stacking between the NCs can direct the formation of these arrays. The structure of the NCs can be also modulated by the change in symmetry of the ligand environment. Though the $Au_{246}(p\text{-}MBT)_{80}$ NCs [p-MBT = p-methylbenzenethiol] have nearly spherical geometry, they do not prefer cubic arrangement; instead, they formed lower symmetry monoclinic arrangement [74]. This monoclinic packing is driven by the intermolecular C–H···π interaction, maximizing the van der Waals interaction between the ligands. It is essential to match the symmetry to maximize the interaction of surface ligands in ordered assembly. In this case, the p-MBT ligands self-organize amongst themselves in a highly ordered manner. Twenty-five of the surface ligands at the pole site are arranged into four pentagonal circles (Figure 4A). Along with this α-rotation, there are β-parallel patterns at the waist site having six p-MBTs arranged into three alternating parallel pairs. The highly ordered surface p-MBTs are found to interact between the C–H bonds from the phenyl moiety or the $-CH_3$ groups and the π-electrons. This intermolecular C−H···π interaction specifically stabilizes the 25 p-MBT ligands at the pole as well as ligands of the β-parallels (Figure 4B) [74]. Another exciting phenomenon was discovered by Tang's group where Au NCs self-assembled into nanocubes with uniform body-centered cubic (BCC) packing and generated unique properties different from the individual NCs [75]. The chirality of Au NCs was changed by simply changing the chirality of the capping ligand from R to S-Tol-BINAP. Au_3[R-Tol-BINAP]$_3$Cl and Au_3[S-Tol-BINAP]$_3$Cl NCs [(*R*)- or (*S*)-2,2′-bis(di-*p*-tolylphosphino)-1,1′-binaphthyl] were prepared, and the Au atoms experienced strong aurophilic interactions amongst each other (Figure 4C) [75]. The DLS study reveals that the clusters were in a dispersed state when 40% n-hexane was used, and after that, aggregates started forming. With 60% n-hexane, the aggregates were irregular and amorphous, becoming uniform nanocubes with 70% n-hexane. From the crystal structural analysis of the self-assembled nanocubes, it was found that each Au cluster faced six nearby Au clusters where three Au clusters were involved in C−H···π interaction with the central Au one forming both outward and inward ring pairs. Another three Au clusters also experience C−H···π interactions, and the energy was ~1.5 to 2.5 kcal/mol, which is much higher than the molecular thermal energy (0.57 kcal/mol).

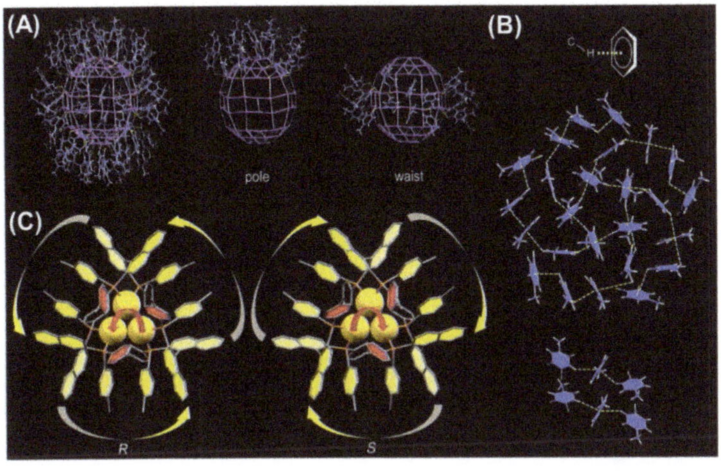

Figure 4. (**A**) The arrangement of the ligands on the surface and (**B**) representation of the C–H···π interactions stabilizing the surface patterns of $Au_{246}(p\text{-}MBT)_{80}$ NCs. Adapted with permission from Ref. [74]. Copyright 2016, American Association for the Advancement of Science. (**C**) Structural representation of Au_3[(R)-Tol-BINAP]$_3$Cl and Au_3[(S)-Tol-BINAP]$_3$Cl. Adapted from Ref. [75] with permission. Copyright 2017 Wiley-VCH Verlag GmbH & Co. KGaA, Weinheim.

Self-assembly of Cu_7 NCs from rods to platelets to ribbon-like structures can be directed with three different positional isomers of the surface capping ligand (Figure 5A) [76]. From the matrix-assisted laser desorption ionization mass spectrometric study, it is clear that the steric hindrance of the ligand is responsible for morphological evolution, keeping the individual building block (Cu_7 NCs) constant for all the assemblies (Figure 5B). Although the composition is identical, the building blocks experience the different extent of agostic interaction (Cu···H–C) for three different isomers of the dimethylbenzenethiol (DMBT) ligand depending on the position of the $-CH_3$ group. The interplay between the extent of agostic interaction and π–π stacking between the NCs directs the shape from rod to ribbon. Figure 5C depicts the schematic representation of ribbon like assemblies from Cu_7 building blocks where the extent of the agostic interaction is the maximum. With these structural changes, optical properties and excited state relaxation dynamics of the red phosphorescent assemblies are also differently correlated with their degree of compactness (Figure 5D,E).

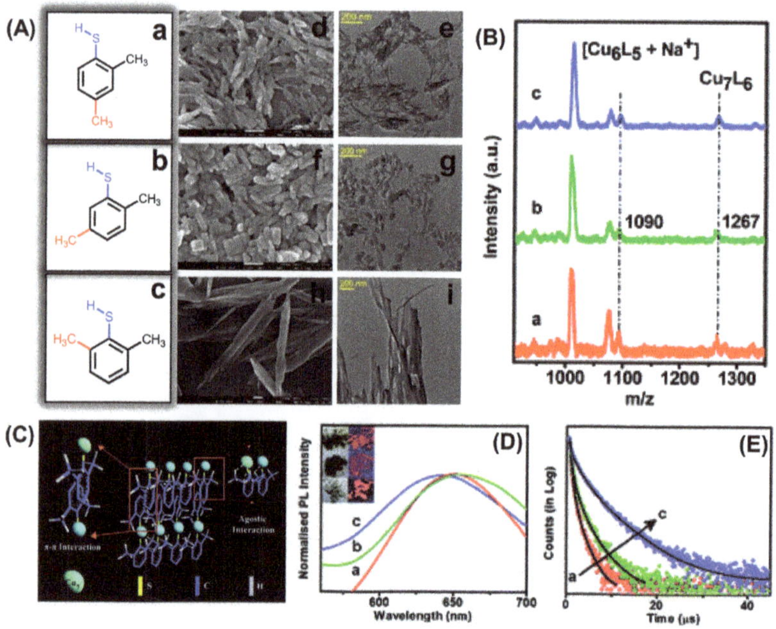

Figure 5. (**A**) The TEM (scale bar is 200 nm) and SEM (scale bar is 100 nm) images of the self-assembled rods, platelets, and ribbon-like structures by varying the isomers of the ligand. (**B**) Structural characterization and (**C**) the evolution process of the assembled NCs. (**D**) Emission spectra and (**E**) PL decay curves of the self-assembled rods (a), platelets (b), and ribbons (c), the inset shows the digital photographs of self-assembled Cu NCs in daylight and under UV (365 nm) excitation. Reprinted with permission from Ref. [76]. Copyright 2021 Royal Society of Chemistry.

Xie et al., obtained customizable shapes of Ag_{44}(p-MBA)$_{30}$ (p-MBA = para-mercaptobenzoicacid) nanocluster super crystals (SCs) by changing the counterions and their concentration of counterions [77]. The geometry changes from a lower symmetry rhombohedral (D_{3d}) to a higher symmetry octahedral (O_h) by altering H^+ with non-H^+ counterion (Cs^+). Again, enhancement of the Cs^+ concentration leads to tailoring the geometry from O_h to concave O_h. The growth kinetics of the SCs manipulate the shape and electrostatic interactions, which follow the charge-screening-assisted nucleation mechanism. Thus, we can conclude that the degree of electrostatic interaction is a primary decisive parameter leading to aggregate formation with the oppositely charged systems. One of the recent studies by Pradeep's group showed that simple hydrogen bonding could also direct the self-assembly of NCs using the previously discussed ligand p-MBA. Gold nanorods were functionalized with p-MBA, and Ag_{44} NCs were allowed to interact with this GNR@p-MBA, which resulted in directed self-assembly (Figure 6A) [78]. Ag_{44} NCs form assembly through H-bonding with carboxylic acid dimerization of the capped surface ligands. Pradeep et al., investigated the factors directing the morphological evolution. This study suggested that the p-MBA molecules functionalized on GNR can interact with the surface capped ligands of Ag_{44} NCs, leading to a decrease in the extent of H-bonding interactions between the ligands of two adjacent NCs. Self-assembly of Ag_{25} NCs capped with p-aminothiophenol (PATP) into uniform and highly ordered lamellar silver nano leaves (NLs) was studied by Wang's group [79]. The main driving forces to direct the self-assembly process are electrostatic or covalent bonding and π–π stacking. A couple of studies confirm that the PATP exists in a quinonoid model where the $-NH_2$ (electron-donating group) and the –SH (electron acceptor group) are connected by a conjugated benzene moiety. The Ag NCs are bound with these –SH and $-NH_2$ groups of the PATP ligand existing in quinonoid form (Figure 6B). The smallest building blocks for this lamellar assembly are $[Ag_{25}(PATP)_6]^{6+}$, and they are interconnected as Ag_{25}-PATP-Ag_{25}. Now, π–π stacking interaction between the adjacent benzene rings of the building blocks leads to the formation of uniformly ordered silver NLs. To minimize the steric repulsion and entropy, head-to-head assembly is formed with the help of dipole–dipole interactions, whereas the π–π stacking having the temperature and solvent effect leads the side-by-side organization process.

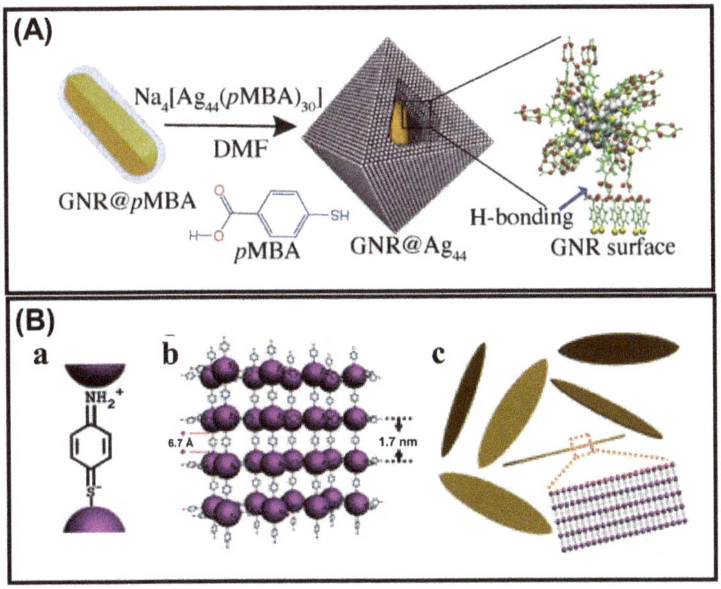

Figure 6. Schematic representation for the (**A**) self-assembly of Ag_{44} NCs on the GNR@p-MBA surface. Adapted with permission from Ref. [78]. Copyright 2018 Wiley-VCH Verlag GmbH &Co. KGaA,Weinheim. (**B**) The quinonoid form of the PATP ligand (a) and formation mechanism of uniform lamellar Ag nano leaves (b,c). Adapted with permission from Ref. [79]. Copyright 2013 American Chemical Society.

3.4. Metallophilic Interaction

The M$\cdots$M bond in self-assembly is attributed to the weak attractive interaction between the electrons in the filled d^{10} or pseudo-filled (d^8) orbital of the metal. The manifestation of metallophilic interaction is crucial for the self-assembly of many NC compounds. Metallophilic interactions have been observed for the late transition metals, more prominently group 11 M(I) salts [80]. Metallophilicity directed assembly has recently turned into an emerging area of research, and several studies are reported based on aurophilic ($Au^{I}\cdots Au^{I}$), argentophilic ($Ag^{I}\cdots Ag^{I}$), and cuprophilic ($Cu^{I}\cdots Cu^{I}$) interactions [41,81,82]. Metallophilic interaction in the shell is a decisive parameter for the luminescence of NCs [83]. Xie et al. designed a colloidal self-assembly of $[Au_{25}(SR)_{18}]^{-}$ NCs into nanoribbons using aurophilic interaction ($Au^{I}\cdots Au^{I}$) (Figure 7) [82]. The self-assembly is initiated by the surface motif reconstruction of NCs from the pristine SR-$[Au^{I}$-$SR]_2$ motifs to longer SR-$[Au^{I}$-$SR]_x$ motifs (x > 2) along with a partial expense of Au^0 species from the Au_{13} core. This is confirmed by XRD patterns of randomly aggregated $[Au_{25}(p\text{-}MBA)_{18}]^{-}$ NCs and XPS of Au 4f (Figure 6). The XPS spectra show an increase in Au^{I} species accompanied by Au^0 reduction in the self-assembled nanoribbons (Figure 7). The Au L_3 edge X-ray absorption near edge structure (XANES) and FT-EXAFS showed a distinct increase in the content of Au-S and $Au_{staple}-Au_{staple}$ characters, which supports the formation of long SR-$[Au^{I}$-$SR]_x$ motifs. The aurophilic interaction is the driving force for the 1D arrangement of Au NCs into nanoribbons. Aurophilic interaction ($Au^{I}\cdots Au^{I}$) plays a crucial role in switching fluorescence to phosphorescence. A small-angle X-ray scattering (SAXS) study confirmed the presence of Au-Au interaction, which drives the self-assembly process and also determines phosphorescence. Bakr et al. reported that argentophilic interactions are crucial for the stability of the silver cluster skeleton [84]. From the single crystal analysis, the Ag–Ag distance is shorter than the van der Waals radius (~3.44 Å), which indicates argentophilic interactions in 0D NCs of the Ag_{16} skeleton. Further, with the employment of a chloride template, 4,4′-bipyridine linker, controlled assembly of 0D, 1D, and 2D NCs based framework can be achieved. The importance of argentophilic interaction in directing the self-assembly is also reflected in a study by Xin et al. [81]. Here, Ag_9 NC constructed a nanofibre structure by interacting with succinic acid. Argentophilic interaction ($Ag^{I}\cdots Ag^{I}$)-directed assembled structure was obtained with aggregation-induced emission. Zhang et al. reported the self-assembly of Cu NCs capped by 1-dodecanethiol with enhanced luminescence. This significant enhancement in PL largely depends on the NC interactions ($Cu^{I}\cdots Cu^{I}$) and the rigidity of the capping ligands. In this case, two different cuprophilic interactions exist, intra and inter cuprophilic interactions between two NCs. In the case of ribbons, the loose arrangement of NCs results in weaker cuprophilic interaction [41]. The control in structural geometry from sheet to ribbon depends on the extent of cuprophilic interactions. Experimental findings and computational studies declare that the metallophilic interactions are weak and are surpassed by strong electrostatic interaction or solvent molecules. Although several studies of self-assembled MNCs have been reported based on metallophilic interaction, the strength and nature of these interactions in MNC assembly are poorly understood.

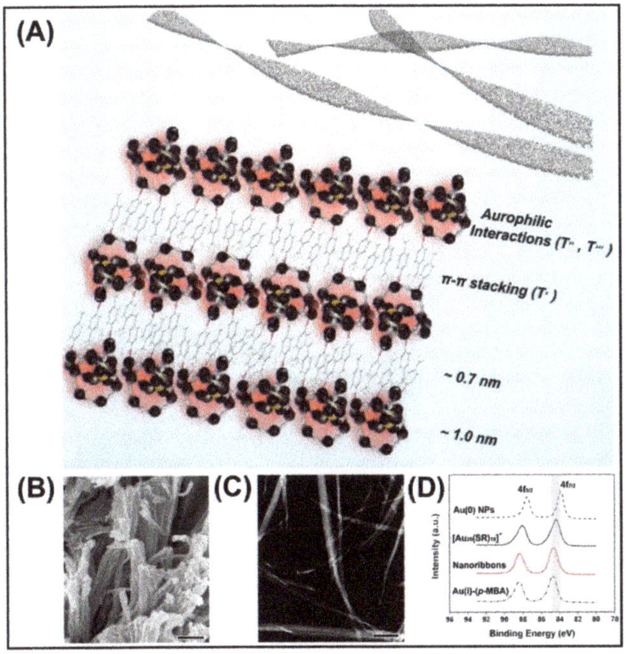

Figure 7. (**A**) Schematic illustration of the aurophilic interaction-directed self-assembly process. (**B**) SEM and (**C**) High-angle dark-field scanning TEM (HAADF-STEM) images of the nanoribbons obtained by the self-assembly of $[Au_{25}(p\text{-}MBA)_{18}]^-$ (scale bars are 300 nm for both). (**D**) Au 4f XPS Study of the self-assembled structures. Reproduced from Ref. [82] with permission. Copyright 2019 Wiley-VCH Verlag GmbH & Co. KGaA, Weinheim.

3.5. Amphiphilicity of NCs

Amphiphiles are natural or synthetic molecules that can form self-assembled micelles, nanotubes, nanofibers, lamellae-like structures, etc. Amphiphiles are composed of hydrophilic and hydrophobic components. Despite having weak forces, it helps the overall structure become more flexible toward the formation of assembly. Hydrogen bonds are essential for the amphiphiles to gain stability in solution, and hydrophobic interaction is the second driving force to form an assembly [85]. The hydrocarbon amphiphiles can be used widely in nanotechnology, self-assembly of noble metal NCs, and also in the surface modification of the NCs. For example, the self-assembly of amphiphilic $Au_{25}(MHA)_{18}$@xCTA NCs (MHA = 6-mercaptohexanoic acid, x = 6–9 and CTA= cetyltrimethylammonium) into stacked bilayers with regular interlayer packing can be observed at the air–liquid interface (Figure 8A) [86]. This was formed by the ion-pairing reaction between CTA^+, hydrophobic cation, and $-COO^-$, derived from the anionic carboxylate group of the hydrophilic NCs. The potential amphiphilicity of the NCs is observed because of the coexistence of hydrophilic MHA and hydrophobic MHA····CTA ligands on the NC surface as well as the flexible chain structure of the surface ligands. Zeng et al. reported the self-assembly of metal NCs guided by the electrostatic repulsive force among surfactant molecules and metal halides. Au NCs form a hierarchical rod/tube-like assembly, whereas giant vesicles and dandelion-like assembly can be found for Pt NCs. Pd NCs and PdS NCs self-assemble into rhombic/hexagonal platelet-like structures (Figure 8B) [87]. Hao et al., in their study, showed an enhancement of fluorescence intensity with the formation of self-assembled amphiphilic Cu NCs [88]. The amphiphilic complex of GSH-capped Cu NCs was prepared by electrostatic interaction and then assembled in an aqueous solution in the presence of a surfactant by rearranging the surface ligands. GSH-Cu NCs are the building blocks that self-assemble in the presence of hydrophilic and hydrophobic moieties, resulting in supramolecular architectures.

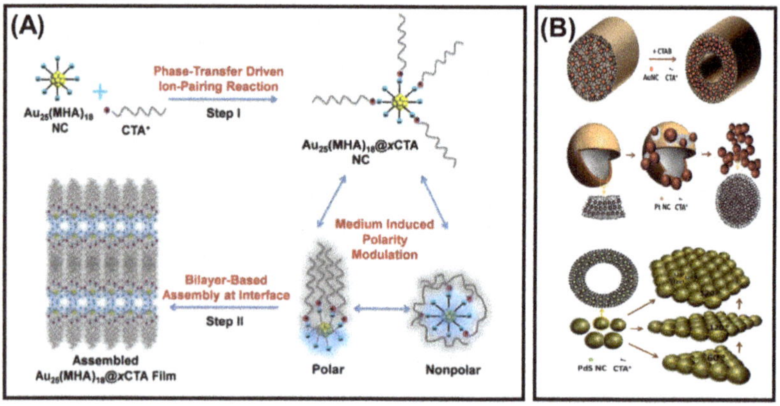

Figure 8. Schematic illustration of the amphiphilicity directed (**A**) self-assembly process of $Au_{25}(MHA)_{18}$@xCTA NCs. Adapted with permission from Ref. [86]. Copyright 2015 American Chemical Society. (**B**) Transformation of Au, Pt, and PdS NCs into rod, dandelion, and platelet-like structures. Adapted with permission from Ref. [87]. Copyright 2014 American Chemical Society.

4. Template-Directed Assembly

In the above section, we have discussed the assembly induced from surface ligands; templates such as DNA, polymers, and macrocycles can also influence the self-assembly process. Templates are a substrate with an active site to bind with the NCs. The variation in templates can result in different assembled morphologies by controlling the interaction in the hybrid assemblies. We will discuss some recent progress on template-directed assembly that entirely depends on template-NC interaction.

4.1. DNA Template-Directed Self Assembly

DNA-templated MNCs are a new generation of functional materials, which have widespread applications in catalysis, sensing, bio-imaging, and therapeutics [51]. This section will demonstrate the structures and optical properties of DNA-functionalized MNCs. DNA is one of the emerging materials for constructing supramolecular assemblies [89]. Highly fluorescent and color-tunable Ag NCs can be synthesized by choosing a proper sequence of DNA templates. Martinez and co-workers synthesized bright emitting Ag NCs by using different arrangements and lengths of single-strand DNA (namely, Ag1, Ag2, Ag3, and Ag4); for example, green fluorescent Ag NCs (λ_{em} = 550 nm) were synthesized by using the Ag1 sequence. Similarly, the emission of the Ag NCs was tuned to 600 nm (orange), 650 nm (red), and 700 nm (NIR) using other DNA sequences Ag2, Ag3, and Ag4, respectively (Figure 9A). The stability of these Ag NCs is mainly dictated by the sequence and the length of DNA [90]. NIR-emitting Ag NCs (Ag4) are most stable under salt conditions compared to the other three NCs because Ag4 has a longer nucleotide length, which prevents the salt-induced aggregation of the NC core.

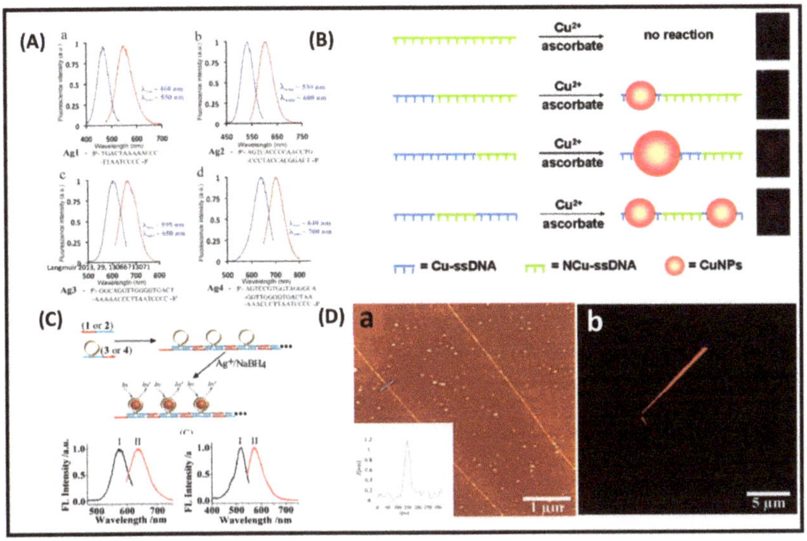

Figure 9. (**A**) Excitation and emission spectra of Ag NCs showing the wavelength-tunable Ag NCs by using different lengths and sequences of DNA. Adapted with permission from ref [90]. Copyright 2010 The Royal Society of Chemistry. (**B**) Cu NPs nanoparticles synthesized by reduction of the ssDNA-Cu^{2+} complex. Cu-ssDNA (blue) represents ssDNA that can serve as a template for Cu NPs, and NCu-ssDNA (green) does not have Cu NP formation capability. The size of Cu NPs depends on the length of Cu-ssDNA. Adapted with permission from ref. [91]. Copyright 2013 John Wiley & Sons, Inc. (**C**) Formation of luminescent Ag NCs by the sequence-specific DNA hairpin structure and excitation (I) and emission (II) spectra of red- and yellow-emitting NCs. (**D**) AFM image of the red emitting Ag NCs (**a**), inset shows the cross-sectional analysis of the self-assembled Ag NCs and confocal microscopy image of wire-like self-assembled Ag NCs with red fluorescence (**b**). Adapted with permission from Ref. [92]. Copyright 2013 American Chemical Society.

Qing et al. synthesized ultrasmall-sized fluorescent copper NCs using poly-thymine as a template [91]. First, they prepared the DNA-Cu^{2+} complex with sequence-specific DNA in MOPS buffer and then reduced it to Cu NCs. Different types of single-stranded DNA were used as templates, namely, single-stranded DNA (ssDNA), poly adenine (poly-A), ploy thymine (poly T), ploy guanine (poly G), and poly cytosine (poly C). Only the Cu NCs templated by poly T emitted fluorescence at 615 nm under the excitation of 340 nm light. In contrast, no such fluorescence was observed for other sequences (poly A, G, and C). The fluorescence intensity of Cu NCs increased as a function of the length of poly T. Moreover, the size of Cu NCs was dictated by the length of poly T of the ssDNA template. Furthermore, they used different types of T-rich domain ssDNA (40 mer). They found that a particular sequence domain produced intense fluorescent Cu NCs. Thus, it was concluded that T polymerization has an important role in forming fluorescent Cu NCs (Figure 9B). NC-based self-assembled structures exhibit more compactness, strong fluorescence, and excellent photostability than bare NCs. Willner and his co-worker reported self-assembly Ag NC-polymerized nanowires using a sequence-specific DNA template [92]. Two different emission colors Ag NCs were prepared, yellow emission (λem = 570 nm) and red emission (λem = 635 nm) (Figure 9C). Self-assembled NCs are seen from AFM and confocal fluorescence microscopy (Figure 9D). The length of the red-emitting self-assembled Ag NCs is in the micrometer order.

4.2. Linker-Directed Assembly

The intrinsic attractive interactions among themselves generally drive regulated self-assembly of molecular NCs, leading to large superstructures. However, small thiolate or non-thiolate ligands that can provide coordination sites to the NCs can control the inter-NC assembly. Ackerson et al. reported a diglyme linker's dynamic assembly of $Au_{20}(PET)_{15}$ NCs [93]. The close spatial proximity between the NCs core via weak diglyme oxygen-Au interaction caused electron delocalization in dimeric Au_{20}-diglyme-Au_{20}. On the flip side, thiolate ligands can covalently attach to the Au core to form a dimeric and trimeric assembly of NCs, which are well-separable by chromatographic techniques [94]. The multimers of plasmonic $Au_{\sim 250}$ NCs exhibit additional transitions in their absorption spectrum due to hybrid LSPR modes. The assembly of NCs via coordinating molecules can be rationally extended to the metal-organic frameworks (MOFs), also referred to as cluster assembled materials (CAMs), with enhanced functionalities and improved photophysical properties. Suitable organic ligands can bind and spatially separate the NCs to form a three-dimensional rigid framework. Luminescent dodecanuclear Ag_{12} NCs form rigid NC-based MOFs by coordinating linear 4,4′-bipyridine (bpy) ligands [95]. The incoming bpy ligands replace the terminal CH_3CN molecules from Ag_{12} NCs and connect the Ag–S NC-based nodes, leading to a porous coordinating framework with yearlong stability in contrast to the few minutes of stable-isolated NCs (Figure 10). Moreover, the MOF exhibits enhanced QY (12.1%) and ultrafast dual-functional fluorescence switching with turn off by O_2 and turn on by volatile organic molecules, which are advantageous in electronic and sensing applications. Solvent-induced transformation of Ag_{12} occurs in the presence of a bpy linker to generate silver CAMs with blue-red dual emission at low temperature [96]. In addition, the mixed linker approach leads to a framework with improved luminescence and thermochromic properties. The chiral $[Au_1Ag_{22}(S\text{-}Adm)_{12}]^{3+}$ superatom serves as the building blocks of a 3D framework with anionic forces [97]. The superatoms are packed into a racemic mixture and enantiomerically pure crystals via intermolecular Ag-F bond formation depending on the linker concentration. The pure enantiomers display strong circularly polarized PL, whereas the racemic crystals show none. In another approach, pre-synthesized GSH-capped Au NCs were incorporated in ZIF-8 by coordination-assisted self-assembly. The spatial distribution of NCs in the framework restricted the ligand rotation and resulted in a 10-fold increase in electroluminescence compared with aggregated Au NCs. The linker-directed assembly of superatomic NCs has laid a foundation for further development in CAMs with diverse functionalities and desirable physical properties.

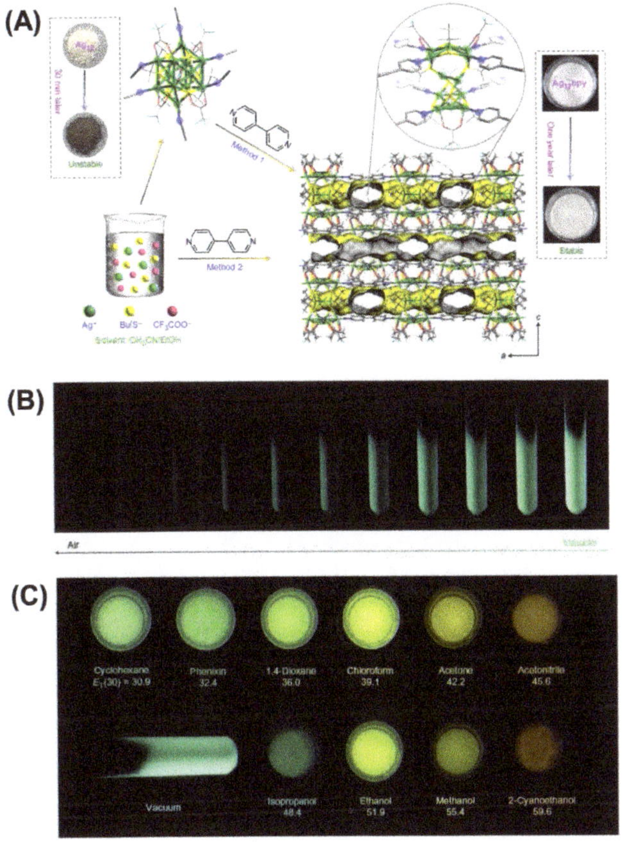

Figure 10. Synthesis and crystal structure of Ag_{12}-bpy (**A**), luminescence quenching response to O_2 (**B**), and vapochromic response of Ag_{12}-bpy (**C**). Reproduced with permission from Ref. [95] Copyright 2017 Macmillan Publishers Limited, part of Springer Nature.

5. Guided Assembly by External Factors

The use of external influences and fields to control the assembly process has long been a powerful method for tailoring the morphology and optical properties of the metal NCs. The self-assembly of metal NCs influenced by external factors such as light and temperature will be discussed here. The self-assembly of metal NCs can be observed by the photoactivation of NCs, with the temperature change, or by coordinating the NCs with organic species.

Light-Triggered Self Assembly

The light-induced self-assembly of metal NCs is an emerging field of research that can be achieved using photoactive ligands such as azobenzene and spiropyran. The critical factor of photoisomerization is extensively studied for these ligands due to the tendency to change their conformation upon irradiation of a particular wavelength of light. In a recent study, self-assembled Cu_3 NCs were prepared using an azo-group-containing ligand, which is highly light-sensitive. The straight self-assembled nanofibers bend upon exposure to UV light instantly (Figure 11A) [98]. This structural transformation is due to the conformational change of the azo-group-containing ligand from trans to cis in light. This isomerization of the peripheral azo-groups in the Cu_3 NCs results in disassembly. Besides the isomerization process, solubility also plays a vital role. The straight nanofibers were fabricated in a binary solvent mixture of toluene and methanol, but the cis conformation was more soluble here. Thus the solvophobic interaction in trans-isomer favors the aggregation process. The light-triggered self-assembly can also be explained by taking an example of phenylethanethiol-tethered Au_{25} NC, which is stapled by the photoactive spiropyran (SP) (Figure 11B) [99]. T. Udayabhaskararao et al. observed that synthesized NCs show photoisomerization in a reversible manner when exposed to UV and visible light due to the dipolar–dipolar interaction (Figure 11C). They demonstrated that the SP-functionalized NCs initially formed in closed rings exhibited no UV absorption peak in the 550–600 nm region. However, exposure to UV light resulted in isomerized into an open ring form, i.e., merocyanine with an absorption band at 587 nm, and the color changed from yellow to purple. The self-assembly of NCs will separate again in a closed ring form if incubated in dark or visible light. Therefore, the assembly–disassembly can be obtained several times for the SP-functionalized NCs; using this strategy, light-induced assembly was found in NCs. Similarly, using the Brust Schiffrin single-phase method, Rival et al. synthesized monothiolated azobenzene-protected Au_{25} NCs showing light-triggered reversible self-assembly—disassembly when irradiated with UV light (365 nm) and visible light (435 nm) [100]. This study shows an attractive disc-like superstructure was obtained due to self-assembly of $[Au_{25}(C_3\text{-AMT})_{18}]^-$ NCs. The photoactivation of NC in the presence of light results in long-range self-assembly and is analyzed using atomic force microscopy AFM, DLS, TEM, and subsequent ET reconstruction. Apart from this, they also synthesized phenylethanethiol (PET)-capped gold NC, with no photoswitching effect. Secondly, the synthesized PET analog $[Au_{25}\ (PET)_{18}]$ NC obtained the same TEM images of individual NCs before and after the light-induced assembly. Here, the attractive dipole–dipole interaction was the driving force for the formation of superstructures of NC upon illumination of light.

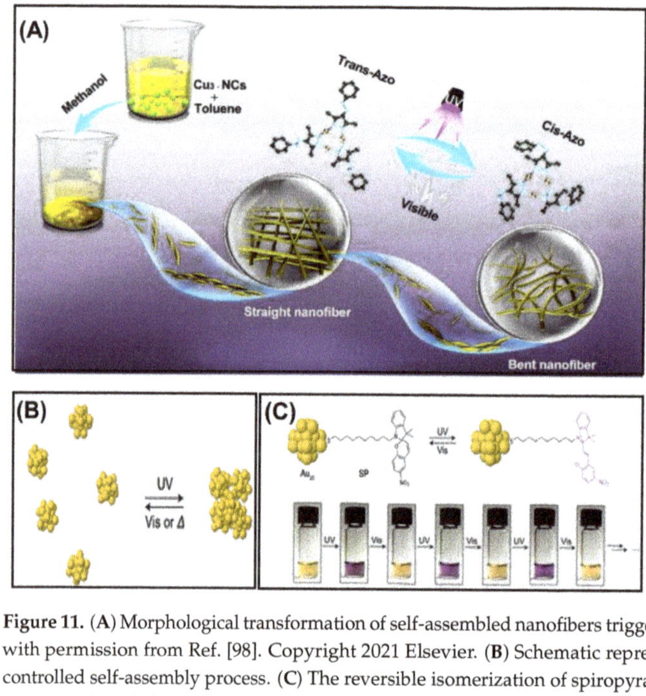

Figure 11. (**A**) Morphological transformation of self-assembled nanofibers triggered by light. Adapted with permission from Ref. [98]. Copyright 2021 Elsevier. (**B**) Schematic representation of the light-controlled self-assembly process. (**C**) The reversible isomerization of spiropyran on Au_{25} NCs in the presence of UV light and digital photographs of $Au_{25}SP_{\sim 3.8}PET_{\sim 14.2}$ in THF solution upon exposure to UV and visible light alternatively. Reprinted with permission from Ref. [99]. Copyright 2016 Wiley-VCH Verlag GmbH & Co. KGaA, Weinheim.

Zhang et al. synthesized thiolated azobenzene-capped Cu NCs with light-triggered self–assembly phenomena in a different study [101]. The formation of long nanoribbons of $Cu_{12}DT_8Ac_4$ NCs was due to the observed permanent dipole moment. The combination of dipolar and van der Waals attraction under UV light formed a spherical superstructure. Here, the light was used as an external stimulus for the reversible self-assembly.

Coordination between the charged MNCs and ions with opposite charges can also direct the self-assembly process. Xie et al. reported an assembly formation via the electrostatic interaction between the negatively charged MNCs in water with the divalent cations, e.g., Cd^{2+} and Zn^{2+}, added externally in the solution [102]. This assembly showed high orderliness and regularity and enhanced PL, implying a strong synergistic effect between the NCs. Another study of zinc-mediated self-assembly was reported by Chattopadhyay et al. [103]. This report investigated programmable assembly formation between the mixed ligand (L-histidine and mercaptopropionic acid)-capped Au_{14} NC and Zn^{2+}. Anisotropic assembly was designed with Ag_{52} and Ag_{76} NCs using $tBuC_6H_4SH$ and $MeOC_6H_4SH$ and a bis-(diphenylphosphino) methane ligand [104]. This study revealed that the ligands' regiospecific distribution and arrangement formed the anisotropic growth. A very recent study reported the formation of chiral assembly from achiral, atomically precise Ag_9 NCs [105]. Here, a second metal ion, Ba^{2+}, facilitated the assembly process, and thus nanotubes with chiral cubic lattice were obtained. These hierarchically organized self-assembled structures formed chiral hydrogels with enhanced luminescence in an aqueous solution. Other external stimuli on which the self-assembly process largely depends are pH and the solvent. As these parts were focused on in many previous articles, we do not discuss them in detail [4,51,52]. We will discuss a couple of examples regarding these factors. Tan et al., in their study, synthesized a self-assembled structure from thiosalicylic acid-capped silver NCs, which showed nanofiber-like morphology. The assembled geometry can be tuned from nanofiber to amorphous morphology with the solvent change from THF to toluene [106]. A diphosphine-protected $[Au_6]^{2+}$ cluster exhibited aggregation in an aqueous organic solvent with an appropriate water content. In a specific range of water, the aligned cluster assembly showed a J-aggregate-like absorption response and PL enhancement [107]. Liu's group used a pH-guided assembly strategy to fabricate protein/Cu NC hybrid nanostructures with stable and bright luminescence properties [108]. The aggregates showed strong AIE behavior at pH 3.0, but the high reversible pH-responsive nature of Cu NCs showed weak luminescence at pH < 1.5 or >4.0, and the aggregates were soluble. These strategies are used to design some new architectures with unique properties for specific applications.

6. Optical Properties of Self-Assembled Nanoclusters: Aggregation-Induced Emission

Luminescent MNCs are highly in demand because of their various application possibilities in bioimaging, biosensing, fabrication of LEDs, etc. However, MNCs with sub 2 nm size have very low QY, limiting their practical applications [109,110]. The luminescence property can be improved by aggregation-induced emission strategy, which Tang's research group first discovered [111]. Self-assembly is an efficient strategy for achieving tunable emission and customizable shape and size of the ordered structures by modulating the spatial distribution between the building blocks. For example, by controlling the inter NC distance between the individual Cu NCs, Zhang et al. tuned the photophysical behavior significantly [38]. The inter-NC distance is controlled by changing the solvents with different dielectric constants. The radiative relaxation from the Cu-centered triplet state influences the charge transfer from ligand to the Cu core, and it changes the emission color. This study reveals that the solvents can regulate the aggregation process effectively along with their luminescent behavior. However, the luminescence property of the MNCs depends both on the metal core and the surface capping ligands (see Section 2). Though many works have been done by restricting the rotational and vibrational motion of the capping ligand to alter the self-assembly-induced emission, very few studies reveal the effect of the metal defects on AIE. Zhang et al. demonstrated the effects of metal defect states in their self-assembly of dodecanethiol (DT)-capped Cu NCs (Figure 12A) [112]. The metal defect states speed up the self-assembly process by coordinating with Cu and changing the surface properties of self-assembled nanosheets. The enhancement in absolute QY originated from a triplet state (T_2) related to metal defects. Figure 12B depicts that T_1 is the original triplet state determined from ligand to metal–metal charge transfer (LMMCT), whereas metal defects largely influence T_2, thus providing a lower energy level. This phenomenon allows the relaxation to occur from the T_2 state to the intermediate state attributed to the metal defects. As a result, they observed a red-shifted emission from 490 to 550 nm with an enhanced QY of 15.4%. Here, the increased Cu(I) facilitates the radiative relaxation process. Based on this strategy, Au(I) was doped onto the Cu NC-self-assembled nanosheets to influence the LMMCT process deliberately [113]. Doping with the Au atom enhances the QY and results in red shifting of the emission. This result is attributed to the Au(I)–Cu (I) metallophilic interaction-directed charge transfer from Cu to Au. Doped Au on Cu self-assembled nanosheets generates a stable Au(I) centered state, which leads the charge transfer from the ligand to Cu and then from Cu to Au. Therefore, a red-shifted emission spectrum is observed at ~600 nm with a longer lifetime, as the Au(I) doping lowers the energy.

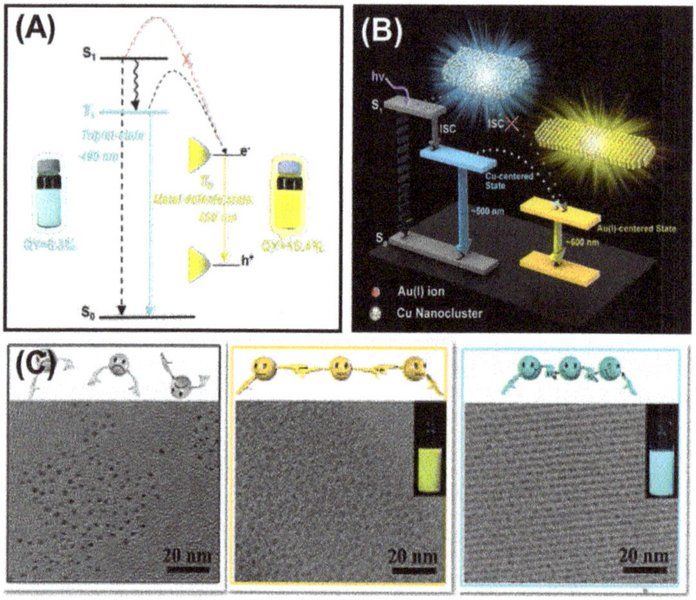

Figure 12. (**A**) Schematic representation of the excited state relaxation dynamics of the Cu NC-self-assembled nanosheets (Adapted with permission from Ref. [112]. Copyright 2017 American Chemical Society) and (**B**) self-assembled nanosheets of Cu NCs with Au doping. Reprinted with permission from Ref. [113]2. Copyright 2017 American Chemical Society. (**C**) TEM images of individual Cu NCs, self-assembled ribbons, and self-assembled sheets of Cu NCs. Inset shows the emission under 365 excitation. Adapted with permission from Ref. [41]. Copyright 2015 American Chemical Society.

In addition to metal defect states, metallophilic interactions are essential to produce self-assembled NCs with significant luminescence enhancement. A controlled and ordered cuprophilic Cu(I)···Cu(I) interaction assisted Zhang's group to report assembly-induced emission [41]. They designed a blue–green emissive assembly, and the emission peak was centered at 490 nm with a QY of 6.5%. The assembly-induced emission greatly depended on inter and intra-NCs cuprophilic interactions and the rigidity of the capping ligands (Figure 12C). Firstly, the cuprophilic interactions influenced the radiative relaxation, and secondly, the non-radiative relaxation decreased upon the restriction of the vibration and rotation of the ligands. Although the highest QY was achieved with self-assembled ribbons, they also prepared loosely bound self-assembled sheets with a reduced QY of 3.6%. This result signifies a strong correlation between the cuprophilic interaction and emission enhancement. Based on this strategy, another self-assembled Ag_9 NC was also prepared, promoting the implementation of AIE [81]. Xin's group prepared $Ag_9(MBA)_9$ NCs, which produce luminescent hydrogels in interaction with succinic acid. Several non-covalent interactions along with argentophilic interaction are responsible for the assembly-induced phosphorescence emission. The phosphorescence is said to originate from a triplet state attributed to argentophilic interaction with LMMCT. On hydrogel formation, the ligand's rotation becomes suppressed, which lowers the energy of the triplet state and yields bright orange–red emission, whereas for solution-state Ag_9 NCs, the surface ligands are free to rotate or vibrate, and thus non-radiative relaxation channels increase. The self-assembled highly phosphorescent nanofibres can be used as a phosphor LED.

Another fundamental question may arise between the M(0) core and M(I)-SR: which is more efficient for the PL enhancement? Several studies have revealed that metalcore solely does not determine the PL properties, as NCs with the same no. of metalcores but different surface capping ligands result in various PL phenomena. Xin et al. discovered an exciting finding based on the AIE strategy where the non-luminescent oligomeric Au-(I)-thiolate complexes became highly luminescent upon aggregation. The luminescence originated from a metal-centered triplet state, and it depended strongly on the degree of aggregation. The increased aurophilic interaction in the denser aggregation enhanced the luminescence by reducing the non-radiative relaxation probability of the excited states [114]. Furthermore, Xie et al. synthesized a weakly emitting gold NC with a QY of 7%; 60% luminescence enhancement occurred by rigidifying the Au(I)-thiolate shell of NC with cations due to aurophilic interaction [115]. Most importantly, the rigidity of the gold shell by binding with bulky groups resulted in a drastic reduction of the non-radiative channels.

The change in core composition also influences the PL properties. For example, Xie et al. developed Au@Ag NCs where Ag(I) acts as a bridge between slight Au(I)-thiolate motifs and generates aggregation-induced emission [116]. The relative QY was found to be 6.8%, with a strong red emission at 667 nm. This luminescence was generated from the metal-centered triplet state affected by the Au(I)/Ag)(I)-thiolate surface state. In another study, an anti-galvanic exchange reaction achieved bimetallic NCs with strong emissions. Bain et al. synthesized MPG capped Au NCs, which exhibited very weak red fluorescence at 735 nm, whereas the bimetallic AuAg NCs showed bright luminescence at 674 nm (r-AuAg) and 574 nm (y-AuAg) with different Ag concentrations (Figure 13A). In this study, aggregation-induced emission strongly depended on the length of surface motifs, and this impacted the photophysical properties of AuAg NCs. Further, the solvent played a vital role in engineering the surface motifs of the NCs, and accordingly, 13-fold PL-enhancement was found for y-AuAg NCs. The TEM images confirmed that circular assembled structures were formed with a 60% water-EtOH mixture (Figure 13B). Therefore, the PL intensity increases with the degree of aggregation. Further, the excited state lifetime study revealed that the Au-thiolate motifs are more stable in a less polar solvent, which causes the enhancement of emission intensity (Figure 13A) [117]. Another solvent polarity-directed controlled assembly of NCs was employed by our group previously to improve the PL QY as well as tune the emission color [118]. The red luminescent $Cu_{34-32}(SG)_{16-13}$ NCs showed very weak emission at ambient temperature but bright-red emission upon freezing (temperature < 0 °C). The restriction of molecular motion in the frozen state may be attributed to the emission enhancement. With increasing EtOH from 0 to 90% (Figure 13C,D), the particles with low surface charge suffered from a loss of stability in the solution. Therefore, the particles came closer due to inter and intra Cu(I)-Cu(I) cuprophilic interaction. The Cu NCs in an aqueous medium undergo fast internal conversion from S_n to S_1 and subsequently to a triplet state (T_1) (Figure 13E). Further, this comes down to a low-lying triplet state T_2 and undergoes radiative relaxation to the ground state (Figure 13E). The enhancement of intensity after aggregation is due to the restriction of the rotational and vibrational motion of the ligands. Here, the radiative decay occurred directly from high-lying T_1, which resulted in blue shifting of the emission maxima from 625 to 597 nm.

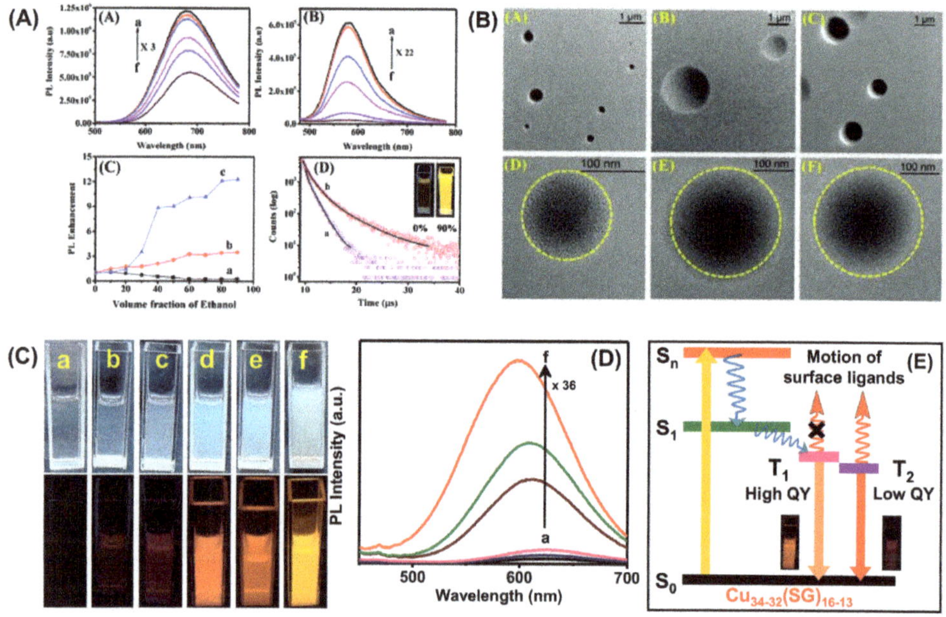

Figure 13. (**A**) Temperature dependent PL spectra of (i) r-AuAg and (ii) y-AuAg NCs, (iii) enhancement of PL for Au and AuAg NCs by changing the volume fraction of ethanol/water (v/v), (a) Au NCs, (b) r-AuAg, and (c) y-AuAg NCs and (iv) time resolved decay curves of y-AuAg NCs with (a) 0% and (b) 60% ethanol (lex = 415 nm). Inset shows the digital picture of y-AuAg NCs in fv = 0% and fv = 90% under UV-light (365 nm) excitation. (**B**) TEM images of NCs in 60% water–ethanol (v/v) mixture: (A,D) Au NCs, (B,E) r-AuAg NCs, and (C,F) y-AuAg NCs. Adapted with permission from Ref. [117]. Copyright 2020 The Royal Society of Chemistry. (**C**) Digital photographs of $Cu_{34-32}(SG)_{16-13}$ NCs with EtOH volume fraction of (a) 0%, (b) 20%, (c) 40%, (d) 60%, (e) 80%, and (f) 90% under daylight and UV (365 nm) illumination. (D) PL spectra of Cu34-32(SG)16-13 NCs in an EtOH volume fraction of (a) 0%, (b) 20%, (c) 40%, (d) 60%, (e) 80%, and (f) 90%. (**E**) Schematic diagram of excited state dynamics and origin of emission in $Cu_{34-32}(SG)_{16-13}$ NCs. Adapted with permission from Ref. [118]. Copyright 2019 American Chemical Society.

The self-assembly of Au_3[R-Tol-BINAP]$_3$Cl drastically changes the optical properties compared to the non-luminescent Au NCs. The self-assembly exhibits an orange emission centered at 583 nm, and a significant stoke shift (~138 nm) suggests phosphorescence, originating from a triplet state LMCT or LMMCT. These assemblies also show strong circularly polarised luminescence (CPL) response. The PL intensity is found to display enhancement with the increase in the n-hexane fraction, and the maximum intensity is found with 70% n-hexane (QY ~3.6%). Structural and morphological characterization was brought into play to decode the factors behind such behavior. Concentration-dependent ^{1}H NMR spectroscopy revealed that the strong C–H$\cdots\pi$ interaction amongst the assembled Au NCs limits the intramolecular rotation of the chiral ligands. Consequently, the non-radiative channels for relaxation are closed, and the radiative decay dynamics increase drastically [75]. The aggregation strategy induces the enhancement of the luminescence property and can change the origin of emission. Konishi et al., in their study, displayed the switching of emission type from fluorescence to phosphorescence upon aggregation of [core + exo]-type $[Au_8]^{4+}$ clusters [119]. Monomeric NCs show fluorescence, whereas their aggregation displays phosphorescence-type emission, primarily due to the closed packing of NC assemblies. The phosphorescence can be observed only at a particular alignment of the NC orientation.

The electronic interaction between NCs in the assembled structures dramatically influences the dynamics at an ultrafast time scale. Dimeric Au_{20}-diglyme-Au_{20} showed different relaxation behavior compared to Au_{20}-diglyme [93]. Circularly polarized pump pulses can differentiate between low-spin monomer and high-spin dimeric species, as Au_{20}-diglyme-Au_{20} exhibits additional picosecond decay due to the spin-fli p relaxation process. Knappenberger, Jr. et al. later investigated the effect of the length of bridging molecules on the electronic transitions of dimeric assemblies [120]. The increase in the dimer-specific electronic relaxation time with the decrease in the length of bridging molecules reveals the presence of intercluster distance-dependent electronic states. An unprecedented appearance of electron–phonon coupling, which is the characteristic of plasmonic NPs, is observed in the one-dimensional assembly of $Ag_4(S\text{-}Adm)_6$ building units [121]. The pump power-dependent exciton relaxation dynamics indicate that the CAMs exhibit molecular-like and plasmonic behavior. Further investigation into the ultrafast dynamics of self-assembled materials will unveil the detailed electronic coupling owing to inter-NC interactions. These discussions mentioned above manifest a solid correlation between the structural and optical properties of the self-assembled NCs. Tailoring the surface motifs to fabricate architectures with customized composition, length, and shape synchronously engineer their optical properties. Indeed, in addition to enhance the emissive properties of NCs through the self-assembly formation, the use of chiral NC leads to strong circular dichroism intensity and a remarkable circularly polarized luminescence response [70].

7. Conclusions and Outlook

In summary, this review article has highlighted MNC-assemblies of different sizes, shapes, and compositions, providing highly luminescent nanostructures. Such self-assembly presents the capacity to reduce solvent accessibility, restrict the floppiness of surface ligands, and increase metallophilic interactions through various supramolecular forces. Different types of nanoscale forces such as dipolar interactions, van der Waals interactions, electrostatic interactions, hydrogen bonding, C–H$\cdots\pi$ interaction, π–π stacking, metallophilic interactions, and amphiphilicity along with the external triggers are responsible for directed self-assembly process of MNCs. The reduction in free energy during the assembly is the main driving force for this process.

A comprehensive mechanistic understanding of the origins of PL in the self-assembly of NCs is mandatory; in particular, understanding the molecular-level details and structure–property correlation will require a more well-established crystal structure of self-assembly of NCs. The quantum chemistry approach (already well explored by the TD-DFT level of theory in individual MNCs) should bridge the gap to address the photophysical properties of MNCs embedded in self-assembly. In particular, a combined quantum chemical and molecular mechanics method (QM/MM) implementation using periodic boundary conditions might be applied to two-dimensional arrays of metal clusters protected by organic ligands. Such hybrid methods could thus address photophysical processes in the assembly of NCs across length- and time-scales. A combined experimental and theoretical approach including QM/MM tools and computational simulation techniques can provide a holistic description of the nature of the interactions present in self-assembled nanoclusters. QM/MM and molecular dynamic simulations should be developed to simulate self-assembled molecular systems, where an explicit description of changes in the electronic structure is necessary.

Although the emphasis of this review is on the structural correlation of the assemblies with the optical properties, the mechanistic understanding behind the assembly formation is also essential. The successful assembly formation into desired morphology depends on maintaining a balance between the attractive and repulsive forces between the NCs.

Concerning the control of self-assembly of MNCs, there is still room for improvement. Indeed, one can take inspiration from recent progress on the fabrication of monodispersed nanoparticles using programmed automated techniques. A possible outcome should be the programmable self-aggregation to activate the AIE process in the assembly of MNCs. The templating motifs with "programmed" surface ligands that drive the assembly formation via different driving forces should be a possibility. Alternatively, the MNC community should learn to create more versatile superstructures with desirable nanoscale optical properties from the metamaterial community. A variety of well-defined MNCs could be obtained through a programmable assembly of NCs. From the scientific perspective, there is a considerable number of reports related to the synthetic chemistry of NCs, whereas the self-assembly of MNCs is not well developed to conclude a relevant mechanism. A wide range of self-assembly processes is reported where NCs are soluble in an organic solvent. In contrast, water-soluble self-assembled NCs are limited due to their lack of stability as crystal structures. However, these water-soluble assembled structures with unique optical properties could be ideal for metal ion sensing, catalysis, bioimaging, and therapeutics [95,122,123]. The assembled structures have been used as biomarkers for cellular imaging and cancer therapy due to their enhanced physicochemical properties, which are advantageous for biological applications [124,125]. MNCs are highly promising in catalysis owing to their tailorable active site structure, which can also tune their catalytic properties. The factors such as size, shape, composition, and isomerization of NCs are also of major concern for new catalysis with high activity and selectivity through site-specific surface tailoring [126–128] The selectivity of a catalyst is highly dependent on the morphology of the MNCs [129]. The wide range of emission color tunability of assembled MNCs makes them eligible as a competitive color conversion material in light-emitting diodes (LEDs) [39,41,130]. The advantage of self-assembled structures in device fabrication is due to their decreased surface-to-volume ratio as well as stable tunable emission and their compact arrangement [32]. These veiled built structures could also bring some eccentric optical properties such as AIE, strong circularly polarized luminescence, enhanced mechanochromic properties, etc. Recent progress on fabricating the NCs using templates or linkers is highly appreciated. Aggregated NCs could be confined in a metal-organic framework [84] or zeolite-like framework to maintain their aggregation, restrict the rotation of their ligands, and further improve their QY [131]. Prescribed assembly of inorganic nanoparticles (NPs) guided by the programmable DNA sequences provides a promising method to fabricate various nano-architectures, mainly using DNA origami frame versatility [132]. Finally, simulations can be used to determine the extent of the assembly process and the governing principles for the assembly mechanism.

Author Contributions: Conceptualization, S.K., D.B., R.A. and A.P.; writing—original draft preparation, S.K., D.B, S.M. and A.D.; writing—review and editing, S.K., D.B., S.M., R.A. and A.P. All authors have read and agreed to the published version of the manuscript.

Funding: S.K. acknowledges University Grants Commission and D.B. thanks INST for the fellowship. S.M. acknowledges Council of Scientific and Industrial Research, and A.D. thanks INST for research support. Science and Engineering Research Board-Department of Science and Technology is acknowledged for financial support.

Institutional Review Board Statement: Not applicable.

Informed Consent Statement: Not applicable.

Data Availability Statement: Not applicable.

Conflicts of Interest: The authors declare no conflict of interest.

List of Abbreviations

Abbreviation	Full Name
MNC	Metal nanocluster
NP	Nanoparticle
PL	Photoluminescence
QY	Quantum yield
NIR	Near-infrared
ISC	Intersystem crossing
vdW	van der Waals
AIE	Aggregation-induced emission
CPL	Circularly polarised luminescence
LMCT	Ligand to metal charge transfer
LMMCT	Ligand to metal–metal charge transfer
NL	Nano leaves
SC	Super crystal
GNR	Gold nanorod
AFM	Atomic force microscopy
DLS	Dynamic light scattering
TEM	Transmission electron microscopy
XRD	X-ray diffraction
XPS	X-ray photo-electron spectroscopy
XANES	X-ray absorption near-edge structure
FT-EXAFS	Fourier transform extended X-ray absorption fine-structure
SAXS	Small-angle X-ray scattering
TD-DFT	Time-dependent density functional theory
BE	Dibenzyl ether
LP	Liquid paraffin
D_{3d}	Rhombohedral
O_h	Octahedral
MOF	Metal-organic framework
CAM	Cluster assembled materials
LED	Light-emitting diode
SG	Glutathione
GSH	Reduced Glutathione
DT	1-dodecanethiol
p-MBT	p-methylbenzenethiol
DMBT	dimethylbenzenethiol
Tol-BINAP	[2,2′-bis(di-*p*-tolylphosphino)-1,1′-binaphthyl]
p-MBA	para-mercaptobenzoicacid
PATP	p-aminothiophenol
MHA	6-mercaptohexanoic acid
CTA	cetyltrimethylammonium
ssDNA	single-stranded DNA
PET	phenylethanethiol
bpy	4,4′-bipyridine
S-Adm	1-adamantanethiol
SP	spiropyran
C_3-AMT	azobenzene-alkyl monothiol
MBA	2- mercaptobenzoic acid
MPG	N-(2- mercaptopropionyl)glycine

References

1. Whitesides, G.M.; Grzybowski, B. Self-Assembly at All Scales. *Science* **2002**, *295*, 2418–2421. [CrossRef]
2. Kotov, N.A.; Weiss, P.S. Self-Assembly of Nanoparticles: A Snapshot. *ACS Nano* **2014**, *8*, 3101–3103. [CrossRef] [PubMed]
3. Miszta, K.; de Graaf, J.; Bertoni, G.; Dorfs, D.; Brescia, R.; Marras, S.; Ceseracciu, L.; Cingolani, R.; van Roij, R.; Dijkstra, M.; et al. Hierarchical self-assembly of suspended branched colloidal nanocrystals into superlattice structures. *Nat. Mater.* **2011**, *10*, 872–876. [CrossRef] [PubMed]

4. Grzelczak, M.; Vermant, J.; Furst, E.M.; Liz-Marzán, L.M. Directed Self-Assembly of Nanoparticles. *ACS Nano* **2010**, *4*, 3591–3605. [CrossRef] [PubMed]
5. Luo, D.; Yan, C.; Wang, T. Interparticle Forces Underlying Nanoparticle Self-Assemblies. *Small* **2015**, *11*, 5984–6008. [CrossRef]
6. Zhou, T.; Zhu, J.; Gong, L.; Nong, L.; Liu, J. Amphiphilic Block Copolymer-Guided in Situ Fabrication of Stable and Highly Controlled Luminescent Copper Nanoassemblies. *J. Am. Chem. Soc.* **2019**, *141*, 2852–2856. [CrossRef]
7. Wang, T.; Zhuang, J.; Lynch, J.; Chen, O.; Wang, Z.; Wang, X.; LaMontagne, D.; Wu, H.; Wang, Z.; Cao, Y.C. Self-assembled colloidal superparticles from nanorods. *Science* **2012**, *338*, 358–363. [CrossRef]
8. Wang, L.; Xu, L.; Kuang, H.; Xu, C.; Kotov, N.A. Dynamic Nanoparticle Assemblies. *Acc. Chem. Res.* **2012**, *45*, 1916–1926. [CrossRef]
9. Boles, M.A.; Engel, M.; Talapin, D.V. Self-Assembly of Colloidal Nanocrystals: From Intricate Structures to Functional Materials. *Chem. Rev.* **2016**, *116*, 11220–11289. [CrossRef]
10. Sugi, K.S.; Bandyopadhyay, P.; Bodiuzzaman, M.; Nag, A.; Hridya, M.; Dar, W.A.; Ghosh, P.; Pradeep, T. Manifestation of Structural Differences of Atomically Precise Cluster-Assembled Solids in Their Mechanical Properties. *Chem. Mater.* **2020**, *32*, 7973–7984. [CrossRef]
11. Hudson, Z.M.; Boott, C.E.; Robinson, M.E.; Rupar, P.A.; Winnik, M.A.; Manners, I. Tailored hierarchical micelle architectures using living crystallization-driven self-assembly in two dimensions. *Nat. Chem.* **2014**, *6*, 893–898. [CrossRef] [PubMed]
12. Zang, L.; Che, Y.; Moore, J.S. One-Dimensional Self-Assembly of Planar π-Conjugated Molecules: Adaptable Building Blocks for Organic Nanodevices. *Acc. Chem. Res.* **2008**, *41*, 1596–1608. [CrossRef] [PubMed]
13. Wu, Z.; Yao, Q.; Zang, S.; Xie, J. Directed Self-Assembly of Ultrasmall Metal Nanoclusters. *ACS Mater. Lett.* **2019**, *1*, 237–248. [CrossRef]
14. Jin, R. Atomically precise metal nanoclusters: Stable sizes and optical properties. *Nanoscale* **2015**, *7*, 1549–1565. [CrossRef] [PubMed]
15. Jin, R.; Zeng, C.; Zhou, M.; Chen, Y. Atomically Precise Colloidal Metal Nanoclusters and Nanoparticles: Fundamentals and Opportunities. *Chem. Rev.* **2016**, *116*, 10346–10413. [CrossRef] [PubMed]
16. Maity, S.; Bain, D.; Bhattacharyya, K.; Das, S.; Bera, R.; Jana, B.; Paramanik, B.; Datta, A.; Patra, A. Ultrafast Relaxation Dynamics of Luminescent Copper Nanoclusters (Cu7L3) and Efficient Electron Transfer to Functionalized Reduced Graphene Oxide. *J. Phys. Chem. C* **2018**, *122*, 13354–13362. [CrossRef]
17. Zhou, D.; Liu, M.; Lin, M.; Bu, X.; Luo, X.; Zhang, H.; Yang, B. Hydrazine-Mediated Construction of Nanocrystal Self-Assembly Materials. *ACS Nano* **2014**, *8*, 10569–10581. [CrossRef] [PubMed]
18. Antoine, R. Supramolecular Gold Chemistry: From Atomically Precise Thiolate-Protected Gold Nanoclusters to Gold-Thiolate Nanostructures. *Nanomaterials* **2020**, *10*, 377. [CrossRef]
19. Zhou, M.; Zeng, C.; Li, Q.; Higaki, T.; Jin, R. Gold Nanoclusters: Bridging Gold Complexes and Plasmonic Nanoparticles in Photophysical Properties. *Nanomaterials* **2019**, *9*, 933. [CrossRef]
20. Maity, S.; Bain, D.; Chakraborty, S.; Kolay, S.; Patra, A. Copper Nanocluster (Cu23 NC)-Based Biomimetic System with Peroxidase Activity. *ACS Sustain. Chem. Eng.* **2020**, *8*, 18335–18344. [CrossRef]
21. Bain, D.; Maity, S.; Patra, A. Opportunities and challenges in energy and electron transfer of nanocluster based hybrid materials and their sensing applications. *Phys. Chem. Chem. Phys.* **2019**, *21*, 5863–5881. [CrossRef] [PubMed]
22. Bonačić-Koutecký, V.; Antoine, R. Enhanced two-photon absorption of ligated silver and gold nanoclusters: Theoretical and experimental assessments. *Nanoscale* **2019**, *11*, 12436–12448. [CrossRef] [PubMed]
23. Antoine, R.; Bonačić-Koutecký, V. Introduction. In *Liganded Silver and Gold Quantum Clusters. Towards a New Class of Nonlinear Optical Nanomaterials*; Springer: Cham, Switzerland, 2018. [CrossRef]
24. Aikens, C.M. Electronic and Geometric Structure, Optical Properties, and Excited State Behavior in Atomically Precise Thiolate-Stabilized Noble Metal Nanoclusters. *Acc. Chem. Res.* **2018**, *51*, 3065–3073. [CrossRef] [PubMed]
25. Xiao, Y.; Wu, Z.; Yao, Q.; Xie, J. Luminescent metal nanoclusters: Biosensing strategies and bioimaging applications. *Aggregate* **2021**, *2*, 114–132. [CrossRef]
26. Yafang, S.; Ziping, Z.; Tong, S.; Lisheng, Q.; Lei, S.; Xueji, Z. Multicolor Luminescent Gold Nanoclusters: From Structure to Biosensing and Bioimaging. *Prog. Chem.* **2021**, *33*, 179–187. [CrossRef]
27. Nonappa. Luminescent gold nanoclusters for bioimaging applications. *Beilstein J. Nanotechnol.* **2020**, *11*, 533–546. [CrossRef]
28. Li, H.; Li, H.; Wan, A. Luminescent gold nanoclusters for in vivo tumor imaging. *Analyst* **2020**, *145*, 348–363. [CrossRef]
29. Bain, D.; Maity, S.; Paramanik, B.; Patra, A. Core-Size Dependent Fluorescent Gold Nanoclusters and Ultrasensitive Detection of Pb2+ Ion. *ACS Sustain. Chem. Eng.* **2018**, *6*, 2334–2343. [CrossRef]
30. Kang, X.; Zhu, M. Tailoring the photoluminescence of atomically precise nanoclusters. *Chem. Soc. Rev.* **2019**, *48*, 2422–2457. [CrossRef]
31. Goswami, N.; Lin, F.; Liu, Y.; Leong, D.T.; Xie, J. Highly Luminescent Thiolated Gold Nanoclusters Impregnated in Nanogel. *Chem. Mater.* **2016**, *28*, 4009–4016. [CrossRef]
32. Liu, Y.; Yao, D.; Zhang, H. Self-Assembly Driven Aggregation-Induced Emission of Copper Nanoclusters: A Novel Technology for Lighting. *ACS Appl. Mater. Interfaces* **2018**, *10*, 12071–12080. [CrossRef] [PubMed]
33. Hikosou, D.; Saita, S.; Miyata, S.; Miyaji, H.; Furuike, T.; Tamura, H.; Kawasaki, H. Aggregation/Self-Assembly-Induced Approach for Efficient AuAg Bimetallic Nanocluster-Based Photosensitizers. *J. Phys. Chem. C* **2018**, *122*, 12494–12501. [CrossRef]

34. Prasad, B.L.V.; Sorensen, C.M.; Klabunde, K.J. Gold nanoparticle superlattices. *Chem. Soc. Rev.* **2008**, *37*, 1871–1883. [CrossRef]
35. Shen, J.; Wang, Z.; Sun, D.; Liu, G.; Yuan, S.; Kurmoo, M.; Xin, X. Self-assembly of water-soluble silver nanoclusters: Superstructure formation and morphological evolution. *Nanoscale* **2017**, *9*, 19191–19200. [CrossRef] [PubMed]
36. Hossain, S.; Imai, Y.; Motohashi, Y.; Chen, Z.; Suzuki, D.; Suzuki, T.; Kataoka, Y.; Hirata, M.; Ono, T.; Kurashige, W.; et al. Understanding and designing one-dimensional assemblies of ligand-protected metal nanoclusters. *Mater. Horiz.* **2020**, *7*, 796–803. [CrossRef]
37. Li, Q.; Russell, J.C.; Luo, T.-Y.; Roy, X.; Rosi, N.L.; Zhu, Y.; Jin, R. Modulating the hierarchical fibrous assembly of Au nanoparticles with atomic precision. *Nat. Commun.* **2018**, *9*, 3871. [CrossRef]
38. Ai, L.; Liu, Z.; Zhou, D.; Liu, J.; Zou, H.; Wu, Z.; Liu, Y.; Zhang, H.; Yang, B. Copper inter-nanoclusters distance-modulated chromism of self-assembly induced emission. *Nanoscale* **2017**, *9*, 18845–18854. [CrossRef] [PubMed]
39. Ai, L.; Jiang, W.; Liu, Z.; Liu, J.; Gao, Y.; Zou, H.; Wu, Z.; Wang, Z.; Liu, Y.; Zhang, H.; et al. Engineering a red emission of copper nanocluster self-assembly architectures by employing aromatic thiols as capping ligands. *Nanoscale* **2017**, *9*, 12618–12627. [CrossRef]
40. Chen, W.-Y.; Huang, C.-C.; Chen, L.-Y.; Chang, H.-T. Self-assembly of hybridized ligands on gold nanodots: Tunable photoluminescence and sensing of nitrite. *Nanoscale* **2014**, *6*, 11078–11083. [CrossRef]
41. Wu, Z.; Liu, J.; Gao, Y.; Liu, H.; Li, T.; Zou, H.; Wang, Z.; Zhang, K.; Wang, Y.; Zhang, H.; et al. Assembly-Induced Enhancement of Cu Nanoclusters Luminescence with Mechanochromic Property. *J. Am. Chem. Soc.* **2015**, *137*, 12906–12913. [CrossRef]
42. Higaki, T.; Li, Y.; Zhao, S.; Li, Q.; Li, S.; Du, X.-S.; Yang, S.; Chai, J.; Jin, R. Atomically Tailored Gold Nanoclusters for Catalytic Application. *Angew. Chem. Int. Ed.* **2019**, *58*, 8291–8302. [CrossRef] [PubMed]
43. Yao, Q.; Yuan, X.; Chen, T.; Leong, D.T.; Xie, J. Engineering Functional Metal Materials at the Atomic Level. *Adv. Mater.* **2018**, *30*, 1802751. [CrossRef] [PubMed]
44. Yahia-Ammar, A.; Sierra, D.; Mérola, F.; Hildebrandt, N.; Le Guével, X. Self-Assembled Gold Nanoclusters for Bright Fluorescence Imaging and Enhanced Drug Delivery. *ACS Nano* **2016**, *10*, 2591–2599. [CrossRef] [PubMed]
45. Yang, T.; Dai, S.; Yang, S.; Chen, L.; Liu, P.; Dong, K.; Zhou, J.; Chen, Y.; Pan, H.; Zhang, S.; et al. Interfacial Clustering-Triggered Fluorescence–Phosphorescence Dual Solvoluminescence of Metal Nanoclusters. *J. Phys. Chem. Lett.* **2017**, *8*, 3980–3985. [CrossRef]
46. Nag, A.; Chakraborty, P.; Thacharon, A.; Paramasivam, G.; Mondal, B.; Bodiuzzaman, M.; Pradeep, T. Atomically Precise Noble Metal Cluster-Assembled Superstructures in Water: Luminescence Enhancement and Sensing. *J. Phys. Chem. C* **2020**, *124*, 22298–22303. [CrossRef]
47. Chen, C.; Li, R.H.; Zhu, B.S.; Wang, K.H.; Yao, J.S.; Yin, Y.C.; Yao, M.M.; Yao, H.B.; Yu, S.H. Highly Luminescent Inks: Aggregation-Induced Emission of Copper-Iodine Hybrid Clusters. *Angew. Chem. Int. Ed.* **2018**, *57*, 7106–7110. [CrossRef]
48. Ghosh, A.; Mohammed, O.F.; Bakr, O.M. Atomic-Level Doping of Metal Clusters. *Acc. Chem. Res.* **2018**, *51*, 3094–3103. [CrossRef] [PubMed]
49. Maity, S.; Bain, D.; Patra, A. An overview on the current understanding of the photophysical properties of metal nanoclusters and their potential applications. *Nanoscale* **2019**, *11*, 22685–22723. [CrossRef]
50. Yao, Q.; Wu, Z.; Liu, Z.; Lin, Y.; Yuan, X.; Xie, J. Molecular reactivity of thiolate-protected noble metal nanoclusters: Synthesis, self-assembly, and applications. *Chem. Sci.* **2021**, *12*, 99–127. [CrossRef]
51. Rival, J.V.; Mymoona, P.; Lakshmi, K.M.; Nonappa; Pradeep, T.; Shibu, E.S. Self-Assembly of Precision Noble Metal Nanoclusters: Hierarchical Structural Complexity, Colloidal Superstructures, and Applications. *Small* **2021**, *17*, 2005718. [CrossRef]
52. Bera, D.; Goswami, N. Driving Forces and Routes for Aggregation-Induced Emission-Based Highly Luminescent Metal Nanocluster Assembly. *J. Phys. Chem. Lett.* **2021**, *12*, 9033–9046. [CrossRef]
53. Yang, T.-Q.; Peng, B.; Shan, B.-Q.; Zong, Y.-X.; Jiang, J.-G.; Wu, P.; Zhang, K. Origin of the Photoluminescence of Metal Nanoclusters: From Metal-Centered Emission to Ligand-Centered Emission. *Nanomaterials* **2020**, *10*, 261. [CrossRef]
54. Li, Y.; Jin, R. Seeing Ligands on Nanoclusters and in Their Assemblies by X-ray Crystallography: Atomically Precise Nanochemistry and Beyond. *J. Am. Chem. Soc.* **2020**, *142*, 13627–13644. [CrossRef]
55. Ebina, A.; Hossain, S.; Horihata, H.; Ozaki, S.; Kato, S.; Kawawaki, T.; Negishi, Y. One-, Two-, and Three-Dimensional Self-Assembly of Atomically Precise Metal Nanoclusters. *Nanomaterials* **2020**, *10*, 1105. [CrossRef]
56. Wang, Q.; Wang, S.; Hu, X.; Li, F.; Ling, D. Controlled synthesis and assembly of ultra-small nanoclusters for biomedical applications. *Biomater. Sci.* **2019**, *7*, 480–489. [CrossRef]
57. Wang, J.; Lin, X.; Shu, T.; Su, L.; Liang, F.; Zhang, X. Self-Assembly of Metal Nanoclusters for Aggregation-Induced Emission. *Int. J. Mol. Sci.* **2019**, *20*, 1891. [CrossRef]
58. Chakraborty, P.; Nag, A.; Chakraborty, A.; Pradeep, T. Approaching Materials with Atomic Precision Using Supramolecular Cluster Assemblies. *Acc. Chem. Res.* **2019**, *52*, 2–11. [CrossRef]
59. Shi, Y.-e.; Ma, J.; Feng, A.; Wang, Z.; Rogach, A.L. Aggregation-induced emission of copper nanoclusters. *Aggregate* **2021**, *2*, e112. [CrossRef]
60. Weerawardene, K.L.D.M.; Aikens, C.M. Theoretical Insights into the Origin of Photoluminescence of Au25(SR)18– Nanoparticles. *J. Am. Chem. Soc.* **2016**, *138*, 11202–11210. [CrossRef] [PubMed]
61. Yuan, C.T.; Lin, C.A.; Lin, T.N.; Chang, W.H.; Shen, J.L.; Cheng, H.W.; Tang, J. Probing the photoluminescence properties of gold nanoclusters by fluorescence lifetime correlation spectroscopy. *J. Chem. Phys.* **2013**, *139*, 234311. [CrossRef] [PubMed]

62. Perić, M.; Sanader Maršić, Ž.; Russier-Antoine, I.; Fakhouri, H.; Bertorelle, F.; Brevet, P.-F.; le Guével, X.; Antoine, R.; Bonačić-Koutecký, V. Ligand shell size effects on one- and two-photon excitation fluorescence of zwitterion functionalized gold nanoclusters. *Phys. Chem. Chem. Phys.* **2019**, *21*, 23916–23921. [CrossRef]
63. Gran, E.R.; Bertorelle, F.; Fakhouri, H.; Antoine, R.; Perić Bakulić, M.; Sanader Maršić, Ž.; Bonačić-Koutecký, V.; Blain, M.; Antel, J.; Maysinger, D. Size and ligand effects of gold nanoclusters in alteration of organellar state and translocation of transcription factors in human primary astrocytes. *Nanoscale* **2021**, *13*, 3173–3183. [CrossRef]
64. Chevrier, D.M.; Thanthirige, V.D.; Luo, Z.; Driscoll, S.; Cho, P.; MacDonald, M.A.; Yao, Q.; Guda, R.; Xie, J.; Johnson, E.R.; et al. Structure and formation of highly luminescent protein-stabilized gold clusters. *Chem. Sci.* **2018**, *9*, 2782–2790. [CrossRef] [PubMed]
65. Bertorelle, F.; Russier-Antoine, I.; Calin, N.; Comby-Zerbino, C.; Bensalah-Ledoux, A.; Guy, S.; Dugourd, P.; Brevet, P.-F.; Sanader, Ž.; Krstić, M.; et al. Au10(SG)10: A Chiral Gold Catenane Nanocluster with Zero Confined Electrons. Optical Properties and First-Principles Theoretical Analysis. *J. Phys. Chem. Lett.* **2017**, *8*, 1979–1985. [CrossRef] [PubMed]
66. Basu, S.; Fakhouri, H.; Moulin, C.; Dolai, S.; Russier-Antoine, I.; Brevet, P.-F.; Antoine, R.; Paul, A. Four orders-of-magnitude enhancement in the two-photon excited photoluminescence of homoleptic gold thiolate nanoclusters following zinc ion-induced aggregation. *Nanoscale* **2021**, *13*, 4439–4443. [CrossRef] [PubMed]
67. Talapin, D.V.; Shevchenko, E.V.; Murray, C.B.; Titov, A.V.; Král, P. Dipole−Dipole Interactions in Nanoparticle Superlattices. *Nano Lett.* **2007**, *7*, 1213–1219. [CrossRef]
68. Wu, Z.; Liu, J.; Li, Y.; Cheng, Z.; Li, T.; Zhang, H.; Lu, Z.; Yang, B. Self-Assembly of Nanoclusters into Mono-, Few-, and Multilayered Sheets via Dipole-Induced Asymmetric van der Waals Attraction. *ACS Nano* **2015**, *9*, 6315–6323. [CrossRef]
69. Bishop, K.J.M.; Wilmer, C.E.; Soh, S.; Grzybowski, B.A. Nanoscale Forces and Their Uses in Self-Assembly. *Small* **2009**, *5*, 1600–1630. [CrossRef]
70. Dabros, T. Electrokinetic and Colloid Transport Phenomena: Jacob H. Masliyah and Subir Bhattacharjee Publisher: Wiley-Interscience, 2006 ISBN: 0471799734. *Can. J. Chem. Eng.* **2006**, *84*, 729. [CrossRef]
71. Wu, Z.; Li, Y.; Liu, J.; Lu, Z.; Zhang, H.; Yang, B. Colloidal Self-Assembly of Catalytic Copper Nanoclusters into Ultrathin Ribbons. *Angew. Chem. Int. Ed.* **2014**, *53*, 12196–12200. [CrossRef]
72. Wu, Z.; Dong, C.; Li, Y.; Hao, H.; Zhang, H.; Lu, Z.; Yang, B. Self-Assembly of Au15 into Single-Cluster-Thick Sheets at the Interface of Two Miscible High-Boiling Solvents. *Angew. Chem. Int. Ed.* **2013**, *52*, 9952–9955. [CrossRef] [PubMed]
73. Li, F.; Josephson, D.P.; Stein, A. Colloidal Assembly: The Road from Particles to Colloidal Molecules and Crystals. *Angew. Chem. Int. Ed.* **2011**, *50*, 360–388. [CrossRef]
74. Zeng, C.; Chen, Y.; Kirschbaum, K.; Lambright, K.J.; Jin, R. Emergence of hierarchical structural complexities in nanoparticles and their assembly. *Science* **2016**, *354*, 1580. [CrossRef]
75. Shi, L.; Zhu, L.; Guo, J.; Zhang, L.; Shi, Y.; Zhang, Y.; Hou, K.; Zheng, Y.; Zhu, Y.; Lv, J.; et al. Self-Assembly of Chiral Gold Clusters into Crystalline Nanocubes of Exceptional Optical Activity. *Angew. Chem. Int. Ed.* **2017**, *56*, 15397–15401. [CrossRef]
76. Kolay, S.; Maity, S.; Bain, D.; Chakraborty, S.; Patra, A. Self-assembly of copper nanoclusters: Isomeric ligand effect on morphological evolution. *Nanoscale Adv.* **2021**, *3*, 5570–5575. [CrossRef]
77. Yao, Q.; Yu, Y.; Yuan, X.; Yu, Y.; Zhao, D.; Xie, J.; Lee, J.Y. Counterion-Assisted Shaping of Nanocluster Supracrystals. *Angew. Chem. Int. Ed.* **2015**, *54*, 184–189. [CrossRef] [PubMed]
78. Chakraborty, A.; Fernandez, A.C.; Som, A.; Mondal, B.; Natarajan, G.; Paramasivam, G.; Lahtinen, T.; Häkkinen, H.; Nonappa; Pradeep, T. Atomically Precise Nanocluster Assemblies Encapsulating Plasmonic Gold Nanorods. *Angew. Chem. Int. Ed.* **2018**, *57*, 6522–6526. [CrossRef]
79. Li, L.; Wang, Q. Spontaneous Self-Assembly of Silver Nanoparticles into Lamellar Structured Silver Nanoleaves. *ACS Nano* **2013**, *7*, 3053–3060. [CrossRef] [PubMed]
80. Pyykkö, P. Strong Closed-Shell Interactions in Inorganic Chemistry. *Chem. Rev.* **1997**, *97*, 597–636. [CrossRef]
81. Sun, P.; Wang, Z.; Bi, Y.; Sun, D.; Zhao, T.; Zhao, F.; Wang, W.; Xin, X. Self-Assembly-Driven Aggregation-Induced Emission of Silver Nanoclusters for Light Conversion and Temperature Sensing. *ACS Appl. Nano Mater.* **2020**, *3*, 2038–2046. [CrossRef]
82. Wu, Z.; Du, Y.; Liu, J.; Yao, Q.; Chen, T.; Cao, Y.; Zhang, H.; Xie, J. Aurophilic Interactions in the Self-Assembly of Gold Nanoclusters into Nanoribbons with Enhanced Luminescence. *Angew. Chem. Int. Ed.* **2019**, *58*, 8139–8144. [CrossRef]
83. Goswami, N.; Yao, Q.; Luo, Z.; Li, J.; Chen, T.; Xie, J. Luminescent Metal Nanoclusters with Aggregation-Induced Emission. *J. Phys. Chem. Lett.* **2016**, *7*, 962–975. [CrossRef] [PubMed]
84. Alhilaly, M.J.; Huang, R.-W.; Naphade, R.; Alamer, B.; Hedhili, M.N.; Emwas, A.-H.; Maity, P.; Yin, J.; Shkurenko, A.; Mohammed, O.F.; et al. Assembly of Atomically Precise Silver Nanoclusters into Nanocluster-Based Frameworks. *J. Am. Chem. Soc.* **2019**, *141*, 9585–9592. [CrossRef] [PubMed]
85. Lombardo, D.; Kiselev, M.A.; Magazù, S.; Calandra, P. Amphiphiles Self-Assembly: Basic Concepts and Future Perspectives of Supramolecular Approaches. *Adv. Condens. Matter Phys.* **2015**, *2015*, 151683. [CrossRef]
86. Yao, Q.; Yuan, X.; Yu, Y.; Yu, Y.; Xie, J.; Lee, J.Y. Introducing Amphiphilicity to Noble Metal Nanoclusters via Phase-Transfer Driven Ion-Pairing Reaction. *J. Am. Chem. Soc.* **2015**, *137*, 2128–2136. [CrossRef] [PubMed]
87. Zhou, Y.; Zeng, H.C. Simultaneous Synthesis and Assembly of Noble Metal Nanoclusters with Variable Micellar Templates. *J. Am. Chem. Soc.* **2014**, *136*, 13805–13817. [CrossRef]

88. Yuan, J.; Liu, Z.; Dong, M.; Wang, L.; Dong, S.; Hao, J. Self-Assembly of Amphiphilic Copper Nanoclusters Driven by Cationic Surfactants. *Langmuir* **2021**, *37*, 6613–6622. [CrossRef]
89. Tao, Y.; Li, M.; Ren, J.; Qu, X. Metal nanoclusters: Novel probes for diagnostic and therapeutic applications. *Chem. Soc. Rev.* **2015**, *44*, 8636–8663. [CrossRef]
90. Sharma, J.; Yeh, H.-C.; Yoo, H.; Werner, J.H.; Martinez, J.S. A complementary palette of fluorescent silver nanoclusters. *Chem. Commun.* **2010**, *46*, 3280–3282. [CrossRef]
91. Qing, Z.; He, X.; He, D.; Wang, K.; Xu, F.; Qing, T.; Yang, X. Poly(thymine)-Templated Selective Formation of Fluorescent Copper Nanoparticles. *Angew. Chem. Int. Ed.* **2013**, *52*, 9719–9722. [CrossRef]
92. Orbach, R.; Guo, W.; Wang, F.; Lioubashevski, O.; Willner, I. Self-Assembly of Luminescent Ag Nanocluster-Functionalized Nanowires. *Langmuir* **2013**, *29*, 13066–13071. [CrossRef] [PubMed]
93. Compel, W.S.; Wong, O.A.; Chen, X.; Yi, C.; Geiss, R.; Häkkinen, H.; Knappenberger, K.L.; Ackerson, C.J. Dynamic Diglyme-Mediated Self-Assembly of Gold Nanoclusters. *ACS Nano* **2015**, *9*, 11690–11698. [CrossRef]
94. Lahtinen, T.; Hulkko, E.; Sokołowska, K.; Tero, T.-R.; Saarnio, V.; Lindgren, J.; Pettersson, M.; Häkkinen, H.; Lehtovaara, L. Covalently linked multimers of gold nanoclusters Au102(p-MBA)44 and Au∼250(p-MBA)n. *Nanoscale* **2016**, *8*, 18665–18674. [CrossRef] [PubMed]
95. Huang, R.-W.; Wei, Y.-S.; Dong, X.-Y.; Wu, X.-H.; Du, C.-X.; Zang, S.-Q.; Mak, T.C.W. Hypersensitive dual-function luminescence switching of a silver-chalcogenolate cluster-based metal–organic framework. *Nat. Chem.* **2017**, *9*, 689–697. [CrossRef] [PubMed]
96. Huang, R.W.; Dong, X.Y.; Yan, B.J.; Du, X.S.; Wei, D.H.; Zang, S.Q.; Mak, T.C.W. Tandem Silver Cluster Isomerism and Mixed Linkers to Modulate the Photoluminescence of Cluster-Assembled Materials. *Angew. Chem. Int. Ed.* **2018**, *57*, 8560–8566. [CrossRef] [PubMed]
97. Chen, S.; Du, W.; Qin, C.; Liu, D.; Tang, L.; Liu, Y.; Wang, S.; Zhu, M. Assembly of the Thiolated [Au(1) Ag(22) (S-Adm)(12)](3+) Superatom Complex into a Framework Material through Direct Linkage by SbF(6) (-) Anions. *Angew. Chem. Int. Ed.* **2020**, *59*, 7542–7547. [CrossRef]
98. Zhou, S.; Sheng, K.; Zhang, N.; Zhang, H.; Li, H.; Sun, P.; Xin, X. Light-triggered reversible supramolecular self-assembly of azo groups-functionalized copper nanoclusters. *J. Mol. Liq.* **2021**, *343*, 117698. [CrossRef]
99. Udayabhaskararao, T.; Kundu, P.K.; Ahrens, J.; Klajn, R. Reversible Photoisomerization of Spiropyran on the Surfaces of Au25 Nanoclusters. *ChemPhysChem* **2016**, *17*, 1805–1809. [CrossRef]
100. Rival, J.V.; Nonappa; Shibu, E.S. Light-Triggered Reversible Supracolloidal Self-Assembly of Precision Gold Nanoclusters. *ACS Appl. Mater. Interfaces* **2020**, *12*, 14569–14577. [CrossRef]
101. Ai, L.; Li, Y.; Wu, Z.; Liu, J.; Gao, Y.; Liu, Y.; Lu, Z.; Zhang, H.; Yang, B. Photoinduced Conversion of Cu Nanoclusters Self-Assembly Architectures from Ribbons to Spheres. *J. Phys. Chem. C* **2016**, *120*, 24427–24436. [CrossRef]
102. Yao, Q.; Luo, Z.; Yuan, X.; Yu, Y.; Zhang, C.; Xie, J.; Lee, J.Y. Assembly of Nanoions via Electrostatic Interactions: Ion-Like Behavior of Charged Noble Metal Nanoclusters. *Sci. Rep.* **2014**, *4*, 3848. [CrossRef] [PubMed]
103. Basu, S.; Paul, A.; Chattopadhyay, A. Zinc mediated crystalline assembly of gold nanoclusters for expedient hydrogen storage and sensing. *J. Mater. Chem. A* **2016**, *4*, 1218–1223. [CrossRef]
104. Liu, J.-W.; Feng, L.; Su, H.-F.; Wang, Z.; Zhao, Q.-Q.; Wang, X.-P.; Tung, C.-H.; Sun, D.; Zheng, L.-S. Anisotropic Assembly of Ag52 and Ag76 Nanoclusters. *J. Am. Chem. Soc.* **2018**, *140*, 1600–1603. [CrossRef] [PubMed]
105. Bi, Y.; Wang, Z.; Liu, T.; Sun, D.; Godbert, N.; Li, H.; Hao, J.; Xin, X. Supramolecular Chirality from Hierarchical Self-Assembly of Atomically Precise Silver Nanoclusters Induced by Secondary Metal Coordination. *ACS Nano* **2021**, *15*, 15910–15919. [CrossRef] [PubMed]
106. Li, Y.; Xi, W.; Hussain, I.; Chen, M.; Tan, B. Facile preparation of silver nanocluster self-assemblies with aggregation-induced emission by equilibrium shifting. *Nanoscale* **2021**, *13*, 14207–14213. [CrossRef]
107. Sugiuchi, M.; Zhang, M.; Hakoishi, Y.; Shichibu, Y.; Horimoto, N.N.; Yamauchi, Y.; Ishida, Y.; Konishi, K. Aggregation-Mode-Dependent Optical Properties of Cationic Gold Clusters: Formation of Ordered Assemblies in Solution and Unique Optical Responses. *J. Phys. Chem. C* **2020**, *124*, 16209–16215. [CrossRef]
108. Su, X.; Liu, J. pH-Guided Self-Assembly of Copper Nanoclusters with Aggregation-Induced Emission. *ACS Appl. Mater. Interfaces* **2017**, *9*, 3902–3910. [CrossRef]
109. Coutiño-Gonzalez, E.; Baekelant, W.; Steele, J.A.; Kim, C.W.; Roeffaers, M.B.J.; Hofkens, J. Silver Clusters in Zeolites: From Self-Assembly to Ground-Breaking Luminescent Properties. *Acc. Chem. Res.* **2017**, *50*, 2353–2361. [CrossRef]
110. Deng, H.-H.; Shi, X.-Q.; Wang, F.-F.; Peng, H.-P.; Liu, A.-L.; Xia, X.-H.; Chen, W. Fabrication of Water-Soluble, Green-Emitting Gold Nanoclusters with a 65% Photoluminescence Quantum Yield via Host–Guest Recognition. *Chem. Mater.* **2017**, *29*, 1362–1369. [CrossRef]
111. Luo, J.; Xie, Z.; Lam, J.W.Y.; Cheng, L.; Chen, H.; Qiu, C.; Kwok, H.S.; Zhan, X.; Liu, Y.; Zhu, D.; et al. Aggregation-induced emission of 1-methyl-1,2,3,4,5-pentaphenylsilole. *Chem. Commun.* **2001**, 1740–1741. [CrossRef]
112. Wu, Z.; Liu, H.; Li, T.; Liu, J.; Yin, J.; Mohammed, O.F.; Bakr, O.M.; Liu, Y.; Yang, B.; Zhang, H. Contribution of Metal Defects in the Assembly Induced Emission of Cu Nanoclusters. *J. Am. Chem. Soc.* **2017**, *139*, 4318–4321. [CrossRef] [PubMed]
113. Liu, J.; Wu, Z.; Tian, Y.; Li, Y.; Ai, L.; Li, T.; Zou, H.; Liu, Y.; Zhang, X.; Zhang, H.; et al. Engineering the Self-Assembly Induced Emission of Cu Nanoclusters by Au(I) Doping. *ACS Appl. Mater. Interfaces* **2017**, *9*, 24899–24907. [CrossRef]

114. Luo, Z.; Yuan, X.; Yu, Y.; Zhang, Q.; Leong, D.T.; Lee, J.Y.; Xie, J. From Aggregation-Induced Emission of Au(I)–Thiolate Complexes to Ultrabright Au(0)@Au(I)–Thiolate Core–Shell Nanoclusters. *J. Am. Chem. Soc.* **2012**, *134*, 16662–16670. [CrossRef] [PubMed]
115. Pyo, K.; Thanthirige, V.D.; Kwak, K.; Pandurangan, P.; Ramakrishna, G.; Lee, D. Ultrabright Luminescence from Gold Nanoclusters: Rigidifying the Au(I)–Thiolate Shell. *J. Am. Chem. Soc.* **2015**, *137*, 8244–8250. [CrossRef]
116. Dou, X.; Yuan, X.; Yu, Y.; Luo, Z.; Yao, Q.; Leong, D.T.; Xie, J. Lighting up thiolated Au@Ag nanoclusters via aggregation-induced emission. *Nanoscale* **2014**, *6*, 157–161. [CrossRef] [PubMed]
117. Bain, D.; Maity, S.; Patra, A. Surface motifs regulated aggregation induced emission in gold–silver nanoclusters. *Chem. Commun.* **2020**, *56*, 9292–9295. [CrossRef]
118. Maity, S.; Bain, D.; Patra, A. Engineering Atomically Precise Copper Nanoclusters with Aggregation Induced Emission. *J. Phys. Chem. C* **2019**, *123*, 2506–2515. [CrossRef]
119. Sugiuchi, M.; Maeba, J.; Okubo, N.; Iwamura, M.; Nozaki, K.; Konishi, K. Aggregation-Induced Fluorescence-to-Phosphorescence Switching of Molecular Gold Clusters. *J. Am. Chem. Soc.* **2017**, *139*, 17731–17734. [CrossRef]
120. Herbert, P.J.; Yi, C.; Compel, W.S.; Ackerson, C.J.; Knappenberger, K.L. Relaxation Dynamics of Electronically Coupled Au20(SC8H9)15-n-glyme-Au20(SC8H9)15 Monolayer-Protected Cluster Dimers. *J. Phys. Chem. C* **2018**, *122*, 19251–19258. [CrossRef]
121. Das, A.K.; Maity, S.; Sengupta, T.; Bista, D.; Reber, A.C.; Patra, A.; Khanna, S.N.; Mandal, S. One-Dimensional Silver-Thiolate Cluster-Assembly: Effect of Argentophilic Interactions on Excited-State Dynamics. *J. Phys. Chem. Lett.* **2021**, *12*, 2154–2159. [CrossRef]
122. Dong, X.-Y.; Si, Y.; Yang, J.-S.; Zhang, C.; Han, Z.; Luo, P.; Wang, Z.-Y.; Zang, S.-Q.; Mak, T.C.W. Ligand engineering to achieve enhanced ratiometric oxygen sensing in a silver cluster-based metal-organic framework. *Nat. Commun.* **2020**, *11*, 3678. [CrossRef] [PubMed]
123. Kuppan, B.; Maitra, U. Instant room temperature synthesis of self-assembled emission-tunable gold nanoclusters: Million-fold emission enhancement and fluorimetric detection of Zn2+. *Nanoscale* **2017**, *9*, 15494–15504. [CrossRef]
124. Wang, M.; Chen, Y.; Cai, W.; Feng, H.; Du, T.; Liu, W.; Jiang, H.; Pasquarelli, A.; Weizmann, Y.; Wang, X. In situ self-assembling Au-DNA complexes for targeted cancer bioimaging and inhibition. *Proc. Natl. Acad. Sci. USA* **2020**, *117*, 308. [CrossRef]
125. Zhang, W.; Lin, D.; Wang, H.; Li, J.; Nienhaus, G.U.; Su, Z.; Wei, G.; Shang, L. Supramolecular Self-Assembly Bioinspired Synthesis of Luminescent Gold Nanocluster-Embedded Peptide Nanofibers for Temperature Sensing and Cellular Imaging. *Bioconjugate Chem.* **2017**, *28*, 2224–2229. [CrossRef] [PubMed]
126. Jin, R.; Li, G.; Sharma, S.; Li, Y.; Du, X. Toward Active-Site Tailoring in Heterogeneous Catalysis by Atomically Precise Metal Nanoclusters with Crystallographic Structures. *Chem. Rev.* **2021**, *121*, 567–648. [CrossRef] [PubMed]
127. Yu, F.; Chen, L.; Li, X.; Shen, X.; Zhao, H.; Duan, C.; Chen, Q. Cu Nanocluster-Loaded TiO_2 Nanosheets for Highly Efficient Generation of CO-Free Hydrogen by Selective Photocatalytic Dehydrogenation of Methanol to Formaldehyde. *ACS Appl. Mater. Interfaces* **2021**, *13*, 18619–18626. [CrossRef] [PubMed]
128. Zhao, M.; Huang, S.; Fu, Q.; Li, W.; Guo, R.; Yao, Q.; Wang, F.; Cui, P.; Tung, C.-H.; Sun, D. Ambient Chemical Fixation of CO_2 Using a Robust Ag27 Cluster-Based Two-Dimensional Metal–Organic Framework. *Angew. Chem. Int. Ed.* **2020**, *59*, 20031–20036. [CrossRef]
129. Zhao, S.; Austin, N.; Li, M.; Song, Y.; House, S.D.; Bernhard, S.; Yang, J.C.; Mpourmpakis, G.; Jin, R. Influence of Atomic-Level Morphology on Catalysis: The Case of Sphere and Rod-Like Gold Nanoclusters for CO_2 Electroreduction. *ACS Catal.* **2018**, *8*, 4996–5001. [CrossRef]
130. Sun, P.; Wang, Z.; Sun, D.; Bai, H.; Zhu, Z.; Bi, Y.; Zhao, T.; Xin, X. pH-guided self-assembly of silver nanoclusters with aggregation-induced emission for rewritable fluorescent platform and white light emitting diode application. *J. Colloid Interface Sci.* **2020**, *567*, 235–242. [CrossRef]
131. Casteleiro, B.; Martinho, J.M.G.; Farinha, J.P.S. Encapsulation of gold nanoclusters: Stabilization and more. *Nanoscale* **2021**, *13*, 17199–17217. [CrossRef]
132. Maye, M.M.; Nykypanchuk, D.; Cuisinier, M.; van der Lelie, D.; Gang, O. Stepwise surface encoding for high-throughput assembly of nanoclusters. *Nat. Mater.* **2009**, *8*, 388–391. [CrossRef] [PubMed]

nanomaterials

Review

Controlling the Chemistry of Nanoclusters: From Atomic Precision to Controlled Assembly

Srestha Basu [1], Anumita Paul [2,*] and Rodolphe Antoine [3,*]

1 Schulich Faculty of Chemistry, Technion—Israel Institute of Technology, Haifa 3200003, Israel; srestha@iitg.ac.in
2 Department of Chemistry, Indian Institute of Technology Guwahati, Guwahati 781039, Assam, India
3 Institut Lumière Matière UMR 5306, Univ Lyon, Université Claude Bernard Lyon 1, CNRS, F-69100 Villeurbanne, France
* Correspondence: anumita@iitg.ac.in (A.P.); rodolphe.antoine@univ-lyon1.fr (R.A.)

Abstract: Metal nanoclusters have gained prominence in nanomaterials sciences, owing to their atomic precision, structural regularity, and unique chemical composition. Additionally, the ligands stabilizing the clusters provide great opportunities for linking the clusters in higher order dimensions, eventually leading to the formation of a repertoire of nanoarchitectures. This makes the chemistry of atomic clusters worth exploring. In this mini review, we aim to focus on the chemistry of nanoclusters. Firstly, we summarize the important strategies developed so far for the synthesis of atomic clusters. For each synthetic strategy, we highlight the chemistry governing the formation of nanoclusters. Next, we discuss the key techniques in the purification and separation of nanoclusters, as the chemical purity of clusters is deemed important for their further chemical processing. Thereafter which we provide an account of the chemical reactions of nanoclusters. Then, we summarize the chemical routes to the spatial organization of atomic clusters, highlighting the importance of assembly formation from an application point of view. Finally, we raise some fundamentally important questions with regard to the chemistry of atomic clusters, which, if addressed, may broaden the scope of research pertaining to atomic clusters.

Keywords: synthesis; purification; reactions; assembly; application; nanoclusters; ligands

1. Introduction

Continuous efforts to achieve superior properties of assembled nanoscale particles have been limited due to the resulting poly-dispersity associated with colloidal routes of synthesis [1]. A solution to this limitation seems to have emerged from the advent of ligand-protected atomic clusters (often called nanoclusters or quantum clusters) [2,3]. In this case, ligands that stabilize the clusters are highly reactive in nature and thus provide facile avenue for "ligand mediated spatial organization of nanoclusters" [4–6]. Furthermore, the most important characteristics of atomic nanoclusters, which distinguish them from other classes of nanomaterials, is their structural integrity [7]. Notwithstanding the significant advancements achieved towards designing chemical routes to a synthesis of nanoclusters, the issue of poly-dispersity—*namely a collection of different ligand protected atomic cluster sizes*—remains a challenge to date. Thus, unlike other forms of nanoscale particles, a dispersion of atomic clusters typically constitutes of structurally and chemically related species. For example, a dispersion of nanoclusters may be purified following conventional separation techniques such as gel electrophoresis, size exclusion based ultrafiltration and normal and reverse phase chromatography [8]. Then, the precise chemical formula of the nanoclusters may be deciphered using mass spectrometry [9]. This has facilitated the use of nanoclusters in a plethora of applications such as bio diagnostics, therapeutics, catalysis and sensing, to mention a few of their uses [10–16].

Additionally, it has recently been demonstrated that the assembly of nanoclusters features superior physicochemical properties vis à vis their non-assembled analogues. This is primarily attributed to the collective properties of constituent nanoclusters in an "assembly" [17]. Thus, as evidenced by the recent upsurge of studies in the literature, [17–19] much attention is being paid towards the fabrication of complex nanostructures by using nanoclusters as building blocks. To this end, several strategies have been developed for the systematic organization of nanoclusters. For example, a straightforward method for assembling atomic clusters into assemblies has been pursued, based on the interaction among the ligands stabilizing the nanoclusters [20]. Similarly, the use of biological molecules such as DNA and proteins as templates for the organization of nanoclusters have also been reported. Notably, supramolecular interactions are also known to be instrumental in the formation of assembly of nanoclusters [21,22]. In this regard, the use of principles of coordination chemistry has been recognized as a possible route for the deterministic ordering of nanoclusters [21]. Formed assemblies of nanoclusters have been demonstrated to have enhanced stability, advanced property and superior application potential vis-a-vis the non-assembled clusters [23].

In this mini review, we aim to focus on the developments made so far with regard to the synthesis and applications of nanoclusters, their chemical reactions and eventual formation of their hierarchical assemblies endowed with superior application potential. This is deemed important to pave the way for the deterministic fabrication of tailor-made assemblies with desired functionalities. The current account is envisioned to not only enrich the scientific community with fundamental insights but also facilitate the advent of advanced nanostructures with controllable chemistry and tuneable properties. First, we focus on the plethora of strategies developed for synthesizing atomically precise clusters comprised of various metal atoms, stabilized by a definite number of ligands. To this end, we provide insight to the mechanism of formation of nanoclusters and thereby highlight the critical roles played by ligands in cluster synthesis. The purification, separation and isolation of nanoclusters is a necessary step between isolated clusters towards controlled self-assembly. We thus emphasise the importance of standard techniques that allow for the thorough purification of such clusters. Next, we discuss the rational strategies developed so far to add functionality to these nanoclusters, based on the principles of coordination and supramolecular chemistry [24]. The added functionality allows the nanoclusters to arrange into higher dimensional structures with superior physicochemical properties vis à vis in comparison to the non-assembled clusters. Thereafter, we justify the importance of assembly formation.

2. Ligand-Protected Metal Nanoclusters. Comparing Gold, Silver, Copper and Nickel

There has been a significant surge in efforts to synthesize atomically precise nanoclusters. The most explored metal nanocluster to date is that of gold. Gold is considered the noblest metal, and gold clusters are usually more stable than other noble metal clusters [25]. Gold nanoclusters have been synthesized using a multitude of stabilizers, including dendrimers, proteins, DNA, peptides and polymers. In this regard, Dickson et al. synthesized dendrimer-protected Au nanoclusters exhibiting a quantum yield as high as 40% [26]. Polyamido amine (4th generation) were used for the reduction and stabilization of Au_8 nanoclusters (Figure 1a) [26]. On the other hand, natural bio macromolecules such as protein-stabilized gold nanoclusters have been found to be biocompatible [27]. Proteins such as Bovine serum albumin (BSA) and Human serum albumin (HSA) have been widely used for the stabilization of a gold nanocluster, comprising 25 Au atoms featuring red luminescence (Figure 1b) [28,29]. Other than these proteins, lysozymes, horseradish peroxidase and insulin have also been reported to stabilize gold nanoclusters [30–32]. DNA-templated gold nanoclusters have also been reported to feature bright luminescence, photo-stability and biocompatibility [33,34]. In addition, small molecules such as tripeptides (glutathione) (Figure 1c), [35–37] mercaptopropionic acid (MPA), [38] mercaptobenzoic acid (MBA), [39] penicillamine, [40] amino acids [41,42], etc., have been used to produce atomically precise gold nanoclusters. Interestingly, microorganisms such as bacteria have also been used as templates for the synthesis of Au NCs [43]. The key to the synthesis of gold nanoclusters is the exploitation of soft–soft interactions between gold and thiolated ligands, although nitrogenous ligands are also reported to be active in the stabilization of Au NCs in some cases.

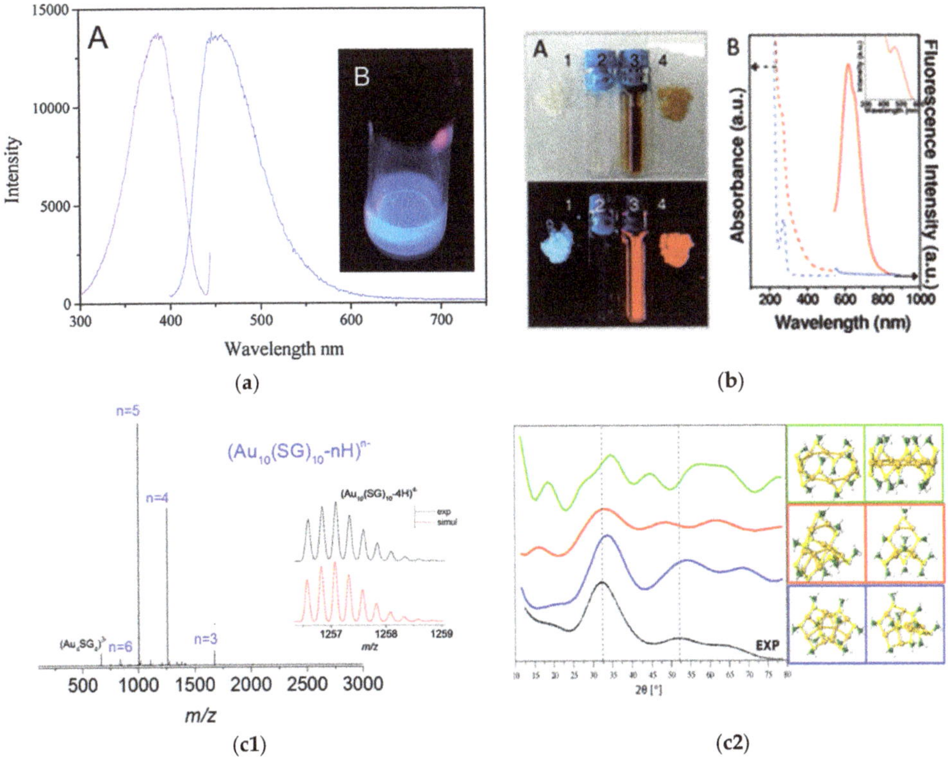

Figure 1. (**a**): Excitation and emission spectra of Au_8 nanodots. Inset shows digital photograph of Au_8 nanodots, upon excitation at 366 nm. Reprinted with permission from [26]. Copyright 2003 American Chemical Society. (**b**) (A) Digital photographs of BSA stabilized Au NCs and control BSA (solution and powder) under visible and UV light. (B) Absorption and emission spectra of BSA stabilized Au NCs and control BSA. Reprinted with permission from [28]. Copyright 2009 American Chemical Society. (**c1**) Electrospray ionization (ESI) mass spectrum of Au_{10} clusters stabilized by glutathione. (**c2**) Experimental and simulated X-ray diffraction patterns of various isomers of Au_{10} clusters stabilized by glutathione. Reprinted with permission from [35]. Copyright 2017 American Chemical Society.

The synthesis of silver nanoclusters has always been considered as more challenging than the synthesis of their gold analogues. This is due to the greater tendency of zero-valent silver to become oxidized to their +1 oxidation state. Thus, much effort has been invested in the synthesis of a well-defined Ag nanocluster and in exploration of their properties at a sub-nanometer scale. Of all the strategies involved in the synthesis of Ag nanoclusters, the case of chemical etching of Ag nanoparticles has been recognized as the most popular one. This is largely due to the possibility of controlled Ag nanoclusters in mild etching environment. In this regard there are several reports where Ag NPs protected by mercapto-succininc acid (MSA) were etched to produce Ag_8 and Ag_7 NCs, which could further be separated using gel electrophoresis [44]. Additionally, red luminescent Ag NCs, comprising 38 Ag atoms, could be produced by etching off Ag NPs capped by citrate ions [45]. Another effective strategy employed for the synthesis of Ag NCs is based on ligand exchange. To this end, glutathione (SG) molecules stabilizing Ag_{35} NCs were exchanged with 4 fluorothiophenol to produce Ag_{44} NCs, stabilized by the latter ligand [46]. Notwithstanding the advantages of the above mentioned synthetic strategies the most effective method to obtain Ag NCs is through direct reduction of silver salts.

Out of all classes of atomic clusters, copper-based nanoclusters were once considered to be extremely challenging. This is due to the reactive nature of copper, leading to the easy transformation of Cu (0) to Cu (II) in the presence of aerial oxygen. In this regard, in 2011, Chen and coworkers successfully reported the synthesis of copper nanoclusters comprising 8 Cu atoms stabilized by four thiolated ligands [47]. The route to the synthesis of the aforementioned clusters was based on mixing of $Cu(NO_3)_2$ and $[N(oct)_4][Br]$ in ethanol, following further treatment with 2-mercato-5N propyl pyrimidyl and reduction with $NaBH_4$. This yielded atomically precise Cu clusters with near homogeneity in composition. On the other hand, highly luminescent Cu nanoclusters were synthesized using a 'green' approach, with protein as template. Specifically, copper precursors were incubated and reduced in the presence of proteins such as lysozymes, which eventually led to the formation of highly fluorescent Cu nanoclusters with a composition ranging from Cu_2 to Cu_9 (Figure 2A) [48]. Cu nanoclusters embedded in BSA [49] and HAS [50] have also been synthesized in an allied fashion. Additionally, other proteins, such as papain, [51] transferrin [52] and trypsin [53] have been used for the synthesis of Cu nanoclusters. A common feature that is typical with regard to the use of proteins in stabilizing Cu nanoclusters is that the synthetic procedure is associated with mixing the metal precursor with protein at ambient temperatures and with a basic pH. The use of an additional reducing agent such as ascorbic acid, $NaBH_4$ is reported to have varied from procedure to procedure. DNA has been used to stabilize Cu nanoclusters featuring bright luminescence [54]. Moreover, small peptides such as L-glutathione, have been used to stabilize ultra-small Cu nanoclusters exhibiting a quantum yield as high as 8.6%. However, by using the same stabilizer, but varying the ratio of metal precursor to stabilizer Cu nanoclusters exhibiting blue luminescence have also been synthesized [55]. Other typical thiolated molecules such as thiosalicylic acid, [56] cysteine, [57] dihydrolipoic acid [58] and mercaptobenzoic acid [59] have been employed to synthesize Cu nanoclusters. When using thiolated molecules for the synthesis of copper nanoclusters, the stabilizer forms a complex with Cu(I) and the latter is reduced to Cu(0), either with the aid of additional reducing agents or by the thiol groups of the stabilizers themselves (Figure 2B). Copper nanoclusters exhibiting temporal, chemical and photo stability have been found suitable for applications ranging from catalysis, fluorescence based sensing of metal ions, disease markers, contaminants, pH, temperature, as vehicles for drug delivery, markers for cell labelling and as electrocatalysts, to name a few [60].

With regard to synthesis of Ni nanoclusters, the typical approach was to use carbonyl ligands as stabilizers. For example, $[Ni_{12-x}(PMe)_x(CO)_{24-3x}]^{2-}$ were synthesized using carbonyl protection [61]. However, carbonyl stabilized Ni clusters are typically negatively charged and thus counter cations are essential for their neutralization. Additionally, post synthetic functionalization of the synthesized clusters is difficult due to the presence of strongly bound carbonyl ligands. Thus, the advent of protected thiolate was deemed important for widening the chemistry of Ni based nanoclusters. To this end, Xu et al. performed a chemical reaction between $NiCl_2.6H_2O$ and $PhCH_2CH_2SH$ in a mixture of THF and methanol followed by reduction with aqueous $NaBH_4$ to obtain Ni_{39} and Ni_{41} clusters [62]. Similarly, $Ni_4(PET)_8$ and $Ni_6(PET)_{12}$ [PET = phenylethane thiol] clusters were synthesized based on the ligand exchange strategy between glutathione-nickel complex and PET. The formed clusters were found to be effective electrocatalysts for water oxidation [63]. A table summarizing the synthetic details and properties of various nanoclusters is provided below (Table 1).

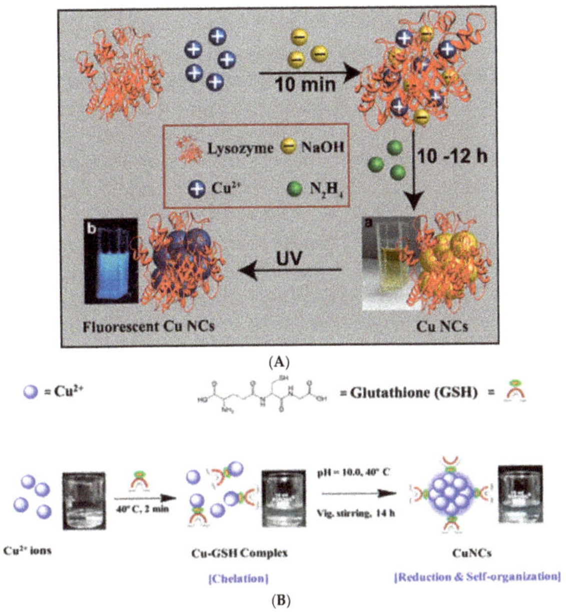

Figure 2. (**A**): Schematic illustration for formation of Cu NCs stabilized by lysozyme. Reprinted with permission from [48]. Copyright 2014 American Chemical Society. (**B**) Schematic illustration for formation of Cu NCs stabilized by glutathione. Reprinted with permission from [55]. Copyright 2015 American Chemical Society.

Table 1. Summary of the synthetic details and properties of various nanoclusters (presented in this review).

Metal	Stabilizer	No. of Metal Atoms Comprising the Clusters	Emission Color	Reference
gold	Polyamido amine (4th generation)	8	blue	26
gold	BSA	25	red	28
gold	HSA	25	red	29
gold	DNA	7	red	33
gold	glutathione	10	Non-emissive	35
gold	MPA + chitosan	20	orange-red (pH dependent)	38
gold	histidine	10	blue	41
gold	MPA + bacteria	Not determined	white	43
silver	produced by Ag NPs protected by mercapto-succininc acid	8, 7	red, blue green	44
silver	produced by etching of Ag NPs capped by citrate ions	38	red	45
copper	lysozymes	2–9	blue	48
copper	glutathione	15	blue	55
copper	cysteine	4	cyan	57
copper	dihydrolipoic acid	4	red	58
nickel	phenylethanethiol	4, 6	Not reported	63

3. Purification, Separation, and Isolation of Nanoclusters

As synthesized NCs are often associated with the presence of NCs with varying compositions, namely, a dispersion of atomic clusters typically constitutes of structurally and chemically related species. This is, however, not desirable, as the properties of NCs are known to vary as a function of the number of atoms constituting the clusters, the structure of the overall clusters and the ligands (number density per cluster and nature) stabilizing the clusters. For example, clusters composed of fewer number of atoms are known to feature luminescence at lower wavelengths as opposed to clusters comprising of greater number of atoms. Similarly, physical properties of clusters such as solubility is dictated by the polarity of the ligands. Additionally, the catalytic activity of nanoclusters is highly dependent on the number density of the ligands stabilizing the clusters. Additionally, the application potential of nanoclusters becomes attenuated owing to the presence of unbound ligands in the reaction mixture. Thus, identifying and deciphering the chemical composition of a particular cluster (amongst a mixture of clusters produced synthetically) is of key importance to understand and modulate their application potential. In order to address this issue, several techniques have emerged for the purification of synthesized nanoclusters. Amongst the several purification techniques employed for the isolation of metal nanoclusters, the most common is the metal ion induced precipitation of nanoclusters. In this regard, Guan et al. successfully isolated BSA-stabilized Au_{25} NCs from unreacted BSA through Zn ion assisted precipitation of the former. In an allied vein, the pH of the reaction mixture containing the NCs can be adjusted to cause the precipitation of NCs. This method has also been used for the purification of BSA-stabilized Au NCs [64]. Precisely, the pH of the solution containing BSA stabilized Au NCs was reduced to reach the isoelectric point of BSA wherein the conformation of BSA was altered, leading to the agglomeration and eventual precipitation of BSA Au NCs. Similarly, methionine-protected Au NCs were separated following a reduction of the pH, leading to the relaxation of charge repulsion of the ligands, consequential aggregation of isolation of the NCs. Furthermore, the tuning of the solvent polarity of the reaction mixture containing the NCs has also been used for purification of NCs. For example, Galchenko et al. could selectively precipitate Au_{25} NCs by varying the ratio of methanol to water [65]. Another well-known strategy for the removal of unbound ligands is through the addition of ionic salts. This is because in the presence of additional ions, the zeta potential of the ligands stabilizing the NCs becomes reduced, which further led to the neutralization of charge and consequential aggregation of NCs. For example, mercaptoundecanoic acid (MUA) protected Au NCs are reported to aggregate and separate in presence of NaCl [66]. Thus, the use of external agents for the separation of NCs from unbound ligands, has emerged as a convenient customizable, rapid and cost effective technique for purification of NCs.

Apart from the aforementioned techniques, the isolation of NCs based on 'size' has also been widely practiced. For example, glutathione-protected Au NCs were separated from unreacted precursors using dialysis membranes with a cutoff of 3.5 k molecular weight [67]. The principal behind this technique is that glutathione molecules having smaller size vis à vis glutathione stabilized Au NCs could feasibly pass through the dialysis membrane, leaving the latter behind. Similarly, lipoic acid stabilized gold nanoclusters were purified from unreacted lipoic using a membrane of 3 kDa cutoff [68]. Furthermore, 14 kDa cutoff membranes were used to remove unreacted reactants from polyvinylpyrrolidone stabilized Cu NCs [69].

In order to isolate NCs of a particular composition from analogous NCs of varying composition, chromatographic methods have been sought after. The principle behind the technique is that clusters with polarity react with the stationary phase to a greater extent as compared to the less polar NCs. This would lead to faster elution of the non-polar NCs and greater retention of the polar NCs in the stationary phase. Using this principle Au_{25} NCs stabilized by phenyl ethane thiol were separated from Au_{25} stabilized by butane thiolate owing to difference of polarity of the respective ligands [70]. Glutathione stabilized clusters of varying composition such as $Au_{10}(SG)_{10}$, $Au_{15}(SG)_{13}$, $Au_{18}(SG)_{14}$, $Au_{22}(SG)_{16}$,

$Au_{25}(SG)_{18}$, $Au_{29}(SG)_{20}$, $Au_{33}(SG)_{22}$, $Au_{39}(SG)_{24}$ were separated using the principle of reversed phase chromatography [71].

Electrophoretic techniques that involve separation, based on the differential mobility of charged species under the influence of an external electric field, have also been utilized for the separation of NCs of varying sizes. To this end, NCs with size difference of only 0.5 nm could be separated using gel electrophoresis. Importantly, the Antoine group recently demonstrated the use of polyacrylamide gel electrophoretic (PAGE) method to isolate $Au_{10}(SG)_{10}$ from $Au_{15}(SG)_{13}$ and $Au_{25}(SG)_{18}$ (see Figure 3) [35].

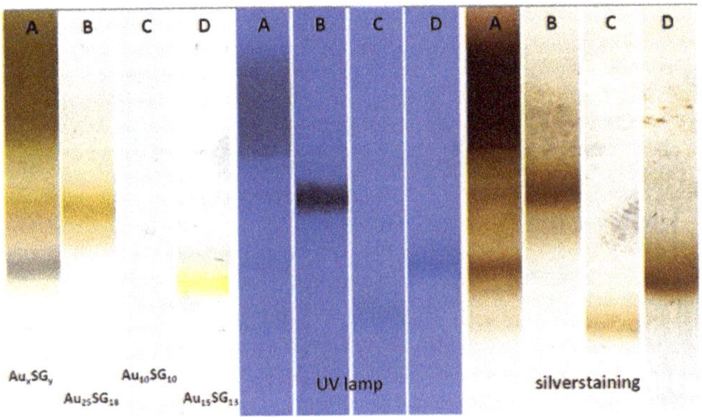

Figure 3. Polyacrylamide gel electrophoretic (PAGE) based separation of $Au_{10}(SG)_{10}$ from $Au_{15}(SG)_{13}$ and $Au_{25}(SG)_{18}$. Reprinted from Supporting Information Ref. [35]. Copyright 2017 American Chemical Society.

The above-described techniques are used for separating, purifying, and isolating a dispersion of atomic clusters, typically constituting different chemically related species. However, dispersions of atomic clusters for a given chemical composition can also comprise different but structurally related species. In such cases, molecular species in terms of 3-dimensional structures and structural dispersity, can be characterized, separated and even isolated based on the coupling of mass spectrometry with ion mobility spectrometry as recently demonstrated by Antoine group and other groups [9,72–78]. Thus, from the above discussion its it is apparent that synthetic and purification techniques of NCs have made great advancements in the last decade. This not only highlights the importance of research pertaining to atomic clusters but also presents them as ideal candidates for further chemical reactions and the advent of newer nanomaterials.

4. Chemical Reactions of Nanoclusters

4.1. Intermolecular Chemical Reactions of Nanoclusters. Example of Zinc Ion Induced Aggregation Strategy

Atomically precise nanoclusters are the closest analogues to molecules in classical chemistry. This is because, akin to molecules, atomic clusters have structural integrity, defined chemical composition and purity. Thus, in the same way that molecules are building blocks of compounds and bulk materials, nanoclusters are ideally suited as building blocks of hierarchical nanomaterials. This can be achieved through 'chemical reactions' of nanoclusters. However, unlike atoms and molecules which provide directed bonds, through covalent bonds for instance, nanoclusters are surrounded by an "isotropic" ligand shell devoid of such directional bonding. Therefore, in order to equip nanoclusters with directional bonding, it is essential that the nanoclusters are stabilized with chemically interacting ligands. This lays the foundation of chemical reactions of nanoclusters, either among themselves or with external chemical agents.

To this end, the Paul group has demonstrated a series of chemical reactions involving gold NCs. As a novel study, gold NCs were stabilized with MUA and were found to feature red luminescence. The Au NCs were then reacted among themselves via (bi)coordination between zinc ions and carboxylate terminals of the MUA molecules stabilizing the clusters. This led to an enhancement in the luminescence of Au NCs. Furthermore, (bi)coordinated zinc ions were coordinatively saturated following reactions with fluorescein molecules. The overall composite, constituting fluorescein and MUA stabilized Au NCs bridged with zinc ions, was rendered with dual red and green fluorescence. The chemical reactivity of the composite was then used to discriminate the biothiols, to the level of a few particles (Figure 4a) [79].

In another report, the chemical reactivity of histidine-stabilized Au NCs was exploited to perform intracellular logic operations. Briefly, the imidazole nitrogen of histidine moieties stabilizing Au NCs were conjugated with zinc ions, wherein the variation of luminescence of the former was suited to form a "tri state" logic operation at the cellular level. Furthermore, the zinc ion-coordinated histidine-stabilized Au NCs, when reacted with sulfide ions, were observed to quench the luminescence of the Au NCs, which could be used to construct an "on-off" intracellular switch. Finally, collective reactions among zinc ions, histidine stabilized Au NCs and sulfide ions, were used to form the basis of the "INHIBIT" gate within HeLa cells [80].

Additionally, zinc ions induced the aggregation of Au NCs, featuring bright green luminescence under visible light excitation, which were found to react differentially with amino acids, based on the pKa of the latter. This formed the basis for the discrimination of amino acids under visible light excitation using zinc ion-induced aggregates of Au NCs (Zn Au NCs) as a probe. The carboxylate ends of mercaptopropionic acid (MPA) stabilizing the clusters were coordinated with zinc ions, leading to the formation of green luminescent aggregates of Au NCs. The surface of the formed aggregates was further covered with MPA molecules. The pKa of MPA is reported to be 4.34. Thus, upon their reaction with amino acids with a pKa less than that of MPA, the surface MPA molecules became protonated, thereby blocking ligand to metal charge transfer, and quenching of the luminescence of Zn Au NCs. On the other hand, amino acids with a pKa greater than that of MPA did not react with the surface MPA molecules, thereby causing no discernible effect on the luminescence of Zn Au NCs. Thus, based on the differential reactivity of the MPA molecules present on the surface of Zn Au NCs towards amino acids with varying pKa, the latter could be distinguished easily (Figure 4b) [81].

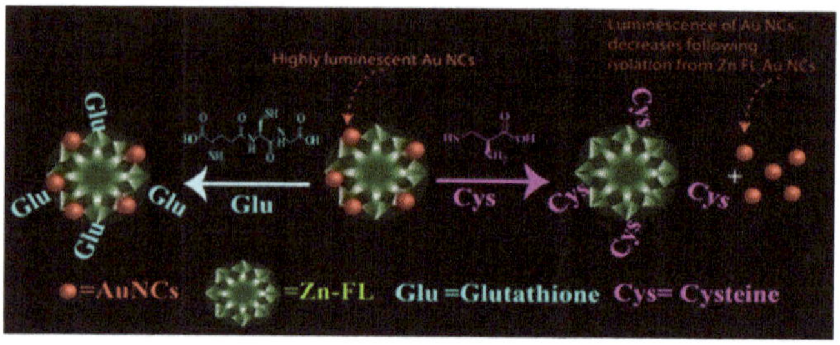

(a)

Figure 4. *Cont.*

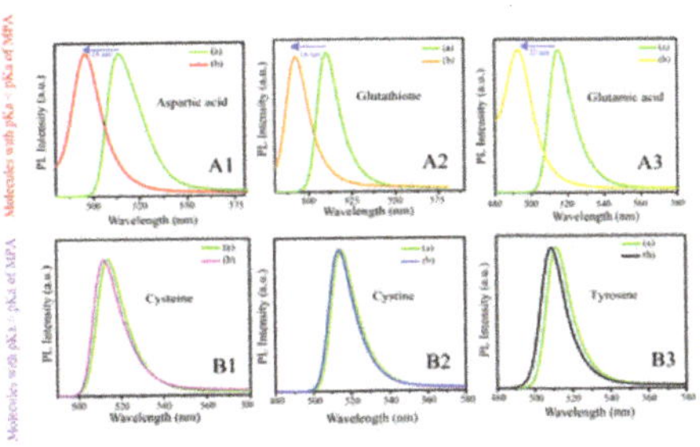

(b)

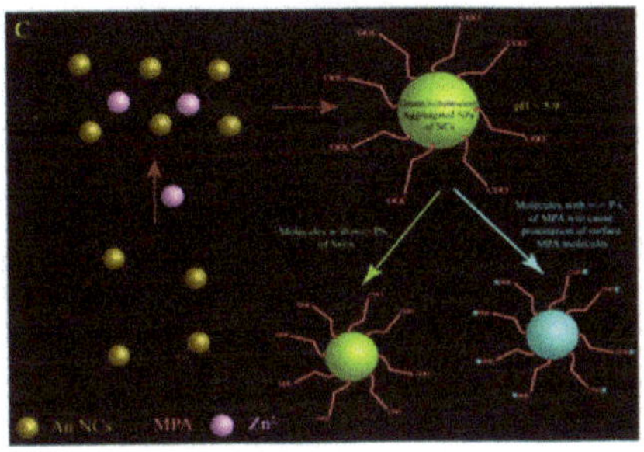

(c)

Figure 4. (**a**): Schematic illustration depicting possible mode of reaction among Au NCs-fluorescein composite and cysteine/glutathione. Reprinted with permission from [79]. Copyright 2018 American Chemical Society. (**b**) Variation in emission spectra of Zn Au NCs upon interaction with aspartic acid, glutathione, glutamic acid, cysteine, cystine and tyrosine. (**c**) Schematic illustration depicting possible mode of reaction among Zn Au NCs and aforementioned molecules. Reprinted with permission from [81]. Copyright 2019 American Chemical Society.

To a similar end, the reactivity of the MPA molecules residing on the surface of allied Zn Au NCs probes, were also used to discriminate between geometrical isomers. However, unlike the case of amino acids, the proton-transfer reaction in the current report was further governed by intra molecular hydrogen bonding interactions among the analytes (Figure 4c) [81]. Interestingly, in a different study, Antoine and coworkers, measured the two photon excited fluorescence cross sections of the Zn Au NCs aggregates. Importantly, a four-fold enhancement in two-photon excited fluorescence of Zn Au NCs was observed vis à vis non assembled clusters. In the same study, for the first time ever, the molar mass of individual Au NCs recorded by time-of-flight mass spectrometry indicated the presence of $Au_{10}MPA_{10}$ catenane nanoclusters, while the entire mass distribution of aggregates of Au NCs was measured using charge-detection mass spectrometry [82].

4.2. Intramolecular Chemical Reactions of Nanoclusters. Example of Oxygen and Ligand and Metal Exchange Reactions

To focus on a different aspect, the luminescence of copper nanoclusters was modulated based on a reaction between the ligands (stabilizing the clusters) and aerial oxygen in presence of light. Briefly, Cu NCs stabilized with cysteine molecules, were found to feature red luminescence. The clusters, however, upon exposure to aerial oxygen in presence of light, were observed to become non luminescent. This was attributed to the desorption of cysteine molecules from the surface of Cu NCs, leading to the transformation of the latter into non luminescent aggregates. The cysteine molecules, in turn, were proposed to have transformed to S-nitrosothiolates following reaction with reactive nitrogen species [83].

Additionally, ligand exchange reaction was performed between histidine stabilized gold nanoclusters and cysteine. The idea behind the study was that cysteine having thiol groups would replace the histidine molecules stabilizing the Au NCs, owing to greater aurophilicity of thiol groups in cysteine as compared to nitrogenous groups present in histidine moieties. This eventually led to alteration in luminescence of the clusters, as ligands stabilizing the clusters are well known for playing key role in the luminescence properties of the clusters. The ligand exchange reaction, involving Au NCs was well validated using X-ray photoelectron spectroscopy (XPS) (Figure 5a) [84].

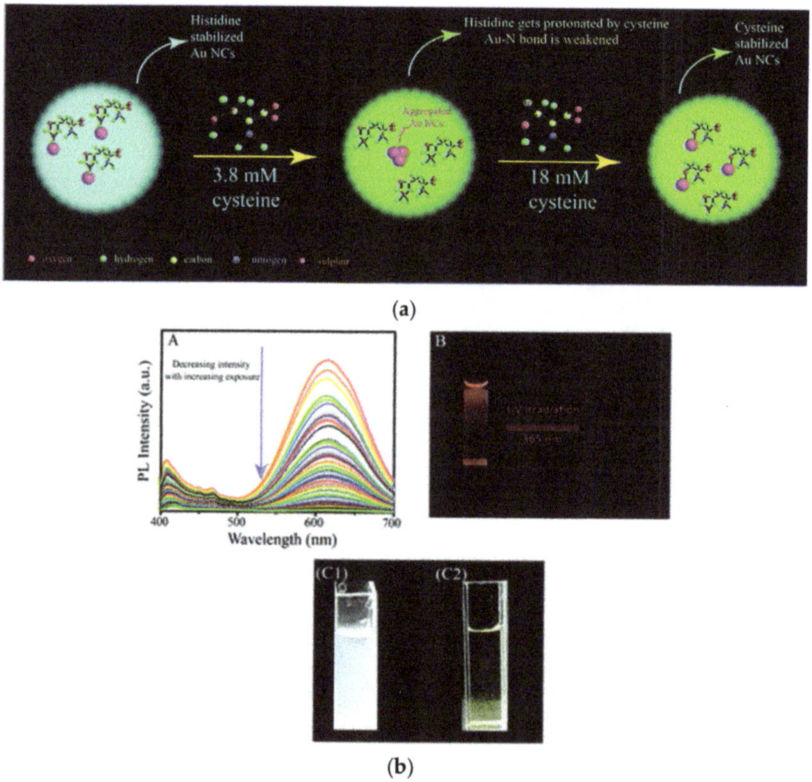

Figure 5. (**a**) Schematic illustration depicting ligand exchange reaction between histidine stabilized Au NCs and cysteine. Reprinted with permission from [84]. Copyright Royal Society of Chemistry 2020 (**b**) Experimental results showing quenching in luminescence of cystine stabilized Cu NCs upon reaction with oxygen in presence of light. Reprinted with permission from [83]. Copyright Royal Society of Chemistry 2019.

For the first time, Antoine and co-workers performed a ligand exchange reaction of glutathione stabilized Au NCs-$Au_{15}(SG)_{13}$ with a single aminooxy ligand to yield $Au_{15}(SG)_{12}(aminooxy)_1$. A detained mass spectrometric analysis revealed the successful ligand exchange of $Au_{15}(SG)_{13}$ by one aminooxy ligand. The ligand exchanged product served as a non-linear optical probe for the detection of protein carbonylation. This study was the first on the ligand exchange reaction for non-linear optical measurement based detection and quantification of protein carbonylation [85].

Furthermore, ligand exchange reactions have been reported to endow nanoclusters with achiral NCs with unprecedented chirality following exchange with chiral ligands. Burgi Yoshiki Niihori et al. successfully demonstrated that ligand exchange reactions between clusters and monothiolates commence at the terminal thiol groups of the thiolate protected NCs proceeding via an associative SN^2 type mechanism [86].

Apart than reactions involving ligands present on surface of clusters, metal atom exchange leading to inter cluster reactions have also been pursued. In this regard, Pradeep et al. have demonstrated inter cluster reactions between $Au_{25}(SR)_{18}$ and $Ag_{44}(SR)_{30}$ where RS = alkyl/aryl thiolate [87]. Detailed experimental and theoretical analyses were performed to prove the occurrence of the metal exchange reaction. Additionally, this has been claimed to be the first ever report on inter cluster alloying. On a similar note, bimetallic NCs constituting Ag and Ni/Pd/Pt protected by dithiols were reacted with monothiol protected Au NCs [88]. This led to formation of tri-metallic NCs due to inter-cluster reactions between the starting clusters. In another report, Au_{25} NCs protected by $PhCH_2CH_2S$ were reacted with Ag_{25} protected by 2,4-dimethyl benzene thiol to inter-cluster alloy. A detailed mechanistic investigation revealed that the reaction proceeded through two step metal exchange processes. In the first step, motif–motif exchange reaction leading to formation of ligand shell doped alloys of clusters occurred whereas in second step motif-motif and motif-kernel exchange reactions between the alloyed clusters took place. To this end, Antoine and co-workers reported the systematic doping of Au_{10} NCs stabilized by glutathione molecules. 1–3 silver atoms have been doped into Au_{10} NCs, displacing equivalent number of gold atoms from the clusters, which led to gradual variation in the two-photon excited fluorescence of the clusters [89].

The aforementioned examples of chemical reactions involving metal NCs clearly highlight their resemblance to the chemical reactivity of molecules. The knowledge gained from all these studies can be further extended to design new materials with tailored properties suited for diverse applications. This may start with the spatial organization of atomic NCs in higher dimensions. For example, ions, molecules and allied chemical species organized into three dimensions, i.e., their crystalline forms, are imbued with properties that are starkly different from that of the individual molecules. This is why crystalline forms of molecules are highly desired not only to gain fundamental insight into their chemistry but also for practical applications. Similarly, the constitutional precision of NCs and the ease with which they undergo chemical reactions provide great opportunities to expand the understanding and utility of NCs following assemblage into higher order dimensions.

5. Routes to Self-Assembled Structures of Nanoclusters. from Crystalline Assembly to Directed Assembly of NCs

Atomically precise clusters feature excellent biocompatibility, photo and temporal stability, catalytic activity, chemical reactivity, and tunable functionalities, and are thus envisioned to be model units for the fabrication of devices with an application potential ranging from catalysis optoelectronics energy storage and theranostics. However, pristine nanoclusters suffer from inherent limitations such as low quantum yield (in general), low stability under harsh conditions such as extremes of pH, heat and solvent. Often, clusters tend to aggregate into undefined moieties thereby losing their integral physicochemical characteristics. This restricts their application potentials. A possible solution for this issue is likely to emerge from the systematic organization of these nanoclusters into deterministic structures. This could be advantageous in several fronts, including gaining complete insight into the structures of NCs, delineating their structure property relationship, fine tuning their structures to tailor their properties for a desired application, conferring them with added stability and making them versatile for practical utilities. In this regard, several efforts have been directed to spatially organize NCs into well-defined structures.

As stated in earlier sections, the ligands stabilizing the clusters play a determining role in not only controlling their interactions but also additional chemical agents. For instance, the charges on the ligands keeps the individual cluster units apart from each other and prevent their uncontrolled agglomeration. Likewise, ligands with functional groups capable of hydrogen bonding interact among themselves to confer stability to clusters. On the other hand, ligands with terminal groups capable of chemical coordinating with metal ions, provide a facile platform for a complexation reaction mediated assembly of NCs.

In this regard, Chattopadhyay and coworkers and Paul and coworkers demonstrated, for the first time, that the complexation reaction between ligands stabilizing gold NCs and zinc ions can lead to the formation of three to two dimensional crystalline assemblies of clusters. Interestingly, they have shown MPA and Histidine stabilized Au_{14} NCs upon reaction with Zn ions results in the formation of three-dimensional assembly of the clusters. The three-dimensional structure derived from transmission electron microscopic (TEM) and selected area electron diffraction (SAED) analysis. The formed 3D assembly of Au NCs was further used for storage and sensing of hydrogen gas at ambient conditions of 20 °C and 20 bar (Figure 6a) [90]. In a similar fashion, two dimensional crystalline nanosheets were formed out of reaction between Au_{14} NCs (stabilized with L phenylaniline and MPA) and zinc ions. The two dimensional nanosheets were used for reversible storage of oxygen at pliant conditions of 20 °C and 20 bar pressures [91].

Similarly, MPA and L/D tryptophan stabilized Au NCs were assembled into three-dimensional organization following complexation reaction amongst Zn ions and carboxylate groups of MPA and L/D tryptophan. The resultant assembly was characterized using TEM, SAED and powder XRD, which revealed their crystalline nature. The crystalline assemblies of Au_{14} clusters were then used for fluorimetric chiral recognition and separation of externally added L and D tryptophan. Interestingly, it was observed that when Au_{14} clusters were stabilized with L tryptophan, the resultant crystalline assembly was responsive to L tryptophan only. Similarly, D tryptophan stabilized Au14 clusters constituting the crystalline assembly was responsive to D tryptophan only. The basis of chiral recognition and separation of tryptophan by the crystalline assembly was attributed to the classical three-point vs. two-point interactions between tryptophan analytes and the crystalline assembly. It is important to note here that, as opposed to the assembled clusters, non-assembled Au clusters were observed to be nonresponsive to the chiral behavior of L and D tryptophan [92].

MPA and L tyrosine-stabilized Au NCs were assembled into nanocrystals following coordination of the carboxylate groups of the ligand and zinc ions. The so formed nanocrystals were characterized using analytical techniques such as TEM and SAED. The crystalline assemblies were found to be effective for mitochondria-targeted cancer theranostics with the rare potential of facile renal clearance. The anticancer activity of the zinc mediated assembly was attributed to the generation of reactive oxygen species within HeLa cells [93].

Moreover, MPA and L cysteine stabilized Au_{14} NCs upon complexation reaction with zinc ions led to the formation of allied crystalline assembles, which could also be successfully characterized with TEM and SAED techniques. Importantly, the non-assembled Au_{14} clusters were found to feature photoluminescence intermittency when illuminated at few particle level, whereas the zinc ion-mediated assembly of clusters was found to be non-blinking under identical conditions. This was attributed to the fact that the non-radiative relaxation pathways active in non-assembled clusters, due to the conformational relaxation of the ligands, were blocked due the structural rigidity gained by the clusters upon complexation with zinc ions in the assembled structure. Additionally, in contrast to non-assembled clusters, the crystalline assembly of Au_{14} clusters could be used for storage of CO_2 and sensing of the latter at a few particle level (Figure 6b) [94].

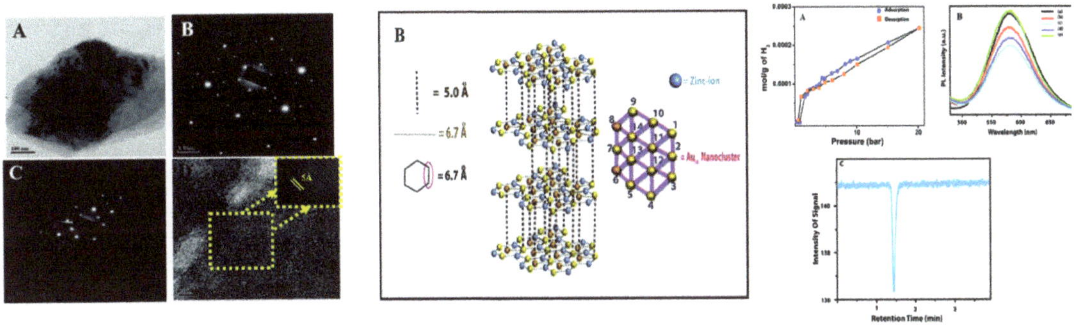

(**a**)

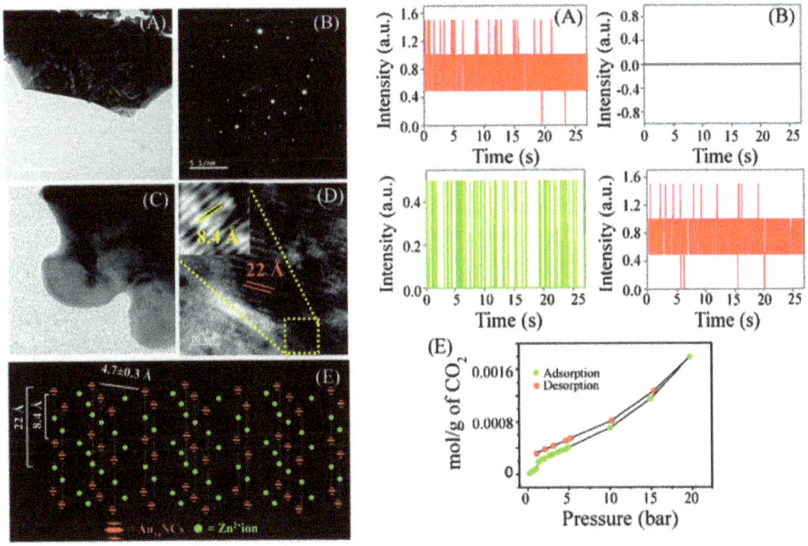

(**b**)

Figure 6. (**a**) Experimental results showing formation zinc mediated three-dimensional crystalline assembly of Au NCs stabilized by MPA and histidine, and use of the assembly in storage and sensing of hydrogen at plaint conditions. Reprinted with permission from [90]. Copyright Royal Society of Chemistry 2016. (**b**) Experimental results showing formation zinc mediated three-dimensional crystalline assembly of Au NCs stabilized by MPA and cysteine, and use of the assembly in storage and sensing of carbon dioxide at a few particle level. Reprinted from Ref. [94] with permission. Copyright Royal Society of Chemistry 2018.

MPA and MBA stabilized Au_{14} nanoclusters assembled into two dimensional organizations, following the complexation reaction with zinc ions was demonstrated to exhibit delayed fluorescence with ultra-long luminescence lifetime of 0.5 ms and quantum yield of ~19%. This marked the advent of the first ever report of crystalline assembly of gold nanoclusters exhibiting lifetime in sub millisecond range. The occurrence of delayed fluorescence was proposed to have originated from the restricted intramolecular motion of the ligands stabilizing the clusters (in assembly), which consequently led to reduced non radiative transitions otherwise occurring in non-assembled Au_{14} clusters [95].

The aforementioned reports of complexation reaction mediated assemblage of Au_{14} nanoclusters with synergistic properties of nanoscale clusters as well as inorganic complexes portends to form a new class of hierarchical nanomaterials concerted with futuristic application potential. Additionally, as opposed to assemblies of clusters, devoid of periodicity, crystalline assemblies are envisioned to have uniform chemical properties and a predictable application potential.

In addition to the principles of coordination chemistry, other strategies have also been adopted for the construction of an assembly of nanoclusters. In this regard, assemblies related to supramolecular chemical interactions are worth mentioning. An important example is provided by Yoon et al., who performed a large scale preparation of super crystalline lattices of $[Na_4Ag_{44}(p\text{-}MBA)_{30}]$. Hydrogen bonding amongst the carboxylate groups of the ligands was recognized as the key to form rhombus-shaped super crystals of $[Na_4Ag_{44}(p\text{-}MBA)_{30}]$ [96]. Single crystal X-ray crystallography revealed that the layers of the super lattice were connected through 24 and 36 intra and inter layer hydrogen bonds, respectively. Yao et al. constructed super structures of $[Ag_{44}(p\text{-}MBA)_{30}]^{4-}$ NCs, following the interaction of the carboxylate ends of the ligands with appropriate counter ions [97]. Zhang et al. fabricated the self-assembly of glutathione stabilized Au NCs, which closely mimics the self-assembly of capsid proteins. In this study, the initial disruption of hydrogen bonding between glutathione and water molecules was achieved following introduction of DMSO. Eventually, hydrogen bonding among glutathione molecules stabilizing the clusters was promoted through the subsequent removal of water molecules, which led to the formation of large spherical assemblies of Au_{22} NCs [98].

In addition to hydrogen bonding, other interactions such as electrostatic interactions, have also been utilized for the fabrication of self-assembled structures of NCs. The key idea behind this approach is the proper balance of charges among the ligands present on the surface of neighboring clusters, such that the clusters should neither undergo uncontrolled agglomeration nor should they remain too apart to hinder self-assembly. In order to construct an assembly of clusters, electrostatic interactions are often used in conjunction with other driving forces such as hydrogen bonding and chemical coordination, to maintain the delicate balance of proximity of the constituent clusters. Moreover, van der Waals interactions based assemblies of NCs are also known. For instance, dodecane thiol stabilized Au_{15} clusters were self-assembled into distinct nanosheets at elevated temperatures [99]. Hydrophobic interactions among the non-polar chains of dodecane thiol were proposed to have played a key role in the assemblage process.

In addition to self-assembly, a directed assembly of NCs have also been pursued. In this regard, templates, either stabilizing the clusters or externally added to the medium, have been proposed to play the determinant role in the assemblage process. To this end, macromolecules such as DNA, [100] macrocycles, [101] synthetic and natural polymers, [102] functionalized graphene oxide [103] have been reported to be instrumental in guiding the assembly of NCs. The chief idea in this approach is to allow concomitant binding of the NCs to the templates as well as spatial organization of the former in a way directed by the latter. For example, polymers imbued with appropriate functionalities can be tailored to interact with NCs in a specific manner, followed by hierarchical organization of the clusters into a geometry guided by the polymer. In this regard, nanospheres exhibiting aggregation induced emission have been fabricated via polyethyleneimine guided self-assembly of Ag(I) NCs [104]. Further, multi-thiolated co-polymers were used as scaffold for synthesis of Cu NCs. This was claimed to not only render the as synthesized NCs resistant towards oxidation and uncontrolled aggregation, but also imbued the clusters with thermo and pH responsive properties of the pristine co-polymers. The so formed polymeric hybrid structure of NCs was further used for sensing of Hg ions [105]. Moreover, natural biopolymers such as bacterial proteins have been used to self-assemble glutathione stabilized NCs. These nanocomposites could further be used as agents for cell imaging and cargo for protein delivery [106]. However, assembly induced by polymers often lead to uncontrolled agglomeration of clusters which render them inappropriate for practical applications.

This issue is often circumvented by the use of DNA as a template for assemblage of NCs. This is primarily attributed to the fact that DNA templates provide the combined options of deterministic structural motif as well as intrinsic binding sites, namely, the nucleobases. As a consequence, several nano-architectures have been constituted following the DNA-guided assembly of atomic clusters. For example, DNA nanoribbons constituting appropriate binding sites have been used as a template for synthesis cum assembly of ultra-small Cu NCs. DNA assembled Cu NCs were reported to feature aggregation induced emission attributed to structural rigidity gained by Cu NCs upon being assembled by DNA [107]. Furthermore, Wang and coworkers assembled Au NCs in the presence of double stranded DNA and used them for cancer theranostics [108]. Similarly, Chattopadhyay and co-workers synthesized Au NCs using double stranded DNA as templates, mimicking the process of polymerase chain reaction. Further luminescence intensity of the clusters could be used to probe and quantify double stranded DNA, i.e., the PCR products [34].

Moreover, macrocyclic molecules have been appropriately tailored to interact with the ligands of the atomic clusters to produce rigid assemblies featuring alluring photo physical properties. For example, Au_{22} NCs stabilized by peptides of sequence Phe-Gly-Gly-Cys were reported to undergo non-covalent interactions with Cucurbiturils, leading to the formation of rigid assemblies exhibiting enhanced quantum yield as opposed to non-assembled clusters [101]. Moreover, Ag_{29} NCs stabilized by 1,3-benzene dithiol and triphenyl phosphine were reported to react with crown ethers, leading to crystallization induced organization of the crown ether molecules into the interstitial sites of the lattice of Ag_{29} NCs [109]. The crystalline co-assembly of Ag_{29} NCs were found to be 3.5 times more luminescent than for the non-assembled clusters. Likewise, an aggregation-induced emission probe, constituting cyclodextrin functionalized Cu NCs, and di(adamantan-1-yl) phosphine, was designed for in situ imaging of membrane associated glycoproteins [110].

Additionally, external stimuli such as pH have been used to trigger assemblies of ultra-small copper nanoclusters. For instance, Cu NCs stabilized by L cysteine were found to undergo reversible aggregation and disaggregation, as a function of pH. Monodispersed Cu NCs were found to form insoluble macroscopic aggregates featuring red luminescence at pH ~ 3, whereas the aggregates were reported to disintegrate into soluble dispersion, showing weak luminescence at pH > 4. Similarly, orange-red emitting Cu NCs were spontaneously self-assembled into nano-spheres, nano-meshes and nanosheets by fine tuning the extent of hydrogen bonding interactions among m-amino thiophenols stabilizing the Cu NCs clusters [111]. A further development was achieved by Chattopadhyay and co-workers, who reported the synthesis of pH responsive Cu NCs, featuring aggregation-induced emission within cancer cells. Cu NCs were reported to exhibit orange-red emission at pH 4.5, whereas the same Cu NCs, following intracellular aggregation, featured bright green luminescence at pH 7.4. Interestingly, the rate constants for intracellular aggregation of Cu NCs were found to be different for different cancer cell lines [112].

6. Concluding Remarks

Atomically precise nanoclusters, once considered to be "just another" luminescent nanomaterial, has now become the heart and soul of research pertaining to nanoscale science and technology. This, as discussed in the earlier sections, is primarily attributed to their chemical and structural integrity. Thus, nanoclusters are being widely explored as model analogues of molecules in classical chemistry. They are often called supertoms or superatomic molecules [113–115]. Significant resources are being invested in designing rational and facile strategies for the synthesis of atomic clusters. Purification techniques are coming up to isolate ultra-pure atomic clusters [8]. Chemical reactions are being performed on atomic clusters to form "products" akin to reactions of molecules and compounds. Research is being directed to elucidate the mechanism of such reactions, gaining insights into the intermediates formed and maximizing the yield of the products. Further, controlled assemblies of atomic clusters are being fabricated using various principles of chemistry ranging from coordination to supramolecular. The as-described assemblies of nanoclusters have found finding applications in wide areas ranging from catalysis to theranostics.

In this mini review, we aimed to summarize the recent advancements made with regard to all these aspects of atomic clusters. Additionally, in the conclusion, we raise questions, which, if addressed in the near future, may further illustrate the versatility of atomic clusters. Firstly, it is important to elucidate the relationship between the number of atoms constituting the clusters and their corresponding properties (optical and chemical). For instance, it is well known that clusters with less atoms feature luminescence at wavelengths lower than the emission wavelengths of clusters with comparatively more atoms. However, an exact relationship between the number of atoms composing the clusters and the resultant emission properties is not fully understood. This is largely because, in addition to the number of atoms constituting the clusters, the ligands stabilizing the clusters also play key roles in determining their luminescence properties [116]. Thus, the emergence of generalized principles governing the properties of clusters is deemed essential. Clearly, quantum chemistry methods elucidating the geometry of the clusters and configuring their optical properties in terms of molecular transitions should be conducted in further depth for the assembly of clusters, and innovative approaches such as the hybrid QM/MM (quantum mechanics/molecular mechanics) approach might address their structure–properties relationships [117–121].

Secondly, with regard to the reactions of atomic clusters, the study of chemical kinetics has remained largely unexplored. Additionally, in classical chemistry, the mystery of "bond breaking and making" in molecules has been, and continues to be, gradually unveiled. However, such mysteries continue to prevail with regard to the chemistry of atomic clusters. These studies, if performed, may open up newer domains to control the yield of reactions, optimize the conditions of reactions, trap the essential intermediates, and identify the transition states of relevant reactions.

Thirdly, with regard to the controlled assembly of clusters, delineation of a definite structure property relationship is deemed important. Notwithstanding the robustness of the strategies for systematic organization of clusters developed so far, little effort has been directed towards deciphering a "one to one" correlation between the structures of assembled clusters and their emerging properties. This is not only important to enhance the ease of assemblage of nanoscale clusters, but also to widen their application potential. Finally, in an assembly of clusters, it is important to understand whether the ultimate properties are an outcome of additive behavior of the individual clusters, or a result of a synergistic behavior between the clusters, ligands, and the templates.

Funding: A.P. acknowledges funding from Department of Electronics and Information Technology, Government of India, (No. 5(9)/2012-NANO (Vol. IV)).

Conflicts of Interest: The authors declare no conflict of interest.

References

1. Overbeek, J.T.G. Monodisperse colloidal systems, fascinating and useful. *Adv. Colloid Interface Sci.* **1982**, *15*, 251–277. [CrossRef]
2. Jin, R.; Zeng, C.; Zhou, M.; Chen, Y. Atomically Precise Colloidal Metal Nanoclusters and Nanoparticles: Fundamentals and Opportunities. *Chem. Rev.* **2016**, *116*, 10346–10413. [CrossRef] [PubMed]
3. Jin, R. Quantum sized, thiolate-protected gold nanoclusters. *Nanoscale* **2010**, *2*, 343–362. [CrossRef] [PubMed]
4. Chakraborty, I.; Pradeep, T. Atomically Precise Clusters of Noble Metals: Emerging Link between Atoms and Nanoparticles. *Chem. Rev.* **2017**, *117*, 8208–8271. [CrossRef]
5. Heuer-Jungemann, A.; Feliu, N.; Bakaimi, I.; Hamaly, M.; Alkilany, A.; Chakraborty, I.; Masood, A.; Casula, M.F.; Kostopoulou, A.; Oh, E.; et al. The Role of Ligands in the Chemical Synthesis and Applications of Inorganic Nanoparticles. *Chem. Rev.* **2019**, *119*, 4819–4880. [CrossRef]
6. Li, Y.; Zhou, M.; Jin, R. Programmable Metal Nanoclusters with Atomic Precision. *Adv. Mater.* **2021**, *33*, 2006591. [CrossRef]
7. Jin, R. Atomically precise metal nanoclusters: Stable sizes and optical properties. *Nanoscale* **2015**, *7*, 1549–1565. [CrossRef]
8. Negishi, Y.; Hashimoto, S.; Ebina, A.; Hamada, K.; Hossain, S.; Kawawaki, T. Atomic-level separation of thiolate-protected metal clusters. *Nanoscale* **2020**, *12*, 8017–8039. [CrossRef]

9. Comby-Zerbino, C.; Dagany, X.; Chirot, F.; Dugourd, P.; Antoine, R. The emergence of mass spectrometry for characterizing nanomaterials. Atomically precise nanoclusters and beyond. *Mater. Adv.* **2021**, *2*, 4896–4913. [CrossRef]
10. Xiao, Y.; Wu, Z.N.; Yao, Q.F.; Xie, J.P. Luminescent metal nanoclusters: Biosensing strategies and bioimaging applications. *Aggregate* **2021**, *2*, 114–132. [CrossRef]
11. Sun, Y.F.; Zhou, Z.P.; Shu, T.; Qian, L.S.; Su, L.; Zhang, X.J. Multicolor Luminescent Gold Nanoclusters: From Structure to Biosensing and Bioimaging. *Prog. Chem.* **2021**, *33*, 179–187. [CrossRef]
12. Jin, R.; Li, G.; Sharma, S.; Li, Y.; Du, X. Toward Active-Site Tailoring in Heterogeneous Catalysis by Atomically Precise Metal Nanoclusters with Crystallographic Structures. *Chem. Rev.* **2021**, *121*, 567–648. [CrossRef] [PubMed]
13. Higaki, T.; Li, Y.; Zhao, S.; Li, Q.; Li, S.; Du, X.-S.; Yang, S.; Chai, J.; Jin, R. Atomically Tailored Gold Nanoclusters for Catalytic Application. *Angew. Chem. Int. Ed.* **2019**, *58*, 8291–8302. [CrossRef] [PubMed]
14. Tao, Y.; Li, M.; Ren, J.; Qu, X. Metal nanoclusters: Novel probes for diagnostic and therapeutic applications. *Chem. Soc. Rev.* **2015**, *44*, 8636–8663. [CrossRef] [PubMed]
15. Combes, G.F.; Vučković, A.-M.; Perić Bakulić, M.; Antoine, R.; Bonačić-Koutecky, V.; Trajković, K. Nanotechnology in Tumor Biomarker Detection: The Potential of Liganded Nanoclusters as Nonlinear Optical Contrast Agents for Molecular Diagnostics of Cancer. *Cancers* **2021**, *13*, 4206. [CrossRef] [PubMed]
16. Porret, E.; Le Guével, X.; Coll, J.L. Gold nanoclusters for biomedical applications: Toward in vivo studies. *J. Mater. Chemistry. B* **2020**, *8*, 2216–2232. [CrossRef] [PubMed]
17. Rival, J.V.; Mymoona, P.; Lakshmi, K.M.; Nonappa; Pradeep, T.; Shibu, E.S. Self-Assembly of Precision Noble Metal Nanoclusters: Hierarchical Structural Complexity, Colloidal Superstructures, and Applications. *Small* **2021**, *17*, 2005718. [CrossRef] [PubMed]
18. Ebina, A.; Hossain, S.; Horihata, H.; Ozaki, S.; Kato, S.; Kawawaki, T.; Negishi, Y. One-, Two-, and Three-Dimensional Self-Assembly of Atomically Precise Metal Nanoclusters. *Nanomaterials* **2020**, *10*, 1105. [CrossRef]
19. Wang, J.X.; Lin, X.F.; Shu, T.; Su, L.; Liang, F.; Zhang, X.J. Self-Assembly of Metal Nanoclusters for Aggregation-Induced Emission. *Int. J. Mol. Sci.* **2019**, *20*, 1891. [CrossRef]
20. Li, Y.; Jin, R. Seeing Ligands on Nanoclusters and in Their Assemblies by X-ray Crystallography: Atomically Precise Nanochemistry and Beyond. *J. Am. Chem. Soc.* **2020**, *142*, 13627–13644. [CrossRef]
21. Kang, X.; Zhu, M. Intra-cluster growth meets inter-cluster assembly: The molecular and supramolecular chemistry of atomically precise nanoclusters. *Coord. Chem. Rev.* **2019**, *394*, 1–38. [CrossRef]
22. Wei, X.; Kang, X.; Zuo, Z.; Song, F.; Wang, S.; Zhu, M. Hierarchical structural complexity in atomically precise nanocluster frameworks. *Natl. Sci. Rev.* **2020**, *8*, nwaa077. [CrossRef]
23. Wu, Z.; Yao, Q.; Zang, S.; Xie, J. Directed Self-Assembly of Ultrasmall Metal Nanoclusters. *ACS Mater. Lett.* **2019**, *1*, 237–248. [CrossRef]
24. Antoine, R. Supramolecular Gold Chemistry: From Atomically Precise Thiolate-Protected Gold Nanoclusters to Gold-Thiolate Nanostructures. *Nanomaterials* **2020**, *10*, 377. [CrossRef] [PubMed]
25. Maity, P.; Xie, S.; Yamauchi, M.; Tsukuda, T. Stabilized gold clusters: From isolation toward controlled synthesis. *Nanoscale* **2012**, *4*, 4027–4037. [CrossRef] [PubMed]
26. Zheng, J.; Petty, J.T.; Dickson, R.M. High Quantum Yield Blue Emission from Water-Soluble Au_8 Nanodots. *J. Am. Chem. Soc.* **2003**, *125*, 7780–7781. [CrossRef] [PubMed]
27. Zare, I.; Chevrier, D.M.; Cifuentes-Rius, A.; Moradi, N.; Xianyu, Y.; Ghosh, S.; Trapiella-Alfonso, L.; Tian, Y.; Shourangiz-Haghighi, A.; Mukherjee, S.; et al. Protein-protected metal nanoclusters as diagnostic and therapeutic platforms for biomedical applications. *Mater. Today* **2021**, *308*, 118323. [CrossRef]
28. Xie, J.; Zheng, Y.; Ying, J.Y. Protein-Directed Synthesis of Highly Fluorescent Gold Nanoclusters. *J. Am. Chem. Soc.* **2009**, *131*, 888–889. [CrossRef] [PubMed]
29. Chan, P.-H.; Chen, Y.-C. Human Serum Albumin Stabilized Gold Nanoclusters as Selective Luminescent Probes for Staphylococcus aureus and Methicillin-Resistant Staphylococcus aureus. *Anal. Chem.* **2012**, *84*, 8952–8956. [CrossRef]
30. Liu, C.-L.; Wu, H.-T.; Hsiao, Y.-H.; Lai, C.-W.; Shih, C.-W.; Peng, Y.-K.; Tang, K.-C.; Chang, H.-W.; Chien, Y.-C.; Hsiao, J.-K.; et al. Insulin-Directed Synthesis of Fluorescent Gold Nanoclusters: Preservation of Insulin Bioactivity and Versatility in Cell Imaging. *Angew. Chem. Int. Ed.* **2011**, *50*, 7056–7060. [CrossRef] [PubMed]
31. Lu, D.; Liu, L.; Li, F.; Shuang, S.; Li, Y.; Choi, M.M.F.; Dong, C. Lysozyme-stabilized gold nanoclusters as a novel fluorescence probe for cyanide recognition. *Spectrochim. Acta Part A Mol. Biomol. Spectrosc.* **2014**, *121*, 77–80. [CrossRef]
32. Wen, F.; Dong, Y.; Feng, L.; Wang, S.; Zhang, S.; Zhang, X. Horseradish Peroxidase Functionalized Fluorescent Gold Nanoclusters for Hydrogen Peroxide Sensing. *Anal. Chem.* **2011**, *83*, 1193–1196. [CrossRef]
33. Chakraborty, S.; Babanova, S.; Rocha, R.C.; Desireddy, A.; Artyushkova, K.; Boncella, A.E.; Atanassov, P.; Martinez, J.S. A Hybrid DNA-Templated Gold Nanocluster for Enhanced Enzymatic Reduction of Oxygen. *J. Am. Chem. Soc.* **2015**, *137*, 11678–11687. [CrossRef] [PubMed]
34. Sahoo, A.K.; Sailapu, S.K.; Dutta, D.; Banerjee, S.; Ghosh, S.S.; Chattopadhyay, A. DNA-Templated Single Thermal Cycle Based Synthesis of Highly Luminescent Au Nanoclusters for Probing Gene Expression. *ACS Sustain. Chem. Eng.* **2018**, *6*, 2142–2151. [CrossRef]

35. Bertorelle, F.; Russier-Antoine, I.; Calin, N.; Comby-Zerbino, C.; Bensalah-Ledoux, A.; Guy, S.; Dugourd, P.; Brevet, P.-F.; Sanader, Ž.; Krstić, M.; et al. $Au_{10}(SG)_{10}$: A Chiral Gold Catenane Nanocluster with Zero Confined Electrons. Optical Properties and First-Principles Theoretical Analysis. *J. Phys. Chem. Lett.* **2017**, *8*, 1979–1985. [CrossRef] [PubMed]
36. Negishi, Y.; Nobusada, K.; Tsukuda, T. Glutathione-Protected Gold Clusters Revisited: Bridging the Gap between Gold(I)−Thiolate Complexes and Thiolate-Protected Gold Nanocrystals. *J. Am. Chem. Soc.* **2005**, *127*, 5261–5270. [CrossRef]
37. Hamouda, R.; Bellina, B.; Bertorelle, F.; Compagnon, I.; Antoine, R.; Broyer, M.; Rayane, D.; Dugourd, P. Electron Emission of Gas-Phase $[Au_{25}(SG)_{18}\text{-}6H]^{7-}$ Gold Cluster and Its Action Spectroscopy. *J. Phys. Chem. Lett.* **2010**, *1*, 3189–3194. [CrossRef]
38. Sahoo, A.K.; Banerjee, S.; Ghosh, S.S.; Chattopadhyay, A. Simultaneous RGB Emitting Au Nanoclusters in Chitosan Nanoparticles for Anticancer Gene Theranostics. *ACS Appl. Mater. Interfaces* **2014**, *6*, 712–724. [CrossRef] [PubMed]
39. Tvedte, L.M.; Ackerson, C.J. Size-Focusing Synthesis of Gold Nanoclusters with p-Mercaptobenzoic Acid. *J. Phys. Chem. A* **2014**, *118*, 8124–8128. [CrossRef]
40. Yang, X.; Gan, L.; Han, L.; Li, D.; Wang, J.; Wang, E. Facile preparation of chiral penicillamine protected gold nanoclusters and their applications in cell imaging. *Chem. Commun.* **2013**, *49*, 2302–2304. [CrossRef] [PubMed]
41. Yang, X.; Shi, M.; Zhou, R.; Chen, X.; Chen, H. Blending of $HAuCl_4$ and histidine in aqueous solution: A simple approach to the Au_{10} cluster. *Nanoscale* **2011**, *3*, 2596–2601. [CrossRef] [PubMed]
42. Chen, Y.; Li, W.; Wang, Y.; Yang, X.; Chen, J.; Jiang, Y.; Yu, C.; Lin, Q. Cysteine-directed fluorescent gold nanoclusters for the sensing of pyrophosphate and alkaline phosphatase. *J. Mater. Chem. C* **2014**, *2*, 4080–4085. [CrossRef]
43. Goswami, U.; Basu, S.; Paul, A.; Ghosh, S.S.; Chattopadhyay, A. White light emission from gold nanoclusters embedded bacteria. *J. Mater. Chem. C* **2017**, *5*, 12360–12364. [CrossRef]
44. Udaya Bhaskara Rao, T.; Pradeep, T. Luminescent Ag_7 and Ag_8 Clusters by Interfacial Synthesis. *Angew. Chem. Int. Ed.* **2010**, *49*, 3925–3929. [CrossRef]
45. Dhanalakshmi, L.; Udayabhaskararao, T.; Pradeep, T. Conversion of double layer charge-stabilized Ag@citrate colloids to thiol passivated luminescent quantum clusters. *Chem. Commun.* **2012**, *48*, 859–861. [CrossRef] [PubMed]
46. Bootharaju, M.S.; Burlakov, V.M.; Besong, T.M.D.; Joshi, C.P.; AbdulHalim, L.G.; Black, D.M.; Whetten, R.L.; Goriely, A.; Bakr, O.M. Reversible Size Control of Silver Nanoclusters via Ligand-Exchange. *Chem. Mater.* **2015**, *27*, 4289–4297. [CrossRef]
47. Wei, W.; Lu, Y.; Chen, W.; Chen, S. One-Pot Synthesis, Photoluminescence, and Electrocatalytic Properties of Subnanometer-Sized Copper Clusters. *J. Am. Chem. Soc.* **2011**, *133*, 2060–2063. [CrossRef] [PubMed]
48. Ghosh, R.; Sahoo, A.K.; Ghosh, S.S.; Paul, A.; Chattopadhyay, A. Blue-Emitting Copper Nanoclusters Synthesized in the Presence of Lysozyme as Candidates for Cell Labeling. *ACS Appl. Mater. Interfaces* **2014**, *6*, 3822–3828. [CrossRef]
49. Singh, R.; Majhi, S.; Sharma, K.; Ali, M.; Sharma, S.; Choudhary, D.; Tripathi, C.S.P.; Guin, D. BSA stabilized copper nanoclusters as a highly sensitive and selective probe for fluorescence sensing of Fe^{3+} ions. *Chem. Phys. Lett.* **2021**, *76*, 139226. [CrossRef]
50. Rajamanikandan, R.; Ilanchelian, M. Red emitting human serum albumin templated copper nanoclusters as effective candidates for highly specific biosensing of bilirubin. *Mater. Sci. Eng. C* **2019**, *98*, 1064–1072. [CrossRef]
51. Miao, H.; Zhong, D.; Zhou, Z.; Yang, X. Papain-templated Cu nanoclusters: Assaying and exhibiting dramatic antibacterial activity cooperating with H_2O_2. *Nanoscale* **2015**, *7*, 19066–19072. [CrossRef]
52. Goswami, U.; Dutta, A.; Raza, A.; Kandimalla, R.; Kalita, S.; Ghosh, S.S.; Chattopadhyay, A. Transferrin–Copper Nanocluster–Doxorubicin Nanoparticles as Targeted Theranostic Cancer Nanodrug. *ACS Appl. Mater. Interfaces* **2018**, *10*, 3282–3294. [CrossRef] [PubMed]
53. Feng, J.; Chen, Y.; Han, Y.; Liu, J.; Ma, S.; Zhang, H.; Chen, X. pH-Regulated Synthesis of Trypsin-Templated Copper Nanoclusters with Blue and Yellow Fluorescent Emission. *ACS Omega* **2017**, *2*, 9109–9117. [CrossRef] [PubMed]
54. Jia, X.; Li, J.; Han, L.; Ren, J.; Yang, X.; Wang, E. DNA-Hosted Copper Nanoclusters for Fluorescent Identification of Single Nucleotide Polymorphisms. *ACS Nano* **2012**, *6*, 3311–3317. [CrossRef]
55. Das, N.K.; Ghosh, S.; Priya, A.; Datta, S.; Mukherjee, S. Luminescent Copper Nanoclusters as a Specific Cell-Imaging Probe and a Selective Metal Ion Sensor. *J. Phys. Chem. C* **2015**, *119*, 24657–24664. [CrossRef]
56. Lin, Y.-S.; Chiu, T.-C.; Hu, C.-C. Fluorescence-tunable copper nanoclusters and their application in hexavalent chromium sensing. *RSC Adv.* **2019**, *9*, 9228–9234. [CrossRef]
57. Yang, X.; Feng, Y.; Zhu, S.; Luo, Y.; Zhuo, Y.; Dou, Y. One-step synthesis and applications of fluorescent Cu nanoclusters stabilized by l-cysteine in aqueous solution. *Anal. Chim. Acta* **2014**, *847*, 49–54. [CrossRef]
58. Ghosh, R.; Goswami, U.; Ghosh, S.S.; Paul, A.; Chattopadhyay, A. Synergistic Anticancer Activity of Fluorescent Copper Nanoclusters and Cisplatin Delivered through a Hydrogel Nanocarrier. *ACS Appl. Mater. Interfaces* **2015**, *7*, 209–222. [CrossRef]
59. Lin, Y.-J.; Chen, P.-C.; Yuan, Z.; Ma, J.-Y.; Chang, H.-T. The isomeric effect of mercaptobenzoic acids on the preparation and fluorescence properties of copper nanoclusters. *Chem. Commun.* **2015**, *51*, 11983–11986. [CrossRef]
60. Shahsavari, S.; Hadian-Ghazvini, S.; Hooriabad Saboor, F.; Menbari Oskouie, I.; Hasany, M.; Simchi, A.; Rogach, A.L. Ligand functionalized copper nanoclusters for versatile applications in catalysis, sensing, bioimaging, and optoelectronics. *Mater. Chem. Front.* **2019**, *3*, 2326–2356. [CrossRef]

61. Rieck, D.F.; Gavney, J.A.; Norman, R.L.; Hayashi, R.K.; Dahl, L.F. Synthesis, chromatographic separation, and stereophysical analysis of the homologous [Ni12-x(PMe)x(CO)24-3x]2- series (x = 2, 3, 4) containing noncentered Ni12-xPx icosahedral cages and the [Ni10(.mu.5-PMe)2(.mu.4-PMe)5(CO)10]2- dianion containing a structurally unprecedented heptacapped pentagonal prismatic metal cage: Structural, spectroscopic, and electrochemical consequences due to replacement of Ni(CO)3 fragments with electronically equivalent (isolobal) PMe fragments. *J. Am. Chem. Soc.* **1992**, *114*, 10369–10379. [CrossRef]
62. Ji, J.; Wang, G.; Wang, T.; You, X.; Xu, X. Thiolate-protected Ni39 and Ni41 nanoclusters: Synthesis, self-assembly and magnetic properties. *Nanoscale* **2014**, *6*, 9185–9191. [CrossRef]
63. Joya, K.S.; Sinatra, L.; AbdulHalim, L.G.; Joshi, C.P.; Hedhili, M.N.; Bakr, O.M.; Hussain, I. Atomically monodisperse nickel nanoclusters as highly active electrocatalysts for water oxidation. *Nanoscale* **2016**, *8*, 9695–9703. [CrossRef] [PubMed]
64. Guan, G.; Zhang, S.-Y.; Cai, Y.; Liu, S.; Bharathi, M.S.; Low, M.; Yu, Y.; Xie, J.; Zheng, Y.; Zhang, Y.-W.; et al. Convenient purification of gold clusters by co-precipitation for improved sensing of hydrogen peroxide, mercury ions and pesticides. *Chem. Commun.* **2014**, *50*, 5703–5705. [CrossRef]
65. Galchenko, M.; Schuster, R.; Black, A.; Riedner, M.; Klinke, C. Preparation of high-yield and ultra-pure Au25 nanoclusters: Towards their implementation in real-world applications. *Nanoscale* **2019**, *11*, 1988–1994. [CrossRef]
66. Sun, J.; Yue, Y.; Wang, P.; He, H.; Jin, Y. Facile and rapid synthesis of water-soluble fluorescent gold nanoclusters for sensitive and selective detection of Ag+. *J. Mater. Chem. C* **2013**, *1*, 908–913. [CrossRef]
67. Tang, Y.; Xu, J.; Xiong, C.; Xiao, Y.; Zhang, X.; Wang, S. Enhanced electrochemiluminescence of gold nanoclusters via silver doping and their application for ultrasensitive detection of dopamine. *Analyst* **2019**, *144*, 2643–2648. [CrossRef]
68. Pramanik, G.; Humpolickova, J.; Valenta, J.; Kundu, P.; Bals, S.; Bour, P.; Dracinsky, M.; Cigler, P. Gold nanoclusters with bright near-infrared photoluminescence. *Nanoscale* **2018**, *10*, 3792–3798. [CrossRef]
69. Li, Y.; Feng, L.; Yan, W.; Hussain, I.; Su, L.; Tan, B. PVP-templated highly luminescent copper nanoclusters for sensing trinitrophenol and living cell imaging. *Nanoscale* **2019**, *11*, 1286–1294. [CrossRef] [PubMed]
70. Ghosh, A.; Hassinen, J.; Pulkkinen, P.; Tenhu, H.; Ras, R.H.A.; Pradeep, T. Simple and Efficient Separation of Atomically Precise Noble Metal Clusters. *Anal. Chem.* **2014**, *86*, 12185–12190. [CrossRef]
71. Niihori, Y.; Kikuchi, Y.; Shima, D.; Uchida, C.; Sharma, S.; Hossain, S.; Kurashige, W.; Negishi, Y. Separation of Glutathionate-Protected Gold Clusters by Reversed-Phase Ion-Pair High-Performance Liquid Chromatography. *Ind. Eng. Chem. Res.* **2017**, *56*, 1029–1035. [CrossRef]
72. Comby-Zerbino, C.; Perić, M.; Bertorelle, F.; Chirot, F.; Dugourd, P.; Bonačić-Koutecký, V.; Antoine, R. Catenane Structures of Homoleptic Thioglycolic Acid-Protected Gold Nanoclusters Evidenced by Ion Mobility-Mass Spectrometry and DFT Calculations. *Nanomaterials* **2019**, *9*, 457. [CrossRef]
73. Bertorelle, F.; Russier-Antoine, I.; Comby-Zerbino, C.; Chirot, F.; Dugourd, P.; Brevet, P.-F.; Antoine, R. Isomeric Effect of Mercaptobenzoic Acids on the Synthesis, Stability, and Optical Properties of Au25(MBA)18 Nanoclusters. *ACS Omega* **2018**, *3*, 15635–15642. [CrossRef] [PubMed]
74. Soleilhac, A.; Bertorelle, F.; Comby-Zerbino, C.; Chirot, F.; Calin, N.; Dugourd, P.; Antoine, R. Size Characterization of Glutathione-Protected Gold Nanoclusters in the Solid, Liquid and Gas Phases. *J. Phys. Chem. C* **2017**, *121*, 27733–27740. [CrossRef]
75. Ligare, M.R.; Baker, E.S.; Laskin, J.; Johnson, G.E. Ligand induced structural isomerism in phosphine coordinated gold clusters revealed by ion mobility mass spectrometry. *Chem. Commun.* **2017**, *53*, 7389–7392. [CrossRef] [PubMed]
76. Baksi, A.; Ghosh, A.; Mudedla, S.K.; Chakraborty, P.; Bhat, S.; Mondal, B.; Krishnadas, K.R.; Subramanian, V.; Pradeep, T. Isomerism in Monolayer Protected Silver Cluster Ions: An Ion Mobility-Mass Spectrometry Approach. *J. Phys. Chem. C* **2017**, *121*, 13421–13427. [CrossRef]
77. Angel, L.A.; Majors, L.T.; Dharmaratne, A.C.; Dass, A. Ion mobility mass spectrometry of $Au_{25}(SCH_2CH_2Ph)_{18}$ nanoclusters. *ACS Nano* **2010**, *4*, 4691–4700. [CrossRef] [PubMed]
78. Kalenius, E.; Malola, S.; Matus, M.F.; Kazan, R.; Bürgi, T.; Häkkinen, H. Experimental Confirmation of a Topological Isomer of the Ubiquitous $Au_{25}(SR)_{18}$ Cluster in the Gas Phase. *J. Am. Chem. Soc.* **2021**, *143*, 1273–1277. [CrossRef]
79. Gayen, C.; Basu, S.; Pan, U.N.; Paul, A. Few Particle-Level Chromaticity Index-Based Discrimination of Biothiols Using Chemically Interactive Dual-Emitting Nanoprobe. *ACS Omega* **2018**, *3*, 17220–17226. [CrossRef]
80. Gayen, C.; Goswami, U.; Gogoi, K.; Basu, S.; Paul, A. Crystallization-Induced Emission Enhancement of Nanoclusters and One-Step Conversion of "Nanoclusters to Nanoparticles" as the Basis for Intracellular Logic Operations. *ChemPhysChem* **2019**, *20*, 953–958. [CrossRef]
81. Gayen, C.; Basu, S.; Goswami, U.; Paul, A. Visible Light Excitation-Induced Luminescence from Gold Nanoclusters Following Surface Ligand Complexation with Zn^{2+} for Daylight Sensing and Cellular Imaging. *Langmuir* **2019**, *35*, 9037–9043. [CrossRef] [PubMed]
82. Basu, S.; Fakhouri, H.; Moulin, C.; Dolai, S.; Russier-Antoine, I.; Brevet, P.-F.; Antoine, R.; Paul, A. Four orders-of-magnitude enhancement in the two-photon excited photoluminescence of homoleptic gold thiolate nanoclusters following zinc ion-induced aggregation. *Nanoscale* **2021**, *13*, 4439–4443. [CrossRef]
83. Basu, S.; Nawaj, M.W.; Gayen, C.; Paul, A. Photo induced chemical modification of surface ligands for aggregation and luminescence modulation of copper nanoclusters in the presence of oxygen. *Phys. Chem. Chem. Phys.* **2019**, *21*, 21776–21781. [CrossRef]

84. Basu, S.; Gayen, C.; Dolai, S.; Paul, A. Tailoring the luminescence of atomic clusters via ligand exchange reaction mediated post synthetic modification. *Phys. Chem. Chem. Phys.* **2020**, *22*, 3959–3964. [CrossRef]
85. Combes, G.F.; Fakhouri, H.; Moulin, C.; Girod, M.; Bertorelle, F.; Basu, S.; Ladouce, R.; Bakulić, M.P.; Maršić, Ž.S.; Russier-Antoine, I.; et al. Functionalized Au15 nanoclusters as luminescent probes for protein carbonylation detection. *Commun. Chem.* **2021**, *4*, 69. [CrossRef]
86. Wang, Y.; Bürgi, T. Ligand exchange reactions on thiolate-protected gold nanoclusters. *Nanoscale Adv.* **2021**, *3*, 2710–2727. [CrossRef] [PubMed]
87. Krishnadas, K.R.; Ghosh, A.; Baksi, A.; Chakraborty, I.; Natarajan, G.; Pradeep, T. Intercluster Reactions between $Au_{25}(SR)_{18}$ and $Ag_{44}(SR)_{30}$. *J. Am. Chem. Soc.* **2016**, *138*, 140–148. [CrossRef] [PubMed]
88. Khatun, E.; Chakraborty, P.; Jacob, B.R.; Paramasivam, G.; Bodiuzzaman, M.; Dar, W.A.; Pradeep, T. Intercluster Reactions Resulting in Silver-Rich Trimetallic Nanoclusters. *Chem. Mater.* **2020**, *32*, 611–619. [CrossRef]
89. Basu, S.; Bakulić, M.P.; Fakhouri, H.; Russier-Antoine, I.; Moulin, C.; Brevet, P.-F.; Bonačić-Koutecký, V.; Antoine, R. Rationale Strategy to Tune the Optical Properties of Gold Catenane Nanoclusters by Doping with Silver Atoms. *J. Phys. Chem. C* **2020**, *124*, 19368–19374. [CrossRef]
90. Basu, S.; Paul, A.; Chattopadhyay, A. Zinc mediated crystalline assembly of gold nanoclusters for expedient hydrogen storage and sensing. *J. Mater. Chem. A* **2016**, *4*, 1218–1223. [CrossRef]
91. Paul, M.; Basu, S.; Chattopadhyay, A. Complexation Reaction-Based Two-Dimensional Luminescent Crystalline Assembly of Atomic Clusters for Recyclable Storage of Oxygen. *Langmuir* **2020**, *36*, 754–759. [CrossRef]
92. Basu, S.; Paul, A.; Chattopadhyay, A. Zinc-Coordinated Hierarchical Organization of Ligand-Stabilized Gold Nanoclusters for Chiral Recognition and Separation. *Chem.–Eur. J.* **2017**, *23*, 9137–9143. [CrossRef]
93. Basu, S.; Goswami, U.; Paul, A.; Chattopadhyay, A. Crystalline assembly of gold nanoclusters for mitochondria targeted cancer theranostics. *J. Mater. Chem. B* **2018**, *6*, 1650–1657. [CrossRef]
94. Basu, S.; Bhandari, S.; Pan, U.N.; Paul, A.; Chattopadhyay, A. Crystalline nanoscale assembly of gold clusters for reversible storage and sensing of CO_2 via modulation of photoluminescence intermittency. *J. Mater. Chem. C* **2018**, *6*, 8205–8211. [CrossRef]
95. Basu, S.; Chattopadhyay, A. Room-Temperature Delayed Fluorescence of Gold Nanoclusters in Zinc-Mediated Two-Dimensional Crystalline Assembly. *Langmuir* **2019**, *35*, 5264–5270. [CrossRef] [PubMed]
96. Yoon, B.; Luedtke, W.D.; Barnett, R.N.; Gao, J.; Desireddy, A.; Conn, B.E.; Bigioni, T.; Landman, U. Hydrogen-bonded structure and mechanical chiral response of a silver nanoparticle superlattice. *Nat. Mater.* **2014**, *13*, 807–811. [CrossRef] [PubMed]
97. Yao, Q.; Yu, Y.; Yuan, X.; Yu, Y.; Zhao, D.; Xie, J.; Lee, J.Y. Counterion-Assisted Shaping of Nanocluster Supracrystals. *Angew. Chem. Int. Ed.* **2015**, *54*, 184–189. [CrossRef] [PubMed]
98. Zhang, C.; Zhang, A.; Hou, W.; Li, T.; Wang, K.; Zhang, Q.; de la Fuente, J.M.; Jin, W.; Cui, D. Mimicking Pathogenic Invasion with the Complexes of $Au_{22}(SG)_{18}$-Engineered Assemblies and Folic Acid. *ACS Nano* **2018**, *12*, 4408–4418. [CrossRef] [PubMed]
99. Wu, Z.; Dong, C.; Li, Y.; Hao, H.; Zhang, H.; Lu, Z.; Yang, B. Self-assembly of Au_{15} into single-cluster-thick sheets at the interface of two miscible high-boiling solvents. *Angew. Chem.* **2013**, *52*, 9952–9955. [CrossRef]
100. Ouyang, X.; Wang, M.; Guo, L.; Cui, C.; Liu, T.; Ren, Y.; Zhao, Y.; Ge, Z.; Guo, X.; Xie, G.; et al. DNA Nanoribbon-Templated Self-Assembly of Ultrasmall Fluorescent Copper Nanoclusters with Enhanced Luminescence. *Angew. Chem. Int. Ed.* **2020**, *59*, 11836–11844. [CrossRef]
101. Jiang, T.; Qu, G.; Wang, J.; Ma, X.; Tian, H. Cucurbiturils brighten Au nanoclusters in water. *Chem. Sci.* **2020**, *11*, 3531–3537. [CrossRef]
102. Hembury, M.; Beztsinna, N.; Asadi, H.; van den Dikkenberg, J.B.; Meeldijk, J.D.; Hennink, W.E.; Vermonden, T. Luminescent Gold Nanocluster-Decorated Polymeric Hybrid Particles with Assembly-Induced Emission. *Biomacromolecules* **2018**, *19*, 2841–2848. [CrossRef]
103. Bertorelle, F.; Basu, S.; Fakhouri, H.; Perić Bakulić, M.; Mignon, P.; Russier-Antoine, I.; Brevet, P.-F.; Thomas, S.; Kalarikkal, N.; Antoine, R. Covalent anchoring of atomically precise glutathione-protected gold nanoclusters on graphene oxide nanosheets. *Nano Express* **2020**, *1*, 030005. [CrossRef]
104. Shen, J.; Wang, Z.; Sun, D.; Xia, C.; Yuan, S.; Sun, P.; Xin, X. pH-Responsive Nanovesicles with Enhanced Emission Co-Assembled by Ag(I) Nanoclusters and Polyethyleneimine as a Superior Sensor for Al^{3+}. *ACS Appl. Mater. Interfaces* **2018**, *10*, 3955–3963. [CrossRef]
105. Benavides, J.; Quijada-Garrido, I.; García, O. The synthesis of switch-off fluorescent water-stable copper nanocluster Hg2+ sensors via a simple one-pot approach by an in situ metal reduction strategy in the presence of a thiolated polymer ligand template. *Nanoscale* **2020**, *12*, 944–955. [CrossRef] [PubMed]
106. Ju, E.; Li, T.; Ramos da Silva, S.; Gao, S.-J. Gold Nanocluster-Mediated Efficient Delivery of Cas9 Protein through pH-Induced Assembly-Disassembly for Inactivation of Virus Oncogenes. *ACS Appl. Mater. Interfaces* **2019**, *11*, 34717–34724. [CrossRef] [PubMed]
107. Qing, Z.; He, X.; He, D.; Wang, K.; Xu, F.; Qing, T.; Yang, X. Poly(thymine)-Templated Selective Formation of Fluorescent Copper Nanoparticles. *Angew. Chem. Int. Ed.* **2013**, *52*, 9719–9722. [CrossRef] [PubMed]
108. Wang, M.; Chen, Y.; Cai, W.; Feng, H.; Du, T.; Liu, W.; Jiang, H.; Pasquarelli, A.; Weizmann, Y.; Wang, X. In situ self-assembling Au-DNA complexes for targeted cancer bioimaging and inhibition. *Proc. Natl. Acad. Sci. USA* **2020**, *117*, 308–316. [CrossRef]

109. Chakraborty, P.; Nag, A.; Sugi, K.S.; Ahuja, T.; Varghese, B.; Pradeep, T. Crystallization of a Supramolecular Coassembly of an Atomically Precise Nanoparticle with a Crown Ether. *ACS Mater. Lett.* **2019**, *1*, 534–540. [CrossRef]
110. Huang, Y.; Ji, J.; Zhang, J.; Wang, F.; Lei, J. Host–guest recognition-regulated aggregation-induced emission for in situ imaging of MUC1 protein. *Chem. Commun.* **2020**, *56*, 313–316. [CrossRef]
111. Su, X.; Liu, J. pH-Guided Self-Assembly of Copper Nanoclusters with Aggregation-Induced Emission. *ACS Appl. Mater. Interfaces* **2017**, *9*, 3902–3910. [CrossRef]
112. Dutta, A.; Goswami, U.; Chattopadhyay, A. Probing Cancer Cells through Intracellular Aggregation-Induced Emission Kinetic Rate of Copper Nanoclusters. *ACS Appl. Mater. Interfaces* **2018**, *10*, 19459–19472. [CrossRef]
113. Cheng, L.; Ren, C.; Zhang, X.; Yang, J. New insight into the electronic shell of $Au_{38}(SR)_{24}$: A superatomic molecule. *Nanoscale* **2013**, *5*, 1475–1478. [CrossRef]
114. Häkkinen, H. Electronic shell structures in bare and protected metal nanoclusters. *Adv. Phys. X* **2016**, *1*, 467–491. [CrossRef]
115. Liu, L.; Li, P.; Yuan, L.-F.; Cheng, L.; Yang, J. From isosuperatoms to isosupermolecules: New concepts in cluster science. *Nanoscale* **2016**, *8*, 12787–12792. [CrossRef] [PubMed]
116. Wu, Z.; Jin, R. On the Ligand's Role in the Fluorescence of Gold Nanoclusters. *Nano Lett.* **2010**, *10*, 2568–2573. [CrossRef] [PubMed]
117. Gran, E.R.; Bertorelle, F.; Fakhouri, H.; Antoine, R.; Perić Bakulić, M.; Sanader Maršić, Ž.; Bonačić-Koutecký, V.; Blain, M.; Antel, J.; Maysinger, D. Size and ligand effects of gold nanoclusters in alteration of organellar state and translocation of transcription factors in human primary astrocytes. *Nanoscale* **2021**, *13*, 3173–3183. [CrossRef] [PubMed]
118. Perić, M.; Sanader Maršić, Ž.; Russier-Antoine, I.; Fakhouri, H.; Bertorelle, F.; Brevet, P.-F.; le Guével, X.; Antoine, R.; Bonačić-Koutecký, V. Ligand shell size effects on one- and two-photon excitation fluorescence of zwitterion functionalized gold nanoclusters. *Phys. Chem. Chem. Phys.* **2019**, *21*, 23916–23921. [CrossRef]
119. Bonačić-Koutecký, V.; Antoine, R. Enhanced two-photon absorption of ligated silver and gold nanoclusters: Theoretical and experimental assessments. *Nanoscale* **2019**, *11*, 12436–12448. [CrossRef]
120. Russier-Antoine, I.; Bertorelle, F.; Calin, N.; Sanader, Z.; Krstic, M.; Comby-Zerbino, C.; Dugourd, P.; Brevet, P.-F.; Bonacic-Koutecky, V.; Antoine, R. Ligand-core NLO-phores: A combined experimental and theoretical approach to the two-photon absorption and two-photon excited emission properties of small-ligated silver nanoclusters. *Nanoscale* **2017**, *9*, 1221–1228. [CrossRef]
121. Sanader, Z.; Krstic, M.; Russier-Antoine, I.; Bertorelle, F.; Dugourd, P.; Brevet, P.-F.; Antoine, R.; Bonacic-Koutecky, V. Two-photon absorption of ligand-protected Ag15 nanoclusters. Towards a new class of nonlinear optics nanomaterials. *Phys. Chem. Chem. Phys.* **2016**, *18*, 12404–12408. [CrossRef] [PubMed]

MDPI AG
Grosspeteranlage 5
4052 Basel
Switzerland
Tel.: +41 61 683 77 34

Nanomaterials Editorial Office
E-mail: nanomaterials@mdpi.com
www.mdpi.com/journal/nanomaterials

www.ingramcontent.com/pod-product-compliance
Lightning Source LLC
Chambersburg PA
CBHW061552210926
51850CB00009B/2081